Resonance and Aspect Matched Adaptive Radar

(RAMAR)

Resonance and Aspect Matched Adaptive Radar

(RAMAR)

Terence William Barrett
BSEI, USA

World Scientific

NEW JERSEY · LONDON · SINGAPORE · BEIJING · SHANGHAI · HONG KONG · TAIPEI · CHENNAI

Published by

World Scientific Publishing Co. Pte. Ltd.

5 Toh Tuck Link, Singapore 596224

USA office: 27 Warren Street, Suite 401-402, Hackensack, NJ 07601

UK office: 57 Shelton Street, Covent Garden, London WC2H 9HE

British Library Cataloguing-in-Publication Data
A catalogue record for this book is available from the British Library.

ISBN-13 978-981-4329-89-7
ISBN-10 981-4329-89-4

Typeset by Stallion Press
Email: enquiries@stallionpress.com

PREFACE

This book treats active radar as a branch of spectroscopy by addressing the optimum excitation of a target's resonances. As in conventional spectroscopy, the aim is to identify a system by its spectrum. That statement is limited by some caveats. While in the frequency domain, the resonances, if not their amplitude, are target aspect-independent, in the time domain, the time-of-arrival of the returning signals, constituting an extended target response, will be target aspect-dependent. Furthermore, the target resonances can only be excited if the transmitted radiation elicits so-called Mie, or resonance, scattering, and such scattering can only be defined by the ratio of the wavelength of the incident radiation to the length of the target resonant features — and not the absolute length of either. With those caveats, optimum radiation for exciting a target's resonances is defined in terms of a target's resonant structural features. This book addresses transmitted RF radiation matched optimally to the resonant features of a target, a class of target, or a component of a target, i.e., a RAMAR (Resonance and Aspect Matched Adaptive Radar), which is achieved by MAP (Matched Adaptive Time-Frequency Packet-Signal) signaling.

This book treats targets as linear transducers or linear systems, but not linear time-invariant, or LTI, systems. Rather, the variation in time of signal arrival of returning signal components at the transceiver dictates that the targets be treated as linear time varying, or LTV, systems. That targets are LTV systems has consequences. The well-known, and ubiquitously used, Fourier transform is strictly only appropriate for LTI systems. The more appropriate techniques for LTV systems are time-frequency and wavelet techniques, and these are used throughout the book.

Although the excited RF system response of the reported targets tested is treated as if these systems are linear, and empirical evidence is presented that the targets tested are linear, on the basis of the statistical principle that one cannot prove the null hypothesis, there is always the possibility that there may be targets, unexamined, that do have a nonlinear response. When and if discovered, extended methods treating the system response would then be required.

The major claim of this book is that to elicit an optimized target response the transmitted signal must be matched to the resonant features of the target. However, it will be of engineering interest that in many cases the matching need not be a matching of the carrier frequency of the transmitted signal to the target. Rather, it need only be a matching of the amplitude modulation of the envelope of the carrier frequency of the transmitted signal to the target. This fact means that the choice of carrier frequency (and therefore antenna size) is arbitrary — but only in the absence of absorbing media. If the transmitted and returning signals are also required to penetrate absorbing media, then, indeed, both carrier and envelope must be matched to both media and target.

This book addresses the result of tests conducted with two prototype systems — one at Ka-Band, the other at UHF-Band — and neither of these systems should be considered other than preliminary designs. The designs and the tests conducted were limited by the equipment and funding available.

While the author takes responsibility for all claims scientific, engineering and mathematical in this book, the author wishes to acknowledge that none of the test data could have been collected nor the tests completed, without the collaboration of Raytheon Missile Systems, Tucson, Arizona, personnel. In particular, the author acknowledges and thanks Silvio Cardero, Johann Schleiss, Blake Hilgermann, William Owens and Ralph Tadaki. The author also acknowledges the support and encouragement of Joseph P. Garcia of the United States Naval Surface Weapons Center, Dahlgren, VA, and Timothy Jaynes of the United States Counter Narco-Terrorism Program Office. This work was supported by grants from the Missile Defense Agency and the Counter-Narco-Terrorism Program Office of the United States Department of Defense.

Terence W. Barrett
Vienna, VA, USA

CONTENTS

INTRODUCTION

The radar field is a mature discipline whose focus is electromagnetic systems for the detection of targets or aspects of the natural environment. A transmitted signal, for ease of referral: a TX, is radiated into space, which interacts with those targets or aspects, and a returning received signal, or echo, for ease of referral: an RX, is produced. The RX is captured by a receiver and, following suitable treatment, targets can be detected and other information about the target extracted from the return signal.

Conventional radars operate with TXs with frequencies ranging from the so-called high frequency (HF) range, (3–30 MHz), to the millimeter range (40–300 GHz). Conventionally, a radar system is constructed to produce TXs either at a set frequency, or over a limited frequency bandwidth, as in a linear frequency modulated (LFM) chirp, or as a very short duration pulse, the short duration of which produces a very wide bandwidth or ultrawideband (UWB) signal. In conventional approaches, only in a very general sense are the frequency and time duration attributes of the TX matched or aligned with similar attributes of the target. In contrast, the radar systems addressed by the present book are directly focused on matching as much as possible the attributes of the TX to the attributes of a designated target, class of targets, or medium-and-target. In a special emphasis, the major attributes of a target addressed are the target's scattered RX radio frequency (RF) spectrum, and, when a target's RX response is extended in time, i.e., if an extended target, the timing sequence of the separate returned wave packets constituting the target's total RX.

Conventionally, it is either assumed that the RX of a target is qualitatively the same in response to all frequencies in the TX radar frequency bandwidth of a specific radar system, or the differences are not brought to full attention, and so are largely neglected. However, paradoxically, it is well known that the electromagnetic response of any target is a function of target size and TX frequency. For example, there are three major so-called "scattering regions": (i) the Rayleigh, (ii) the Mie[1] or resonance, and (iii) the optical, regions. It is important to note, that these regions are not defined with respect to specific TX frequency regions. Rather, they are defined relative to the *ratio* of specific target lengths and TX wavelengths. Furthermore, these three (joint target-and-transmitted-signal-wavelength-dependent-) scattering regions have different underlying scattering mechanisms as a function of that ratio. In the case of optical scattering, the TX wavelength is *less than* a target's dimensions, and the RX is due to *parts of* the target, and not due to the whole target or its volume or area. In the case of Rayleigh scattering, the TX wavelength is *greater than* a target's dimensions, and the RX is due to its volume or area. At ratios in between these, Mie or resonance scattering occurs when the TX wavelength is *approximately equal* to both the specific target length and the lengths of subcomponents of the target. In these cases the whole target and its subcomponents oscillate at specific resonant frequencies and, significantly, the RX is somewhat greater in amplitude than the RX in the case of either optical or Rayleigh scattering. The distinction between the physical mechanisms underlying the three scattering regions is noteworthy, because the designs of the radar systems addressed by the present book are focused on techniques of matching, as much as possible, the wavelengths or frequencies of the TX to the size or resonances of designated targets and target subcomponents, in order to achieve Mie or *resonance scattering*.

One quantitative method of analysis describing the extent to which signals can be matched in time and frequency to receiver properties was available very early in the development of radar. The ambiguity function (Woodward, 1953; Cook & Bernfeld, 1967; Vakman, 1968) and its relative, the cross-ambiguity function, that relate properties of the RX signal to receiver properties, are analysis methods closely related to the matched filter concept. However, here we extend the matched filter concept by matching the TX signal in *time and frequency* to the known target RX, and the matched filter concept is expressed in both an optimum *TX-target response* relation as well as an optimum *RX-receiver* relation, rather merely in the conventional optimum *RX-receiver* relation regardless of the TX. In the conventional case of matching the receiver to a general RX, an optimal relation is only achieved by chance, because targets and their RX responses may vary in the extent of optimization, when the TX remains constant.

---

[1]Gustav Adolf Feodor Wilhelm Ludwig Mie (1869–1957).

Conversely, in the case of matching the TX to a known target response, the target transfer response is optimally configured "in the channel" or "in-the-loop" between the transmitter and receiver as a transfer function, and an optimum response, RX, is due to matching the TX to that frequency dependent transfer function. Therefore the radar systems addressed by the present book are focused on both a TX-target response relation matched filtering *as well as* an RX-receiver relation, rather than a RX-receiver response relation matched filtering alone.

In the case of a conventional radar that transmits a signal unmatched to a designated target, there is both waste of transmitted energy at those frequencies unmatched to target resonances (because energy at those frequencies is minimally reflected or returned), or inadvertently matched to undesignated targets (i.e., "clutter") with resonant frequencies other than target resonances. In the latter case the designated target RX is mixed with clutter returns, resulting in an undesirable signal-to-clutter ratio (SCR). However, when the transmit signal is matched to the designated target resonances, and, with the proviso that the resonances of the clutter do not coincide with the resonances of the designated target, an improved SCR is obtained.

The difference between the conventional approach to radar, and the approach taken in this book, is also mirrored in the difference in statistical decision theory between maximum-likelihood (ML) estimation detection, and maximum a posteriori (MAP) estimation detection.[2] In the case of ML estimation, detection is optimal for decisions made on the basis of the magnitude of the RX given the TX, whatever the TX may be. This is similar to conventional radar detection. In the case of MAP estimation, detection is optimal for decisions made on the basis of the magnitude of the RX given that the TX was matched to known target attributes. Without that matching made possible by the availability of *a priori* information, MAP detection coincides with ML detection. The radar systems addressed by the present book implement MAP detection rather than ML detection, and we refer to such systems as MAP radars.

Since this book addresses MAP detection which requires target *a priori* information, how is this *a priori* information obtained? There are two ways. The first, and preferred, way provides class-of-target, and/or target,

---

[2]The abbreviation and mnemonic "MAP" serves two purposes in this book: as referring to (1) Maximum A Posteriori estimation; and to (2) to (Target) **M**atched **A**daptive Time-frequency **P**acket Signal. Both uses implicate *a priori* information (Barrett, 1996).

and/or target subcomponent, frequency response information from either an anechoic chamber or "clutter-free" environment tests. The second, and more difficult, way provides the same information but gathered "on-the-fly" using heuristic transmit and filtering techniques. The obtainment of the (i) class-of-target, and/or (ii) target, and/or (iii) subcomponent of target, frequency response information permits the design of TXs to match the desired class-of-target, and/or target, and/or subcomponent of target. Furthermore, target identification by "question-and-answer" TX-RX protocol sequences are enabled. For example, if the target frequency response is known *a priori* concerning (i) the major class of target sought; (ii) the minor class of target sought; and (iii) a critically identifying component of the specific target sought, then matched signals can be transmitted in a $TX_1 - RX_1 \to TX_2 - RX_2 \to TX_3 - RX_3 \ldots$ sequence to confirm, or not, the identification of (a) the major class of target sought; (b) the minor target sought; and (c) a critically identifying component of the specific target sought. As a specific example: (i) might be all four-wheel vehicles of a certain size; (ii) might be all Humvee vehicles; and (iii) might be the antenna used by only certain Humvee vehicles. Furthermore, in the case of (i) we are concerned with major RX resonances shared by all targets in the class; in the case of (ii), with major resonances of the minor class of targets; and in the case of (iii), with minor resonances identifying specific targets in the minor class.

Due to the above considerations and with a focus of the present book on target major and minor resonances, a target's *a priori* frequency response information essentially amounts to a target's radio frequency (RF) spectrum. Therefore, whereas conventional radar addresses *ranging and detection* of generally an unknown target — usually a point scatterer — and more recently, target imaging, the present book is focused on what might be called: RF spectroscopy, addressing the spectral characteristics of targets and target subcomponents.

In introducing an RF spectroscopy it is necessary to identify particular kinds of TXs and RXs. For ease of reference throughout this book we shall use some simplifying nomenclature. Apart from a target's frequency response, there is the related impulse response. If the target *a priori* information is obtained by means of pulsing the target with a very short monocycle pulse, we shall call that transmitted pulse: a PTX. The target echo, or returned signal elicited by the PTX, we shall call: a PRX. A PTX is identical to a UWB TX signal, and a PRX is identical to a UWB RX signal.

Similarly the target-matched TX we shall call: an MTX, which is the time reversal of a PRX target return echo pulse. The advantage to using an MTX is that it is matched (i) in frequency to the resonances of the target and also (ii) in time to the target's orientation, or aspect angle, that determines when an extended target's resonance responses are received. In the case of (i) resonance matching, we shall show that the return signal from the target, MRX, is target-and-transceiver-platform-position independent, or *aspect independent*, with respect to the target *spectral frequency bands*, while the *amplitude of those frequency bands* may vary with aspect. In the case of the (ii), the sequencing or time of arrival of RX packets, constituting an extended target's subcomponent echo returns elicited by either PRXs or MRXs, is *aspect dependent* on target-transceiver-platform relative orientation. For these reasons, the radars addressed in this book are Resonance and Aspect Matched Adaptive Radars, or RAMARs.

An underlying assumption of a MAP RAMAR is that the target and its subcomponents, act as linear transfer functions of PTXs and MTXs. Hence PTX, MTX, PRX and MRX components are treated as statistically independent. Independent component analysis (ICA) processing is used, below, to indicate that the statistical independence assumption is justified — at least for the targets so far tested (Section 1.1.18).

The duration of a PTX is as short as equipment permits, and the duration of a MTX is dictated by the extended target's response function — and also of short duration. A major disadvantage of using short duration transmit pulses of any kind is that the average energy per TX and RX pulse is low, and in order to detect targets at long range by returned pulse integration, short duration TX signals need to be transmitted at a high repetition rate. On the other hand, in the majority of situations, information about a target's orientation is not required. In these cases a target's *a priori* known frequency response information is sufficient to design a TX matched only to the target's resonances (or RF spectrum), and not also to the target-platform relative orientation using the PRX timing information. In these majority of cases, the designed TX can be as long in time as target range requires and equipment limitations permit, and we refer to these TX signals as DTXs and STXs, below.

Two questions might be asked concerning the method used to obtain target *a priori* information: (1) If the targets tested so far in this program indicate that the target transfer function is linear, why not use an LFM TX signal, rather than a short monocycle PTX? (2) Why not just calculate a target's frequency response from its known dimensions and material composition?

In answer to the first question: As those targets tested so far have largely exhibited linear transfer function responses, LFM TXs, in which "all", or a broad band of frequencies are applied, but applied *sequentially*, might indeed be substituted for a UWB signal, in which, in contrast, the same frequencies are applied "*instantaneously*". However, statistically speaking, one cannot prove the null hypothesis. That is: one cannot extrapolate a conclusion of no differences (from linearity) in an examination of the few (targets), to a conclusion of no differences in the unexamined many (targets), so there always remains the possibility that some yet untested classes of targets are nonlinear, or partially nonlinear.[3]

While the characterization of linear systems is an advanced discipline (cf. Bendat & Piersol, 1993, 2000), the characterization of nonlinear systems has just begun (cf. Bendat, 1990),[4] and is hardly complete. The available techniques characterizing nonlinear systems apply to only a limited number of nonlinear models. The methods to detect nonlinear systems were formalized by Wiener (1958). These methods require the test application of *all* frequencies and *all* phases between those frequencies and if not with an infinite bandwidth, at least within a reasonably wide bandwidth, e.g., testing using white noise (*cf.* Barrett, 1975). The short ultrawideband pulse (PTX) approach that we used in both Ka-band and UHF prototype testing, described below, is a compromise approach. A radiated short pulse is not a Dirac delta function — which cannot be radiated — and is, at best, a monocycle. The substitution of a PTX short pulse for white noise means that as a short pulse is broad band, one can test with "all frequencies" within a reasonably wide bandwidth, but not all phases between those

---

[3] There are many types of nonlinear systems (*cf.* Bendat, 1990), but there is no universally accepted definition of nonlinearity. When referring to a system that is nonlinear, the type of nonlinearity usually meant is one in which the principle of superposition does not apply to a systems's individual frequency responses. There are, however, other definitions and meanings: e.g., an intensity-dependent nonlinearity; nonlinear feedback, etc. Furthermore, below we treat the whole target, with all subcomponent resonances excited, as a linear time variant (LTV) system. A nonlinearity in this case would be the target's complete response not being the summation of subcomponent minor resonances acting in isolation.

[4] The study has begun for third-order nonlinear polynomial systems consisting of linear systems in parallel with finite-memory square-law systems and finite-memory cubic systems.

frequencies — because the phases between frequencies in a short pulse are static and time invariant.

Turning to the second question: Why not just calculate a target's frequency response? — we answer this question referring to empirical tests conducted on designed subcomponent parts systematically joined (*cf.* Section 1.9, below). This question of whether a complex target's RX signal is merely composed of the superposition of the minor resonances of separated subcomponents, that, when joined, compose the target, was answered in the negative. The tests conducted indicate that superposition of separate subcomponent resonances does not apply, and simulation of an empirically untested complex target's RX spectrum is difficult due to this nonlinear addition. However, it is important to distinguish the difference between this nonlinear addition of subcomponent responses in physically forming the target, and a nonlinear addition of frequency responses after that target is physically formed — the latter having been addressed by the first question.

Another question then arises: given that a short pulse provides frequencies across a wide bandwidth, is the bandwidth an *instantaneous* bandwidth? That is, are all the frequencies *instantaneous* frequencies, i.e., all present at the same time? This question actually addresses the issue of frequency dispersion in antennas. The difficulties in designing a wide *instantaneous* bandwidth antenna (Anderson *et al.*, 2003; Schantz, 2005; Ghavami *et al.*, 2007), as opposed to sequentially wide bandwidth antennas, are well-known, and we do not resolve these issues here. Whether a wide instantaneous bandwidth, or a wide sequential bandwidth antenna is used, the pulse method of obtaining a system's frequency response is still, admittedly, a compromise.

Turning now to related previous work: Since this book addresses the optimized match of transmitted signals to designated targets using *a priori* information, it might asked: Have there been other proposals to do the same or similar? The answer is that there have been, but not exactly the same.

In the early years of radar, Ville (1948), Woodward & Davies (1950) and Woodward (1953) addressed optimum radar signals, but the search was for a single optimum signal for all targets. Cook & Bernfield (1967) mention that it might be possible to match a signal to its environment, but acknowledging the difficulties, looked for a signal optimum over a given number of situations and favored the LFM transmitted signal. Gjessig (1978, 1981, 1986; Apel & Gjessig, 1989) proposed a target adapted "matched illumination" radar. But whereas this radar is designed to achieve a target returned signal that is a "delta function" in the frequency domain, the MAP systems addressed by the present book, on the other hand and quite in contrast, are designed to achieve a target returned signal (MRX) that, only under a specific aspect angle of target-and-platform (90°) approximates a "delta function" of the target's impulse response (PRX) yet in the *time* — not in the *frequency* — domain. Thus, the systems addressed in the present book specifically address target resonances that are neglected in the matched illumination radars which address constructive interference by all the reflecting facets of the target.

Furthermore, this book provides evidence that in many, if not most, instances, the target acts as a linear *time varying* or dependent (LTV) system, rather than a linear *time invariant* or independent (LTI) system. Thus the systems addressed in the present book differ from those systems treated as LTI systems (Bell, 1988, 1993; Haykin, 2006; Guerci, 2010). Bell, although exploiting target resonance effects and addressing waveform design, assumes a static target, i.e., an LTI system. Guerci (2010) likewise addresses LTI targets, even although this form of analysis cannot be extrapolated to real world non-static LTV targets. Although a so-called knowledge based radar (Haykin, 2006) uses *a priori* information, that information appears to be collected from three generic conventional radars of limited TX pulse capability and which cannot provide an estimate of target's frequency response. In summary, the closest surveillance/signaling systems to the MAP systems addressed in this book are not in the field of radar, but rather in the fields of photonics and acoustics.

The radar systems described in this book require *a priori* information concerning the time and frequency characteristics of the target's transfer function, and although such information can be obtained with more difficulty, on-the-fly, here we concentrate on the collection of that information by pulsing the target in a clutter-free environment to obtain *empirical knowledge* of a target's transfer function. The information in the returned signal is then used to design other kinds of transmitted signals with optimum properties. As an aid to exposition, from now forward we shall use the following abbreviations and mnemonics to describe these signals, some of which have been already introduced:

- A *MAP* signal: a target-matched adaptive time-frequency wave packet or signal using *a priori* information and permitting maximum *a posteriori* estimation detection according to Bayesian statistics.
- *PTX*: a transmitted (*TX*) *short duration* packet/signal, modulating an arbitrary carrier, generally of 1 or 2 nanosecond duration, and that is

used to obtain *a priori* information concerning the target in a specified aspect to the transmitter. A *PTX* is equivalent to an UWB transmitted pulse and functions as a "δ function" surrogate or approximation.[5]

- *PRX*: the received (*RX*) return target echo, modulating an arbitrary carrier that is elicited by a *PTX*. A *PRX* is equivalent to an UWB return echo signal, and approximates the target impulse response at a specified target aspect to the transmitter.

- *MTX*: a *MAP* transmitted (*TX*) *short duration* packet/signal, envelope or amplitude modulation of an arbitrary carrier. An *MTX* envelope is a *PRX* but time reversed resulting in a matching of the envelope of the *MTX* transmitted signal to the designated target.

- *MRX*: the envelope of the received (*RX*) return target echo, modulating an arbitrary carrier, that is elicited by an *MTX*.

- *DTX*: a *PRX*-derived transmitted (*TX*) *arbitrarily long duration* transmitted signal, modulating an arbitrary carrier, and which addresses some selected resonance or collection of resonances identified in the *PRX* spectrum.

- *DRX*: the envelope of the received (*RX*) return target echo, modulating an arbitrary carrier, that is elicited by a *DTX*.

- *STX*: A transmitted (*TX*) *arbitrarily long duration* *TX* signal that is a collection or bundle of *DTX*s.

- *SRX*: the received (*RX*) return target echo, modulating an arbitrary carrier, that is elicited by an *STX*.

1. *A Priori* and *A Posteriori* Information Captures

The generic protocol for the PRX-MRX relation where there is access to a clutter-free environment involves two "captures" of target information:

(1) The *A Priori* Capture. A PTX is transmitted in a clutter and multipath-free environment with the target in a known orientation or aspect to the transceiver. With respect to this transmitted signal, an extended target acts as a temporal scattering matrix dispersing the resonance response of individual subcomponents in wave packets (Fig. 1) — the target resonances being temporally dispersed as a function of target aspect angle (Figs. 3–5), and an extended target

---

[5]A true δ function is of infinitely short duration and of infinite bandwidth, and even an approximation could not be propagated. What is meant in this context is a very short pulse — usually a monocycle.

A. Test Pulse: PTX-PRX

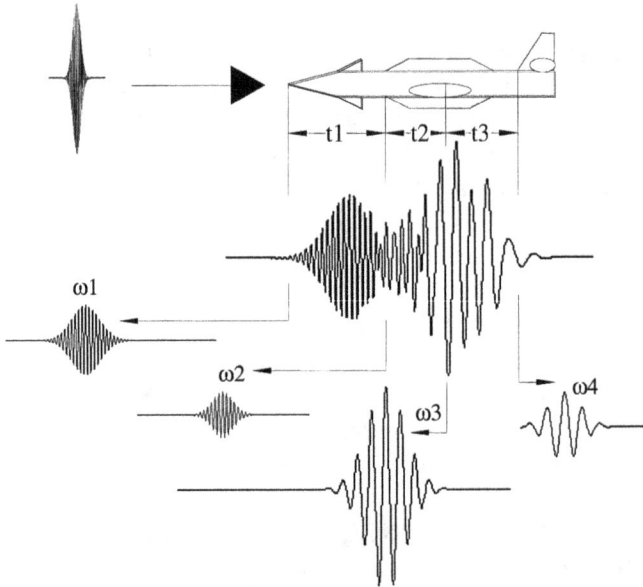

Fig. 1   The *A Priori* Capture. A short duration pulse or wave packet, *PTX* — ideally, shorter in length than any feature of interest on the target — is used to obtain the scattering matrix of a target which is here arbitrarily assigned 4 subcomponent resonances in this figure and is arbitrarily oriented head-on to the transmitter. The *PTX* is used as an approximation to a "δ function". A *PTX* is also an ultrawideband UWB signal. The target does not act as a point scatterer, but decomposes into its individual scattering components — i.e., is an extended target. Each of these individual components is scattered back to the transmitter (1) with specific frequency and amplitude modulation packets, the spectral characteristics of which are *aspect independent*; and (2) at separate phasing or time intervals of arrival that are *aspect dependent*. This is a feature of extended targets. The impulse response of the target can be used for target imaging, but in the case of interest is here used as *a priori* information for the *A Posteriori* Capture. The total response is the target's impulse response and is referred to as *PRX* which is the target's Green's function, $G(t, f; s)$ — t, time, f, frequency, s, spatial position. The target's whole body response is not addressed in this illustration.

response being composed of a collection of subcomponent-related wave packets, as well as a whole body response. The PRX encodes both (i) resonances, the frequency band spectral positions of which are aspect independent, as well as (ii) their temporal dispersion, which is aspect dependent (Fig. 1);

(2) The *A Posteriori* Capture. An MTX is constructed by time reversing the PRX and transmitting (Fig. 2). With respect to this transmitted

B. Matched Pulse: MTX-MRX

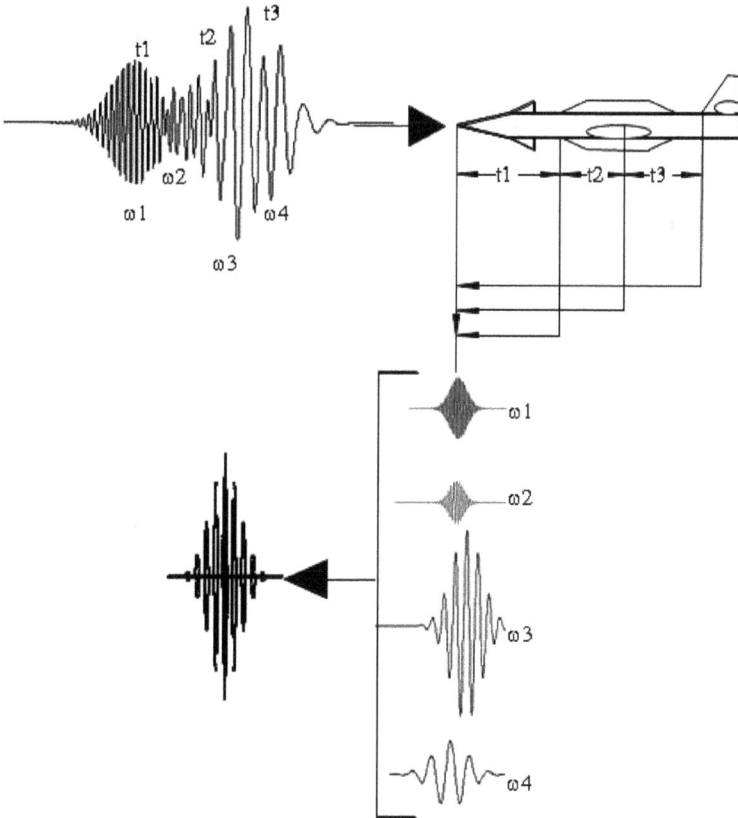

Fig. 2   The *A Posteriori* Capture. A *MAP TX* pulse, or *MTX*, is constructed by time reversing the *PRX* and is thereby matched to the amplitude and frequency modulation of the target at a specific target aspect angle, i.e., the *MTX* is the complex conjugate of the impulse response of the target at set aspect angle, or $G^{-1}(t, f; s)$. The *MTX* signal excites each target subcomponent (1) with the appropriate resonance frequency of the subcomponent; (2) with the appropriate relative amplitude modulation, and also, in the instance shown, (3) at the appropriate timing of TX packets for the target's aspect angle. This technique permits selective enhancement of component parts of the target, if required, and both the avoidance of clutter returns — if the clutter does not share resonant frequencies with the target — as well as confining *TX* energies to those frequencies to which the target is responsive. The returned signal, *MRX*, can be a short duration signal, i.e., acts as a loose approximation to a $\bar{\delta}$ function, if the target orientation is "head-on" ($0°$ aspect agreeing with the *A Priori* Capture). $MRX = G(t, f; s)G^{-1}(t, f; s) = \delta(t, f; s)$. The target's whole body response is also not addressed in this illustration.

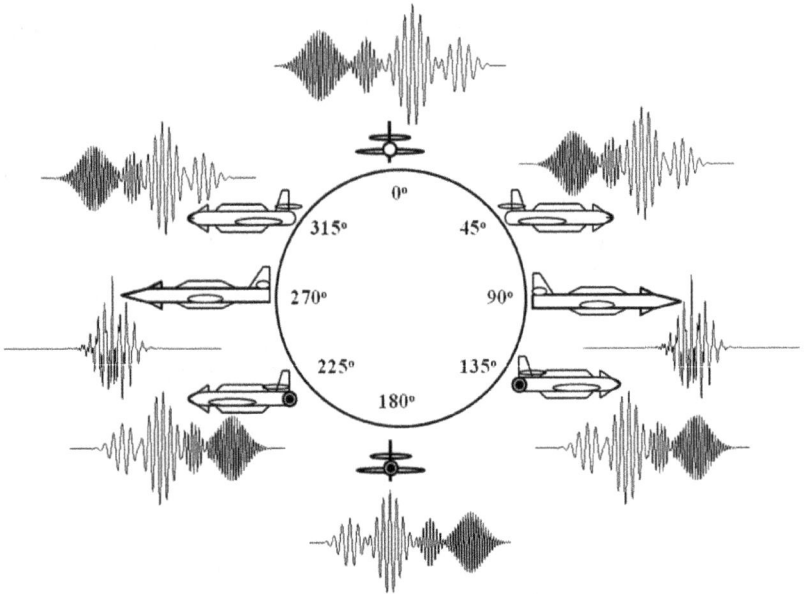

Fig. 3   Target resonances or spectral properties, are aspect independent, but the time of return, or time of arrival, of target signal resonances are platform and target aspect dependent. Here is shown a target with 4 arbitrary resonances or wave packets, that are time sequenced in return according to the target aspect angle. The target RX spectrum is aspect independent with respect to its harmonic components, but aspect dependence is indicated in the amplitude of those components and the timing of the individual RX packets arrival (Fig. 4, below).

signal, an extended target acts as a temporal compressor matrix, the target resonance reflections being almost simultaneously received and adding, if the target is at the identical orientation as in the *A Priori* Capture.

If the target is at a changed orientation, the spectral positioning of the resonance bands identified in a first capture do not change, but there can be changes in (a) the amplitude of the MRX resonances; and (b) the temporal compression (Figs. 3–5). For example, if the target (TGT) is position at an angle, e.g., 00°, or head-on, with respect to the TX-RX transceiver, or $TGT(00)$, and the PRX and MRX received signals for this orientation are $PRX(00)$ and $MRX(00)$, a maximum compressed returned MRX is obtained for:

(1st capture) $PTX(00) \rightarrow TGT(00) \rightarrow PRX(00)$,

(2nd capture) $MTX(00) \rightarrow TGT(00) \rightarrow MRX(00)$.

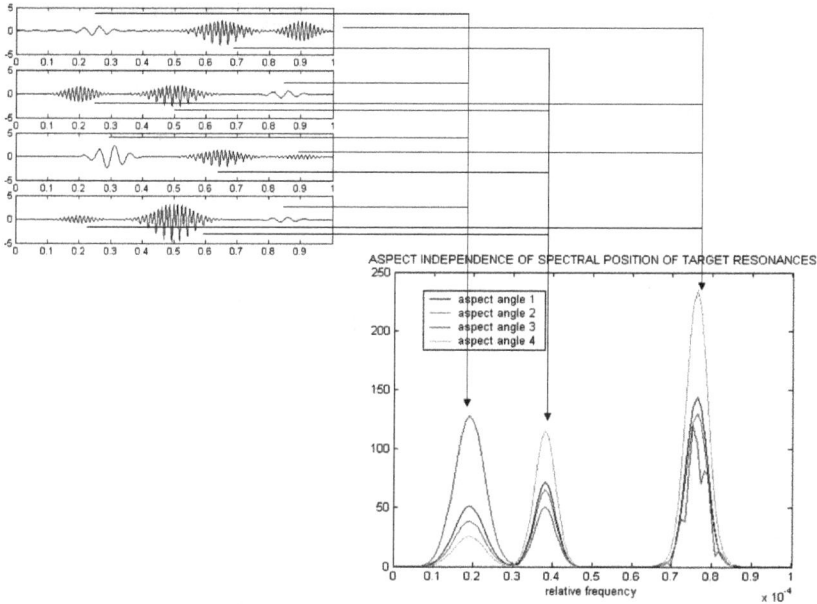

ASPECT INDEPENDENCE OF SPECTRAL POSITION OF TARGET RESONANCES

Fig. 4 Here a target is arbitrarily represented with 3 wave packet subcomponent resonances. As the target changes its aspect angle, the 3 resonance RX signals arrive back at the transmitting platform at different times of arrival — upper figure. However, this aspect dependence is only reflected in the spectrum by differences in amplitude of the 3 spectral bands. Spectral band occupancy, but not band amplitude, is an aspect independent target signature.

However if the target aspect angle is changed to 90°, e.g., $TGT(90)$, but the $MTX(00)$ is transmitted, i.e., the situation is:

(1st capture) $PTX(00) \rightarrow TGT(00) \rightarrow PRX(00)$,

(2nd capture) $MTX(90) \rightarrow TGT(00) \rightarrow MRX(00/90)$.

In this latter case the MTX will still have addressed the target resonances, but (a) the amplitude of the resonances at target aspect angle 90° may have changed from those at target aspect angle 00°. Therefore, (b) the temporal compression will be less than optimum. Yet the MTX will still only have transmitted energy at frequencies matched to target resonances, and in many instances clutter resonances will still not have been excited.

As illustration of a (target) matched adaptive time-frequency packet-signal (MAP) system we consider 3 simplified model MTX signals using an assortment of 4 minor resonances, bandwidths, and differences in time

A: Target as a Resonance Dispersion Matrix, and
B: After TX Equalization to Target: Correlator-Compressor.

A.

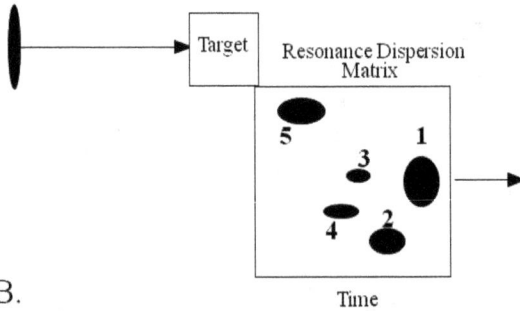

B.   Time
Precoding (Equalization)

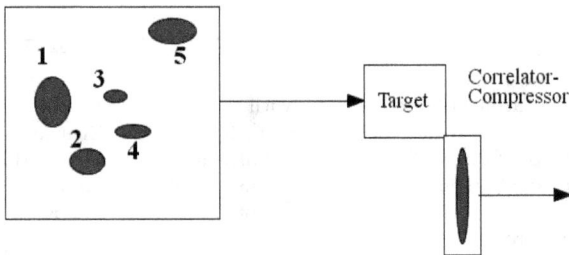

Fig. 5   In the *A Priori* Capture (Fig. 1), the target linear distribution operator, L, acts as a frequency dispersion matrix to the *PTX* according to: $\delta = L(t, f; s) * G(t, f; s)$. In the *A Posteriori* Capture (Fig. 2), the target linear distribution operator, L, acts as a correlator-compressor to this different input, according to $G^{-1}(t, f; s) = L(t, f; s) * \delta(t, f; s)$. Here, $PTX = \delta$; $PRX = G(t, f; s)$; $MTX = G^{-1}(t, f; s)$; $MRX = G(t, f; s)G^{-1}(t, f; s) = \delta(t, f; s)$, where, in all cases, the δ function is loosely defined.

of *MRX* arrival at the receiver in the case of an extended target, and for 2 different target aspect angles: $00°$ (Head-on — Fig. 7A) and $90°$ (Side-on — Fig. 7B). In Fig. 7, the 3 model *MTX*s are shown in the left columns — top to bottom — and the *MRX*s — autocorrelations/selfconvolutions of the *MTX*s — in the right columns.

Time-frequency spectra for the *MRX*s (right columns) of Fig. 7 are shown in Fig. 8. Figure 8A shows the plots for the case of target aspect angle $00°$ (Head-on) and Fig. 8B for the case of target aspect angle $90°$ (Side-on). In Fig. 8A it can be seen that the time-frequency spectrum for *MRX* B-00 in which the four minor resonances are all at $f = 75\,\text{Hz}$ is

1.3 GHz 99% BW Envelope on Miscellaneous Carriers

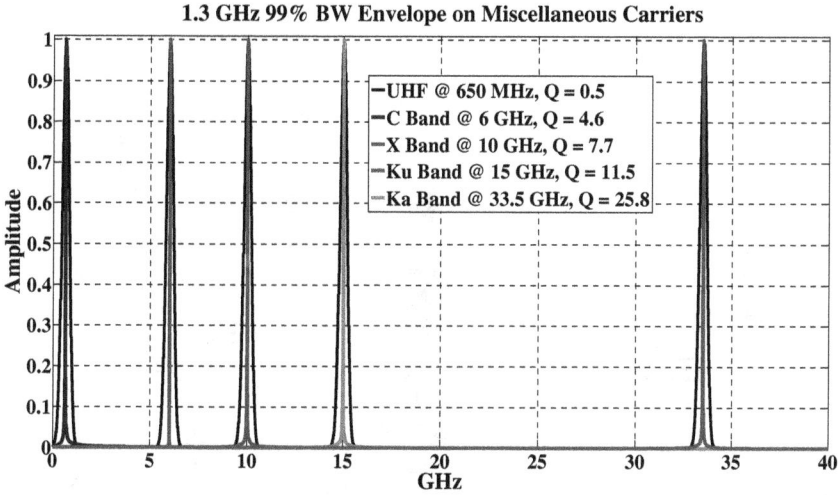

Legend:
- −UHF @ 650 MHz, Q = 0.5
- −C Band @ 6 GHz, Q = 4.6
- −X Band @ 10 GHz, Q = 7.7
- −Ku Band @ 15 GHz, Q = 11.5
- −Ka Band @ 33.5 GHz, Q = 25.8

Fig. 6 Target scattering can be a function of transmitted pulse envelope modulation and independent of the carrier. Here are shown 5 possible carriers for the same target TX/RX resonance bandwidth. If the channel medium between platform and target permits transmission without unacceptable penalties, the MTX can be matched to the target with envelope modulation, rather than carrier modulation. This has the important consequence that as the TX envelope matching is carrier independent, the size of the antenna can be small if the carrier chosen is of high frequency. In the example shown here, 5 possible antennas can be used, for 5 different carriers for matching the same TX envelope modulation to the same target (1.3 GHz, 99% bandwidth), with the difference in size between the Ka band and the UHF band antenna being of orders of magnitude. Of most importance, the required Q (center (carrier) frequency/bandwidth) increases as the center frequency increases: $Q = 0.5$ (UHF), 4.6 (C band), 7.7 (X band), 11.5 (Ku band), and 25.8 (Ka band).

spread in time. There is also time spread for *MRX* C-00 ($f_1 = 60$; $f_2 = 60$; $f_3 = 30$; $f_4 = 30$ Hz) and minimum spread for MRX A-00. In Fig. 8B it can be seen that all *MRX* minor resonances arrive at the receiver at approximately the same time and thus the time-frequency spectra show minimum spread in time ($f_1 = 100$; $f_2 = 75$; $f_3 = 50$; $f_4 = 25$). We see, therefore, that in the time domain the *MRX*s show aspect dependence in the uniqueness of the spread in time of the individual packets constituting the total response of the target. We can also remark on the symmetrical nature of all MRX returns, and a typical MRX obtained in empirical testing is shown in Fig. 8C, agreeing with the model approach.

When the frequency marginals are calculated for Fig. 8, as shown in Fig. 9, target aspect independence is indicated. These model signals are illustrative. It should be cautioned that at different aspect angles the minor

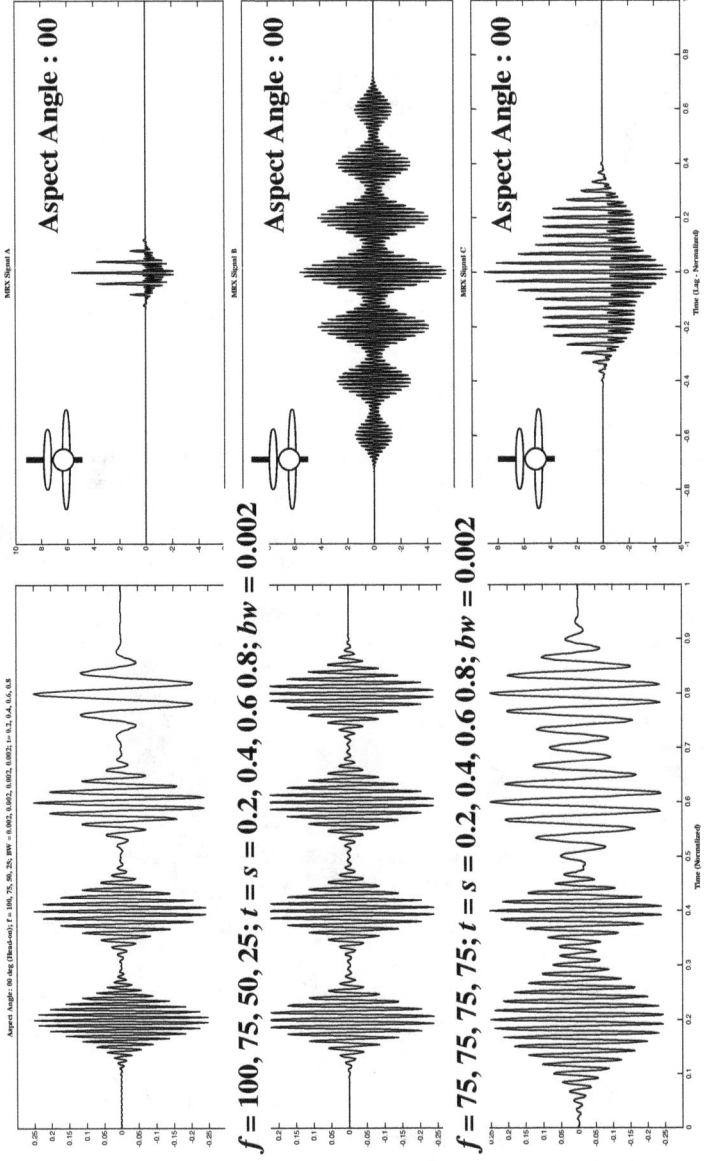

Fig. 7A (Continued)

resonance features of a target may be partially or completely hidden, in which case the amplitude of the target spectral bands will change or even be zero. Yet while the amplitude of a spectral band may vary with aspect, the resonance frequencies will not. With this proviso, we retrun to the scenario of the MTX-MRX spectra shown in Fig. 4, and the dynamic of MTX design based on the time reversal of a PRX providing a MRX that has a target-specific spectrum permitting the interpretation of the empirical results to be shown in succeeding chapters.

As indicated above, although in the frequency domain there is target aspect independent information in the MRX signal, in the time domain there is extended target aspect dependent information in the differences in time of arrival of the sequences of packets in the MTX. However, if information concerning target orientation is not required, then DTX and STX signals can be used of any temporal length.

In the case of DTXs and STXs information concerning target orientation is discarded and target orientation is considered of no consequence. The discarded information is in the time variance of the target's frequency response with the target considered as a linear time variant (LTV) system and we now discuss further the distinction between an LTV system, and a linear time invariant (LTI) system.

←——

Fig. 7A   With reference to the TX/RX and target situation shown in Fig. 2, here the receiver-target aspect angle is at "Target Head-on" ($00°$) aspect angle. There are 3 hypothetical MTX signals shown on the right (top to bottom) that are matched to 3 targets. Each target has 4 subcomponent resonances. Each subcomponent is spread out along the target so that due to the "Target Head-on" aspect angle, the 4 returned MRX minor resonance signals from the target arrive back at the receiver at different times — Left 3 figures. The MTXs convolve with the targets providing the MRXs shown in the Right 3 figures. The details of the normalized frequencies of the minor resonances (f's), normalized bandwidths (bw's) and normalized time of arrival at the receiver differences (Δt's) for the 3 hypothetical MRXs are:

$(A - 00)$ $f_1 = 100; f_2 = 75; f_3 = 50; f_4 = 25;$
$bw_1 = 0.002; bw_2 = 0.002; bw_3 = 0.002; bw_4 = 0.002;$
$\Delta t_1 = 0.2; \Delta t_2 = 0.4; \Delta t_3 = 0.6; \Delta t_4 = 0.8;$
$(B - 00)$ $f_1 = 75; f_2 = 75; f_3 = 75; f_4 = 75;$
$bw_1 = 0.002; bw_2 = 0.002; bw_3 = 0.002; bw_4 = 0.002;$
$\Delta t_1 = 0.2; \Delta t_2 = 0.4; \Delta t_3 = 0.6; \Delta t_4 = 0.8;$
$(C - 00)$ $f_1 = 60; f_2 = 60; f_3 = 30; f_4 = 30;$
$bw_1 = 0.008; bw_2 = 0.002; bw_3 = 0.006; bw_4 = 0.005;$
$\Delta t_1 = 0.2; \Delta t_2 = 0.4; \Delta t_3 = 0.6; \Delta t_4 = 0.8;$

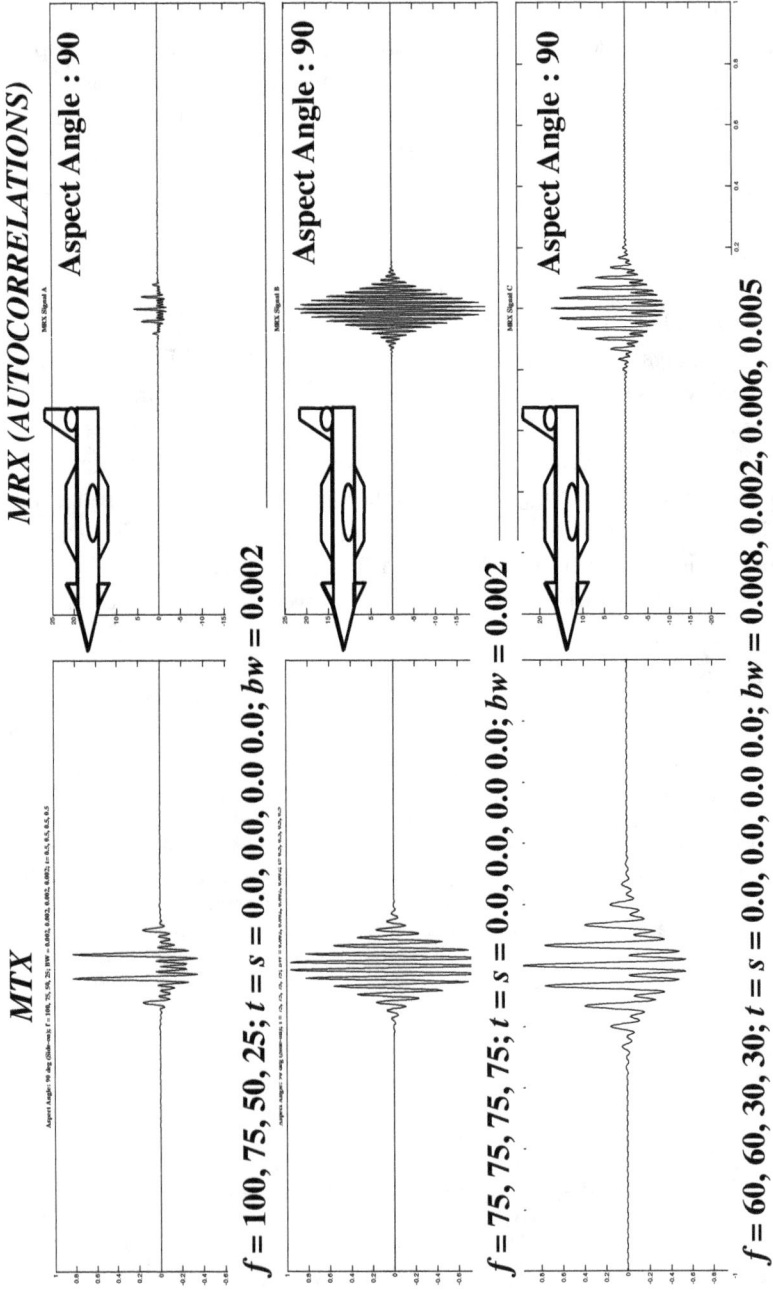

MRX (AUTOCORRELATIONS)

Aspect Angle : 90

Aspect Angle : 90

Aspect Angle : 90

MTX

$f = 100, 75, 50, 25; t = s = 0.0, 0.0, 0.0, 0.0\ 0.0; bw = 0.002$

$f = 75, 75, 75, 75; t = s = 0.0, 0.0, 0.0, 0.0\ 0.0; bw = 0.002$

$f = 60, 60, 30, 30; t = s = 0.0, 0.0, 0.0, 0.0\ 0.0; bw = 0.008, 0.002, 0.006, 0.005$

Fig. 7B   (*Continued*)

2.   LTI versus LTV Systems

If we take a difference equation approach to a system's filtering, then an i'th LTI system, is described by:

$$y_i(n) = \alpha_0 x(n) + \alpha_1 x(n-1) + \alpha_2 x(n-2) + \cdots$$

where each $\alpha_i x(n)$ refers to an RX wavepacket in the present target metaphor and where the coefficients α's can be real or complex, but are *shift invariant*. Shift invariance translates in the present target metaphor, to *time of arrival invariance* — which we do not see in empirical testing.

In contrast, a difference equation for an i'th LTV system is described by:

$$u_i(n) = \beta_0(n)x(n) + \beta_1(n-1)x(n-1) + \beta_2(n-2)x(n-2) + \cdots$$

where the β coefficients are *shift variant*, which translates in the present target model, to aspect angle variance and *time variance* — which we in fact see in empirical testing. A mixed LTV-LTI system will then be:

$$z(n) = \beta_0(n)y_0(n) + \beta_1(n-1)y_1(n-1) + \beta_2(n-2)y_2(n-2) + \cdots$$

This describes targets in which target minor features are y_i LTI systems, but the total composite target is a z LTV system. There is also the possibility

←───

Fig. 7B   Here the receiver-target aspect angle is at "Target Side-on" (90°) aspect. As in Fig. 7A, 3 hypothetical MTX signals are shown on the right (top to bottom) that are matched to the same three 3 targets but at the "Target Side-on" aspect. Each target has the same 4 subcomponent resonances and each subcomponent is again spread out along the target but now due to the "Target Side-on" aspect angle, the 4 returned MRX minor resonance signals from the target arrive back at the receiver at the *same* time — Left 3 figures. The MTXs convolve with the targets providing the MRXs shown in the Right 3 figures. The details of the normalized frequencies of the minor resonances (f's) — same as for Fig. 7A, normalized bandwidths (bw's) — same as for Fig. 7A and normalized time of arrival at the receiver differences (Δt's) — *different* from Fig. 7A — for the 3 hypothetical MRXs are:

$(A-90)$ $f_1 = 100; f_2 = 75; f_3 = 50; f_4 = 25;$
$bw_1 = 0.002; bw_2 = 0.002; bw_3 = 0.002; bw_4 = 0.002;$
$\Delta t_1 = 0.5; \Delta t_2 = 0.5; \Delta t_3 = 0.5; \Delta t_4 = 0.5;$

$(B-90)$ $f_1 = 75; f_2 = 75; f_3 = 75; f_4 = 75;$
$bw_1 = 0.002; bw_2 = 0.002; bw_3 = 0.002; bw_4 = 0.002;$
$\Delta t_1 = 0.5; \Delta t_2 = 0.5; \Delta t_3 = 0.5; \Delta t_4 = 0.5;$

$(C-90)$ $f_1 = 60; f_2 = 60; f_3 = 30; f_4 = 30;$
$bw_1 = 0.008; bw_2 = 0.002; bw_3 = 0.006; bw_4 = 0.005;$
$\Delta t_1 = 0.5; \Delta t_2 - 0.5; \Delta t_3 = 0.5; \Delta t_4 = 0.5;$

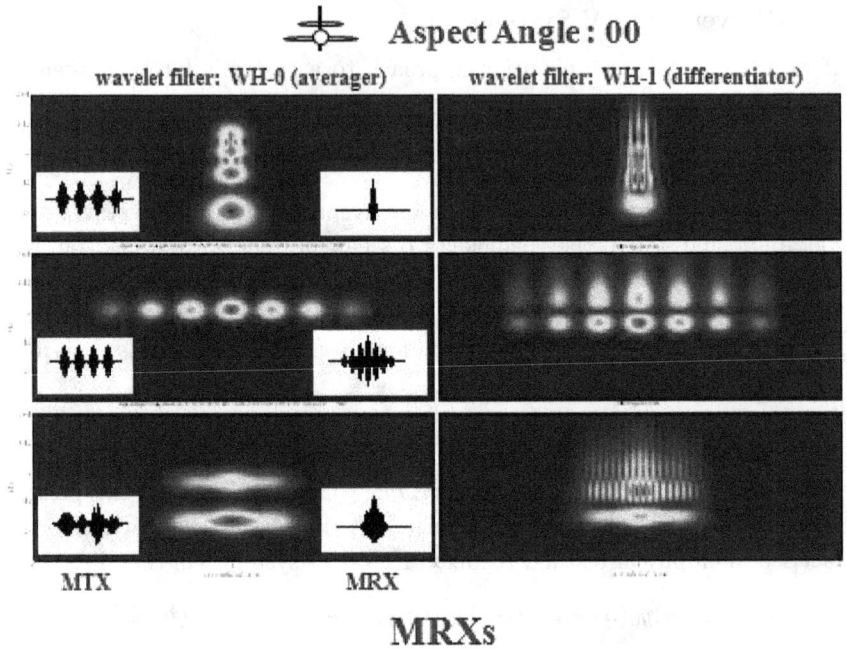

Fig. 8A   Time-frequency spectra for target aspect angle 00° (Head-on) *MRX*s: A-00, B-00, C-00. Left column: using WH-0 wavelet scaling function (averaging) filter set. Right column: using WH-1 wavelet (differentiating) filter set. Noticeably, the time-frequency spectrum for *MRX* B-00 in which the four minor resonances are all at $f = 75\,\text{Hz}$ is spread in time. There is also time spread for *MRX* C-00 and minimum spread for *MRX* A-00.

of further mixtures, e.g.:

$$z_T(n) = \beta_0(n)u_0(n) + \beta_1(n-1)y_1(n-1) + \beta_2(n-2)z_2(n-2) + \cdots$$

where each component may be LTI *or* LTV and with subcomponents which also may be LTI *or* LTV. Because in the present instance any variance is due to changes in transceiver-target spatial orientation (aspect angle) that determine the relative time of arrival of RX packets at the receiver, if the aspect angle is held constant, or eliminated from consideration (by use of *DTX*s or *STX*s), these LTV or mixed LTV-LTI system will reduce to a conventional LTI system.

     Now, the differences in MRX time of subcomponent signal arrivals shown in Figs. 7A & B and 8A, B & C for the model MTX signals A-00, B-00, C-00, A-90, B-90 and C-90, reflect targets that are LTV due to these differences in arrival time being a function of target aspect angle.

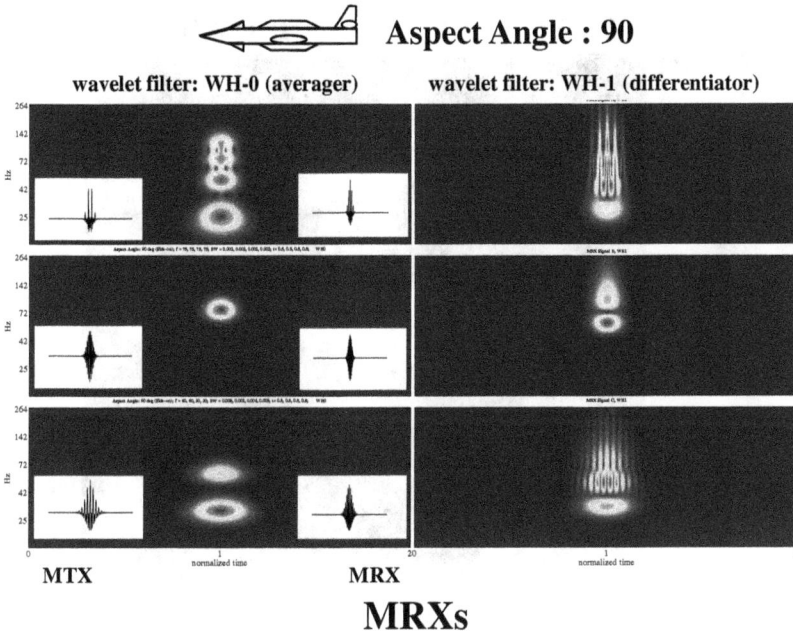

Aspect Angle : 90

wavelet filter: WH-0 (averager)   **wavelet filter: WH-1 (differentiator)**

MTX   MRX

Mrxs

Fig. 8B   Time-frequency spectra for target aspect angle 90° (Side-on). *MRX*s: A-90, B-90, C-90. Left column: using WH-0 wavelet scaling function (averaging) filter set. Right column: using WH-1 wavelet (differentiating) filter set. As to be expected, with target aspect angle 90° all MRX minor resonances arrive at the receiver at approximately the same time and thus the time-frequency spectra show minimum spread in time.

Equally, the differences in arrival time are a function of the spatial distance of each resonating target subcomponent from the receiver. Changes in relative spatial transceiver-target orientation produce changes in relative time of minor signal arrival intervals, i.e., in the β's, but not the α's. These differences are evident in the time domain, but not evident in the frequency domain. Thus they are important when detection of target aspect angle is required and the radar is in an MTX-MRX short pulse mode. However, when knowledge of target aspect angle is of no importance and the radar is in a DTX-DRX or STX-SRX mode, the target can be treated as a conventional LTI system.

However, in the MTX-MRX short pulse mode the distinction between an LTV and an LTI system has consequences. A variable network is defined as one in which one or more element-values are dependent in a specified way upon a combination of three variables: time, input and output. Whereas in LTI systems a transfer or system function is defined as the Fourier transform

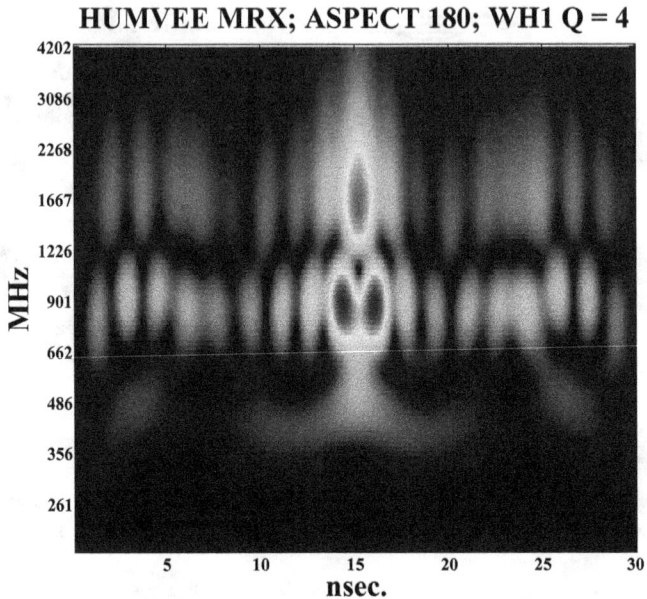

Fig. 8C   Time-Frequency spectrum of an MRX — Target: Humvee, Aspect Angle: 180°, Filter: WH1. This spectrum clearly indicates the typical symmetrical spread of MRXs.

of the response to a unit impulse and hence is a function of frequency and phase but independent of time, in the case of LTV systems a transfer or system function is defined as a function of frequency and phase with time as a parameter (Zadeh, 1950a, b). In the limiting case of time independence (which, in the present instance is aspect angle independence), an LTV transfer function reduces to that of a LTI transfer function and, with a specific time defined, a LTV transfer function is an instantaneous transfer function. We turn now to the relation of matched filtering and LTV systems.

With a single channel time series, a matched filter is conventionally derived to maximize the output signal-to-noise ratio (SNR) at a specified time (Trees, 1968; Papoulis, 1984). In the case of beam-forming or multichannel signal processing, maximizing the SNR results in the minimum mean-square error beam former (Trees, 2002). In a further development, Chambers *et al.* (2004) showed that for multiple radiators identified as a complex Green's function and with a multiple set of excitations identified as a filter set, an SNR is defined as a function of that Green's function, and the matched filter is the integral over time of the product of the Green's function and its complex conjugate. Moreover, the complex Green's function obtained by transmitting a pulse at time t_0 and time reversing the result

Wavelet filter: WH-0 (averager)

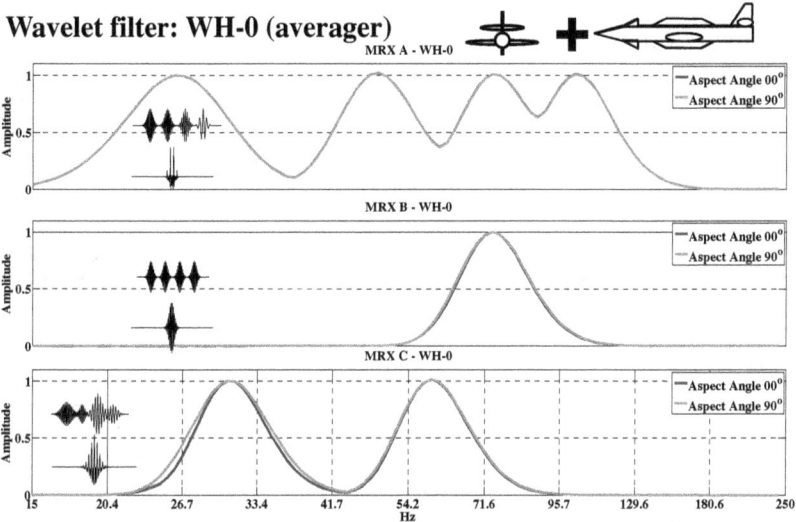

MRXs Overlaid Frequency Marginals: Aspect Angles : 00 & 90

Fig. 9A   Overlap of the marginal spectra for the left columns (using averager scaling function WH-0) of the time-frequency plots of Fig. 7A (target aspect angle 00°) and of Fig. 7B (target aspect angle 90°) indicating MRX aspect independence in the frequency domain.

(i.e., to obtain $MTX = PRX^{-1}$) is equal to the complex conjugate of the complex Green's function at a later time (Kuperman *et al.*, 1998). In other words, in the time domain time, time reversal and matched filtering are equivalent for LTV systems. In the frequency domain, time reversal is phase conjugation and there is also a similar equivalence to matched filtering.

It is well known that LTV matched filters are used in the detection of signals in dispersive media and adaptive equalizers are used in wireless communications to cancel channel characteristics that vary over time. There are also many techniques for dealing with the response of LTI systems. However, these LTI techniques are not strictly valid for LTV systems. In contrast, the discrete wavelet transform is valid for LTV systems, because it is naturally time variant due to the use of the decimation operation, which permits the local (as opposed to global) time-frequency analysis of signals and systems. With this local wavelet analysis approach an input signal is expanded into a linear combination of orthogonal basis signals, and an LTV filter is constructed by replacing each basis signal with a new basis signal. Other methods involve the matching pursuit algorithm (Moll & Fritzen, 2010), or a time-varying autoregressive process, to construct an LTV filter. In the empirically testing examined later, the target's impulse

Wavelet filter: WH-1 (differentiator)

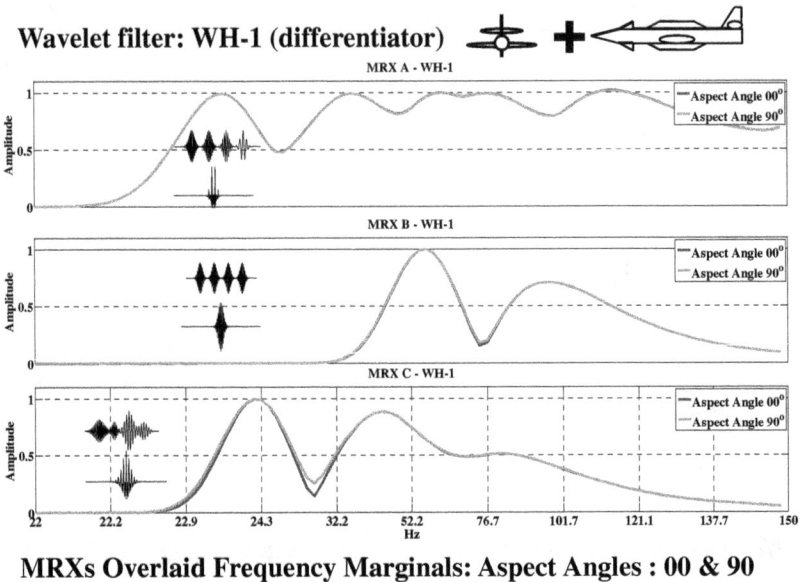

MRXs Overlaid Frequency Marginals: Aspect Angles : 00 & 90

Fig. 9B   Overlap of the marginal spectra for the right columns (using differentiating wavelet WH-1) of the time-frequency plots of Fig. 7A (target aspect angle 00°) and of Fig. 7B (target aspect angle 90°) indicating, again, MRX aspect independence in the frequency domain.

response (PRX) at a known aspect angle is obtained empirically, and the matched signal (MTX) is constructed from the time reversal of that impulse response.

Just as the concept of LTI transfer functions can be extended to LTV, a Green's function and a time-reversal mirror concept can be extended from time invariance to time variance (Chambers *et al.*, 2004), producing an optimal matched filter (Tanter *et al.*, 2000). The steps demonstrating this extension (from LTI to LTV systems) are as follows, and we generalize them here to MAP radar systems.

Fink and Dorme showed that time reversal of fields in an array added coherently at the focus of the array (Fink, 1992; Dorme & Fink, 1995). If $G(p_1, t_1; p_2, t_2)$ is the Green's function describing the target impulse response with the transmitter at position p_1, the target at position, p_2, signal transmission to target at time t_1 and signal reception from target at time t_2, then time invariance implies $G(p_1, t_1; p_2, t_2) = G(p_1, t_1 - \tau; p_2, t_2 - \tau)$ and reciprocity implies $G(p_1, t_1; p_2, t_2)^{-1} = G(p_2, t_2; p_1, t_1)$ — or, in the present instance, $\text{PRX}^{-1} = \text{MRX}$.

The general observation is that the target response is represented by:

$$\boldsymbol{RX}(t) = \sum_{n=1}^{N} \int_0^t h_n(t - t') \, \boldsymbol{TX}_n(t') dt',$$

where $\{\boldsymbol{TX}_n(t) : n = 1, 2, \ldots, N\}$ is the set of input signals (e.g., subcomponents of PTX, MTX), and $\{h_n(t) : n = 1, 2, \ldots, N\}$ is the set of target filters producing n minor and major target resonances.

The MRX time response is then:

$$MRX = G * G^{-1}$$

or

$$MRX = PRX * PRX^{-1} = PRX * MTX$$

where: $G = h_n(t - t') = PRX$ and $G^{-1} = PRX^{-1} = MTX$ as in Figs. 1 and 2, above);

Setting constant the inter-packet time of arrival influences (due to aspect angle influences), if the time reversed signal $PRX^{-1} = MTX$ is associated with the complex Green's function $G(\boldsymbol{TX}_n, t; \boldsymbol{RX}_n, t')$ where $\boldsymbol{RX}_n$ specifies the return signal from a local resonator due to a target subcomponent, n, and $\boldsymbol{TX}_n$ specifies a transmitted signal matched to a local resonance n, in the case of the spatio-temporal matched filter for an LTI system the SNR is maximum when:

$$SNR(\boldsymbol{TX}_s t_s)$$

$$= \frac{\left(\sum_{n=1}^{N} \int_0^{t_s} |G(\boldsymbol{TX}_s, t_s; \boldsymbol{RX}_n, t)|^2 dt\right)\left(\sum_{n=1}^{N} \int_0^{t_s} |G^*(\boldsymbol{TX}_s, t_s; \boldsymbol{RX}_n, t)|^2 dt\right)}{\sum_{n=1}^{N} \int_0^{t_s} |G^*(\boldsymbol{TX}_s, t_s; \boldsymbol{RX}_n, t)|^2 dt}$$

$$= \sum_{n=1}^{N} \int_0^{t_s} |G(\boldsymbol{TX}_s, t_s; \boldsymbol{RX}_n, t)|^2 dt$$

which shows that the SNR is the squared ratio of the energy of a specific transmitted signal $\boldsymbol{TX}_s$ at a specific time t_s to the energy in the total received signal $\boldsymbol{RX}$.

For an LTI system, the MRX or return signal from a target in response to a matched transmitted signal, MTX, is:

$$\boldsymbol{RX}_{MF}(t) = \sum_{n=1}^{N} \int_0^t G(\boldsymbol{TX}_s, t; \boldsymbol{RX}_n, t') G^*(\boldsymbol{TX}_s, t_s; \boldsymbol{RX}_n, t') dt'$$

In the case of a time reversal mirror and an LTV system, and for the case of (i) an impulse being transmitted at time $t = 0$, (ii) the target response being received, time reversed, and (iii) retransmitted, the returned signal, MRX, is:

$$RX_t = \sum_{n=1}^{N} \int_0^t G(TX_s, t; RX_n, t')$$
$$\times\, G(RX_n, t_s - t'; TX_s, 0)dt', \quad 0 < t < t_s.$$

Moreover, if the Green's function satisfies:

$$G(RX_n, t_s - t; TX_s, 0) = G^*(TX_s, t_s; RX_n, t,), \quad 0 < t < t_s,$$

i.e., if the time reversed response of a minor resonance to a specific transmitted signal at a specific time (aspect angle) is equal to the complex conjugate of the impulse response of that minor resonance for that specific time (aspect angle), then under these conditions $RX_t = RX_{MF}(t) = MRX$ and the LTV response is also a matched filter response.

Thus the difference between an LTV and LTI system is in the constancy of specific TX times, t_s (of LTI systems) relating to the minor resonances and that are a function of the target aspect angle or relative times of arrival of extended target wave packets. Elimination of t_s by constancy of time, i.e., constancy of aspect angle, reduces total target LTV condition to that of a minor resonance featured LTI system. Furthermore,

(1) It is well known that the SNR is maximized for an RX signal, when the impulse response of the optimum filter (i.e., in this instance, the target) is a reversed and a time delayed copy of the TX signal. But more can be said due to the self-convolution operation being commutative. In terms of, e.g., a matrix representation, the TX signal and the filter (represented by the target) can switch places. Therefore,

(2) With the switch made, we can state that the SNR is maximized for an RX signal, when the TX signal (as opposed to the filter) is a reversed and a time delayed copy of the impulse response of the filter (as opposed to the TX signal).

This book addresses the condition (2).

3. Signal Envelope Match vs Carrier Match

A major observation in empirical testing described later is that the matching of a TX pulse to the target's transfer characteristics need not be a match of the carrier of the signal to the characteristics, but can also be a match of only the amplitude modulation of a carrier — provided that the channel medium through which the transmit signal passes is transparent to the carrier (Fig. 6). The restriction to carrier transparent channel conditions in this instance is necessary because of the need of, e.g., foliage penetrating radars, which require that the carrier penetrate foliage, and ground/wall penetrating radars, which require that the carrier penetrate ground/walls. Both these cases require *both* carrier and envelope to be at medium-penetrating frequencies. But even in these cases, if a medium-penetrating carrier is chosen, the transmitted signal match to a designated target-and-medium need only be required of the carrier envelope, and the carrier need only be medium-penetrating.

This circumstance has major system benefits in that the size of the transmit-receive antenna is determined by the carrier and less by its bandwidth set by the carrier's envelope modulation, as antenna size scales with wavelength. Thus, absent medium-penetration requirements, the transmit-receive antenna can be at a high frequency and of smaller size — as it is for the Ka-band prototype results reported here — while the target-matched envelope modulation of the carrier can be at much lower frequencies, as opposed to transmittance of those same envelope lower frequencies without a carrier, this latter configuration requiring a large antenna. The advantages of this circumstance for antenna size reduction are therefore considerable (see Fig. 6).

4. Target Modeling and Identification by Coherence Functions

Addressing resonance (Mie) scattering with MAP captures provides target aspect-independent information in the frequency domain, and target aspect-dependent information in the time domain. A system designed to capture and exploit this information is a resonance- and aspect-matched adaptive radar (RAMAR). A RAMAR addresses target/subcomponent resonances and therefore offers identification of targets by RX resonance spectral profiles. Taking resonance spectral profile identification farther: target identification can be based on the degree of coherence-match to a target's transfer function treated as LTV, and there are multiple methods that model a linear target transfer function. One such method is coherence function profiling, which we describe here.

Coherence is a measure of the consistency of phase relations between two time series (Marple, 1987). For purposes of exposition, a model target

Magnitude Response (dB)

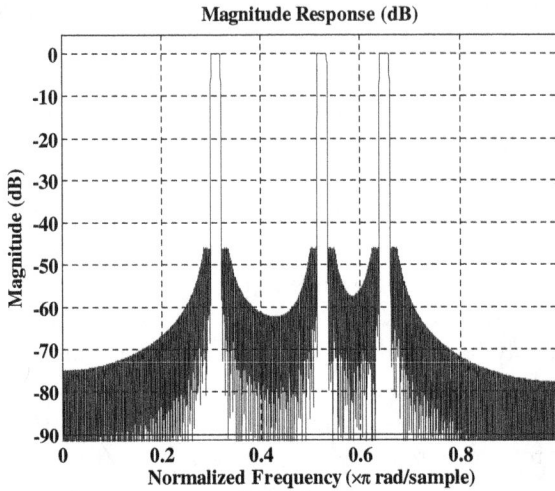

Fig. 10A   The frequency responses of a synthetic target (composed of FIR filters), with resonances (passbands) at $f_c = 310$, 525 & 650 MHz and BW = 20 MHz for each band. $F_s = 2$ GHz.

can be created that has, e.g., three resonances — at 310, 525 & 650 MHz — and whose filter frequency response and impulse response autocorrelation are shown in Figs. 10A & B. Suppose now that DTX signals (i.e., single frequency or constant wavelength (cw) signals) compose the following two STX bundles:

1. An "on-resonance" STX composed of DTX frequencies at the target resonances 310, 525 & 650 MHz (Fig. 11A).
2. An "off-resonance" STX composed of DTX frequencies 290, 545 & 630 MHz (Fig. 11B).

Then the SRXs elicited by either of these STX long duration bundles of DTX signals are obtained by convolving the STXs with the target's transfer function (Fig. 12). The DTXs, together with the SRXs provide the "on-resonance" and "off-resonance" cases, represented by matrices, both composed of 3 DTXs (i.e., three DTX frequencies = the STX bundle) and 1 SRX. These 2 matrices are each composed of 3 system inputs and 1 output.

Coherence functions are defined by (Thomson, 1982; Xu *et al.*, 1999):

$$\gamma_{xy}^2(f) = \frac{\left| \sum_{k=0}^{K-1} x_k(f) y_k^*(f) \right|^2}{\left| \sum_{k=0}^{K-1} x_k(f) \right|^2 \left| \sum_{k=0}^{K-1} y_k(f) \right|^2},$$

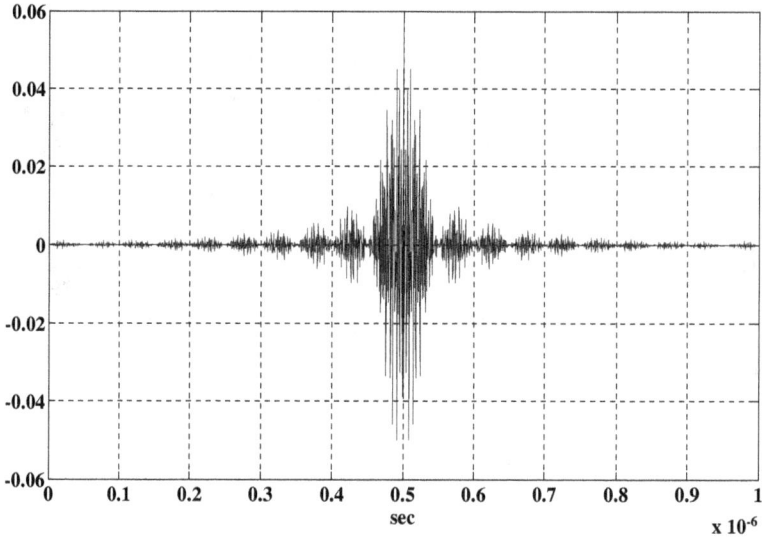

Fig. 10B The autocorrelation of the impulse response of the synthetic target, with resonances (passbands) at 310, 525 & 650 MHz and BW = 20 MHz for each band; F_s = 2 GHz.

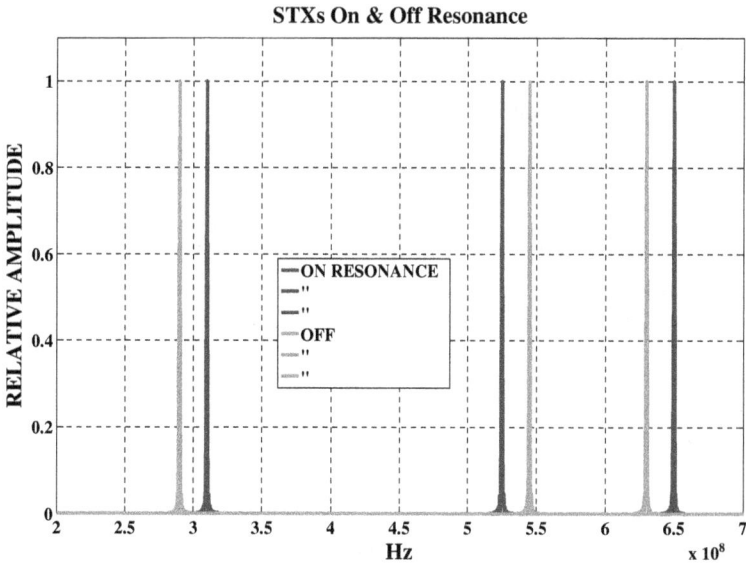

Fig. 11 The "On resonance" STX composed of DTXs at 310, 525 & 650 MHz and the "Off resonance" STX composed of DTXs at 290, 545 & 630 MHz. Both 2000 data points in duration and F_s = 2e9.

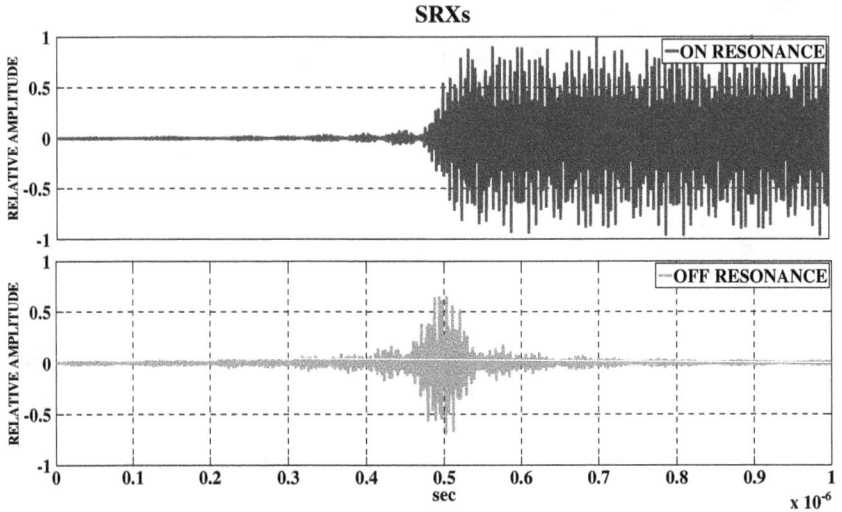

Fig. 12   The SRXs obtained by convolving the on-resonance and off-resonance STXs (Figs. 11A & 11B) with the target's frequency response (Fig. 10A).

where the x's and the y's are power spectra obtained by taking the Fourier transform of the autocorrelation, and k ranges from 1 to 4, the first 3 being the STX's and the 4th being the DRX.

The results of coherence function measurements for the on-resonance and off-resonance model cases are shown in Figs. 13–16. Examination of these figures shows that the coherence function method, in principle, is a promising approach to MAP target identification.

In Fig. 13 are shown coherence functions for SRX frequencies and target resonances, with target resonances at 310, 525 & 650 MHz; STX (on resonance) frequencies at 310, 525 & 650 MHz; and DTX (off resonance) frequencies at 290, 545 & 630 MHz. In (A) DTX$_1$ = 310 MHz (on

---→

Fig. 13   Coherence Functions for DRX frequencies and target resonances.
Target resonances: 310, 525 & 650 MHz;
DTX (on resonance) frequencies: 310, 525 & 650 MHz;
DTX (off resonance) frequencies: 290, 545 & 630 MHz.
In (A) DTX$_1$ = 310 MHz (on resonance), and In (B) DTX$_1$ = 290 MHz (off resonance).
Notice in (A) that γ_{12}^2 & γ_{13}^2 are minimum at all frequencies (lack of coherence), as expected, but γ_{14}^2 is high at 310 MHz (maximum coherence) and minimum at 525 & 650 MHz (lack of coherence), also as expected.
Notice in (B) that γ_{12}^2, γ_{13}^2 & γ_{14}^2 are all minimum (lack of coherence) and broad band, as expected.

Coherence Function: TX On Resonance: $f_1 = 310e6$; $f_2 = 525e6$; $f_3 = 650e6$

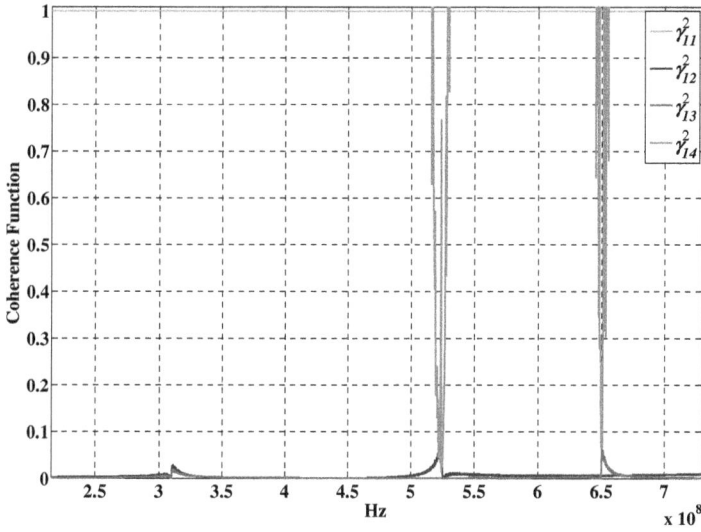

(A)

Coherence Function: TX Off Resonance: $f_1 = 290e6$; $f_2 = 545e6$; $f_3 = 630e6$

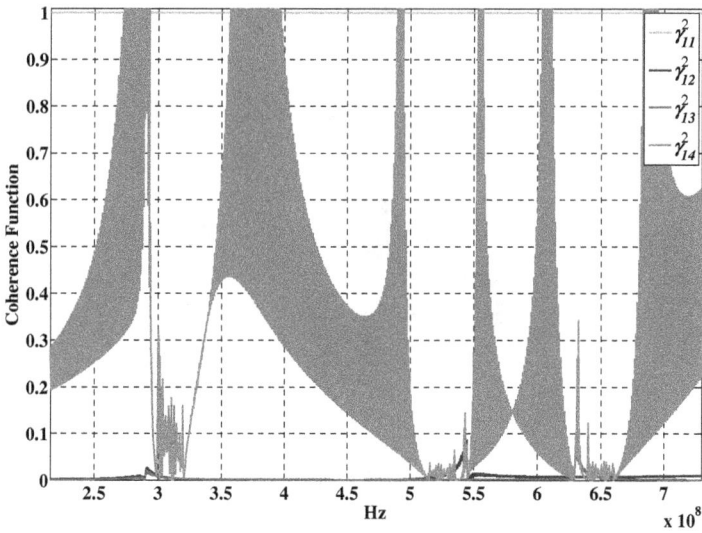

(B)

Fig. 13

Coherence Function: TX On Resonance: $f_1 = 310e6$; $f_2 = 525e6$; $f_3 = 650e6$

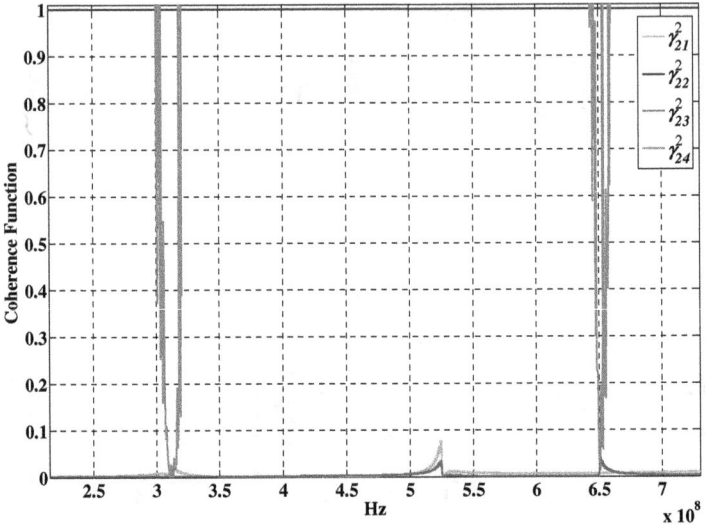

(A)

Coherence Function: TX Off Resonance: $f_1 = 290e6$; $f_2 = 545e6$; $f_3 = 630e6$

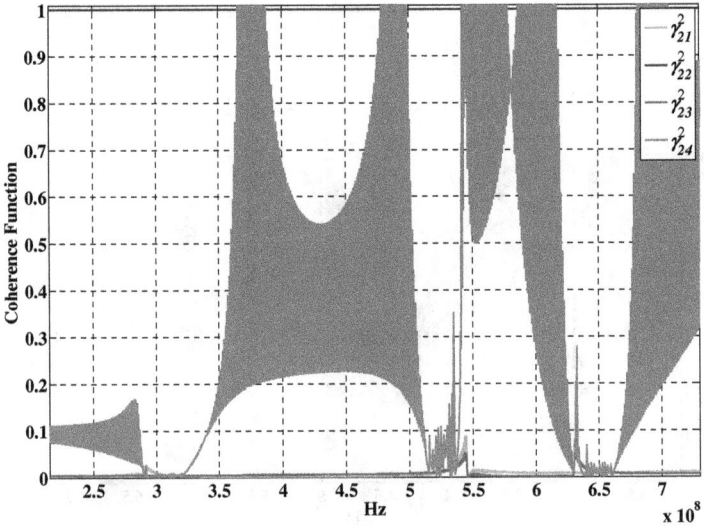

(B)

Fig. 14   (*Continued*)

resonance), and in (B) DTX$_1$ = 290 MHz (off resonance). It can be seen that in (A) that γ_{12}^2 & γ_{13}^2 are minimum at all frequencies (lack of coherence), as expected, but γ_{14}^2 is high at 310 MHz (maximum coherence) and minimum at 525 & 650 MHz (lack of coherence), also as expected. It can be seen in (B) that γ_{12}^2, γ_{13}^2 & γ_{14}^2 are all minimum (lack of coherence) and broad band, as expected.

In Fig. 14 (A) DTX$_2$ = 525 MHz (on resonance), and in (B) DTX$_2$ = 545 MHz (off resonance). It can be seen that in (A) that γ_{21}^2 & γ_{23}^2 are minimum at all frequencies (lack of coherence), as expected, but γ_{24}^2 is high at 525 MHz (maximum coherence) and minimum at 310 & 650 MHz (lack of coherence), also as expected. It can be seen in (B) that γ_{21}^2, γ_{23}^2 & γ_{24}^2 are all minimum (lack of coherence) and broad band, as expected.

In Fig. 15 (A) DTX$_3$ = 650 MHz (on resonance), and In (B) DTX$_3$ = 630 MHz (off resonance). It can be seen in (A) that γ_{31}^2 & γ_{32}^2 are minimum at all frequencies (lack of coherence), as expected, but γ_{34}^2 is high at 650 MHz (maximum coherence) and minimum at 310 & 525 MHz (lack of coherence), also as expected. It can be seen in (B) that γ_{31}^2, γ_{32}^2 & γ_{34}^2 are all minimum (lack of coherence) and broad band, as expected.

In Fig. 16 in the case of both (A) (on resonance), and (B) (off resonance), SRX = 310, 525 & 650 MHz. It can be seen in (A) that γ_{41}^2 is minimum at 525 & 650 MHz (lack of coherence), γ_{42}^2 is minimum at 310 & 650 MHz (lack of coherence), γ_{43}^2 is minimum at 310 & 525 MHz (lack of coherence), but γ_{44}^2 is at all frequencies (maximum coherence) as expected; It can be seen in (B) that γ_{41}^2, γ_{42}^2 & γ_{43}^2 are all minimum at 310, 525 & 650 MHz (lack of coherence), but γ_{44}^2 is at all frequencies (maximum coherence) as expected;

Given an LTV target transfer function, this exercise indicates that target identification can be based quite accurately on the degree of coherence-match.

Fig. 14   Coherence Functions for SRX frequencies and target resonances.
Target resonances: 310, 525 & 650 MHz;
DTX (on resonance) frequencies: 310, 525 & 650 MHz;
DTX (off resonance) frequencies: 290, 545 & 630 MHz.
In (A) DTX$_2$ = 525 MHz (on resonance), and In (B) DTX$_2$ = 545 MHz (off resonance). Notice in (A) that γ_{21}^2 & γ_{23}^2 are minimum at all frequencies (lack of coherence), as expected, but γ_{24}^2 is high at 525 MHz (maximum coherence) and minimum at 310 & 650 MHz (lack of coherence), also as expected. Notice in (B) that γ_{21}^2, γ_{23}^2 & γ_{24}^2 are all minimum (lack of coherence) and broad band, as expected.

Coherence Function: TX On Resonance: $f_1 = 310e6$; $f_2 = 525e6$; $f_3 = 650e6$

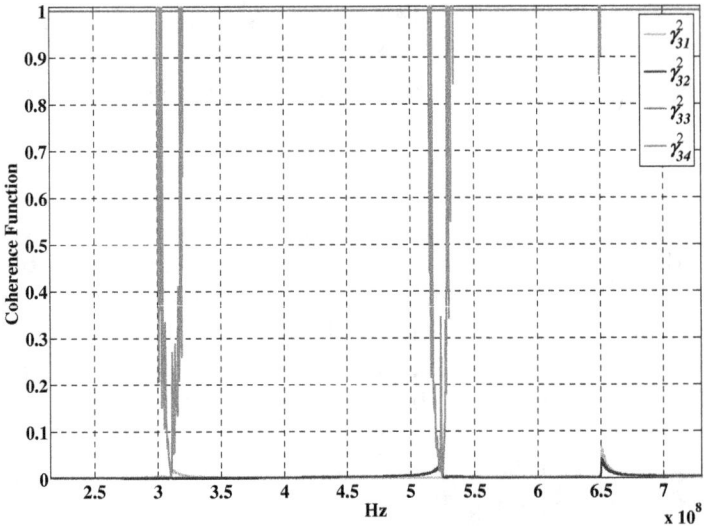

(A)

Coherence Function: TX Off Resonance: $f_1 = 290e6$; $f_2 = 545e6$; $f_3 = 630e6$

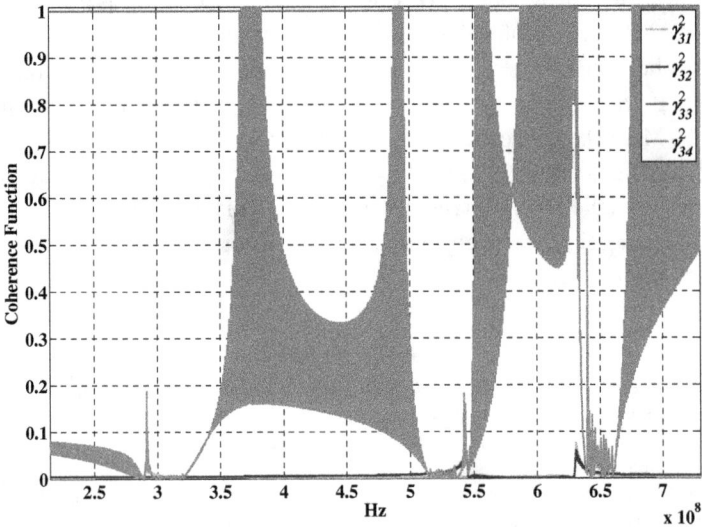

(B)

Fig. 15   (*Continued*)

5. The WH Transform & WHWFs

As the appropriate analysis method for both LTI and LTV system signals is time-frequency analysis, a local wavelet decomposition of RX signals was used to analyze test results. The specific wavelet analysis method used provides a solution to the well-known "energy confinement problem", or time-bandwidth product limiting problem, addressed by Slepian and collaborators (Landau & Pollak, 1961, 1962; Slepian, 1964, 1978; Slepian & Pollak, 1961). We provide a short introduction here, and an expanded version in the following chapters.

Slepian *et al.*'s solution to the energy confinement problem was the prolate spheroidal wavefunction (PSWF) series, that can be recursively, but not analytically, generated. An alternative solution is the parabolic cylinder or Weber-Hermite wave function (WHWF) series (Barrett, 1972, 1973a,b), that are analytic.

The WHWFs are related to *Weber's equation* (Weber, 1869):

$$\frac{d^2\psi_n(x)}{dx^2} + \left(n + \frac{1}{2} - \frac{1}{4}x^2\right)\psi_n(x) = 0,$$

for which the *general Weber equation*, or *parabolic cylinder differential equation*, is (Abramowitz & Stegun, 1972):

$$\frac{d^2\psi_n(x)}{dx^2} + (ax^2 + bx + c)\psi_n(x) = 0,$$

The solutions of this equation are the parabolic cylinder or Weber-Hermite wave functions (WHWFs):

$$\psi_n(x) = 2^{-n/2}\exp[-x^2/4]H_n(x/\sqrt{2}), \quad n = 0, 1, 2, \ldots,$$

where the H_n are Hermite polynomials. When n is an integer, the Weber-Hermite functions become proportional to the Hermite polynomials. Other

---

Fig. 15 Coherence Functions for SRX frequencies and target resonances.
Target resonances: 310, 525 & 650 MHz;
DTX (on resonance) frequencies: 310, 525 & 650 MHz;
DTX (off resonance) frequencies: 290, 545 & 630 MHz.
In (A) $DTX_3 = 650$ MHz (on resonance), and In (B) $DTX_3 = 630$ MHz (off resonance).
Notice in (A) that γ_{31}^2 & γ_{32}^2 are minimum at all frequencies (lack of coherence), as expected, but γ_{34}^2 is high at 650 MHz (maximum coherence) and minimum at 310 & 525 MHz (lack of coherence), also as expected.
Notice in (B) that γ_{31}^2, γ_{32}^2 & γ_{34}^2 are all minimum (lack of coherence) and broad band, as expected.

Coherence Function: TX On Resonance: $f_1 = 310e6$; $f_2 = 525e6$; $f_3 = 650e6$

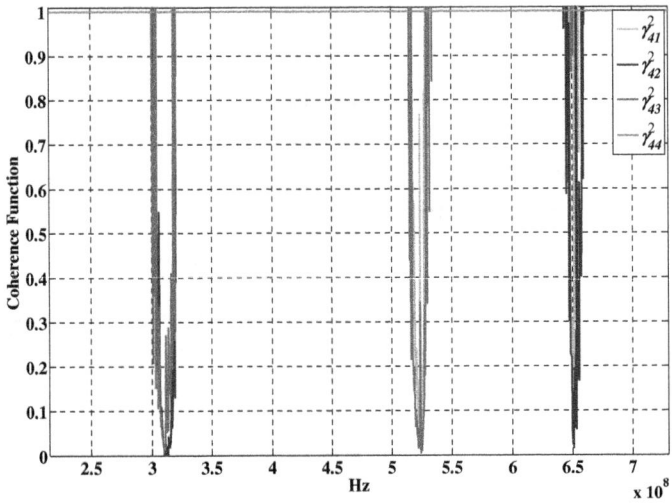

(A)

Coherence Function: TX Off Resonance: $f_1 = 290e6$; $f_2 = 545e6$; $f_3 = 630e6$

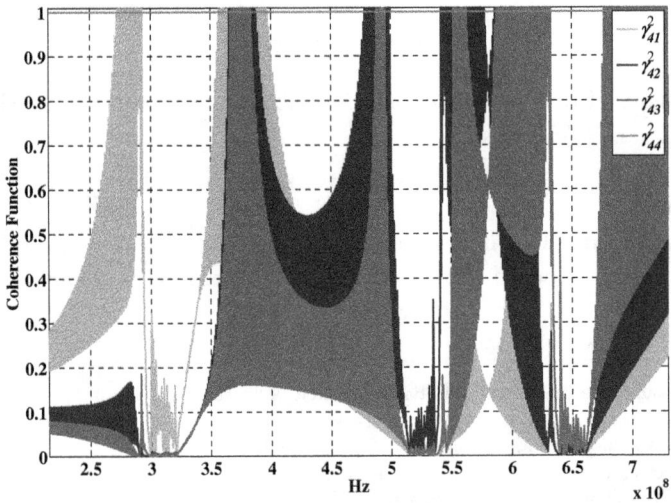

(B)

Fig. 16   (*Continued*)

names are used for the Weber-Hermite wave functions, e.g., Hermite-Gaussian functions. I prefer the designation Weber-Hermite wave function because (a) Hermite-Gaussian implicates Gaussian in all polynomials, $n = 0, 1, 2, \ldots$ but the Gaussian is only just the first, for the case $n = 0$; (b) Weber's equation is more general than Hermite's equation; (c) the name "Weber-Hermite" follows the *Mathematical Encyclopedia* (Hazewinkel, 2002), usage; and (d) other texts, e.g., (Morse & Feshback, 1953, vol. 2, p. 1642; Jones, 1964, p. 86), have also used the name "Weber-Hermite".

The WHWFs are given a physical representation as follows. The one-dimensional wave equation is:

$$-\frac{1}{2m}\frac{\partial^2\psi}{\partial x^2} + V(x)\psi = E\psi$$

with spring potential:

$$V(x) = \frac{1}{2}kx^2 = \frac{1}{2}m\omega^2 x^2,$$

where $\omega = \sqrt{\frac{k}{m}}$ is the angular frequency, k is the stiffness constant, m is the mass, and x is the field deflection of the oscillator. The wave equation can be written in dimensionless form by defining the independent variables $\xi = \alpha x$ and an eigenvalue, λ, and requiring:

$$\alpha^4 = mk, \quad \lambda = 2E\left(\frac{m}{k}\right)^{1/2} = \frac{2E}{\omega}.$$

The dimensionless form is then:

$$\frac{\partial^2\psi}{\partial\xi^2} + (\lambda - \xi^2)\psi = 0.$$

Fig. 16   Coherence Functions for SRX frequencies and target resonances.
Target resonances: 310, 525 & 650 MHz;
DTX (on resonance) frequencies: 310, 525 & 650 MHz;
DTX (off resonance) frequencies: 290, 545 & 630 MHz.
For both (A) (on resonance), and (B) (off resonance), SRX = 310, 525 & 650 MHz.
Notice in (A) that γ_{41}^2 is minimum at 525 & 650 MHz (lack of coherence),
γ_{42}^2 is minimum at 310 & 650 MHz (lack of coherence),
γ_{43}^2 is minimum at 310 & 525 MHz (lack of coherence),
but γ_{44}^2 is at all frequencies (maximum coherence) as expected;
Notice in (B) that γ_{41}^2, γ_{42}^2 & γ_{43}^2 are all minimum at 310, 525 & 650 MHz (lack of coherence),
but γ_{44}^2 is high at all frequencies (maximum coherence) as expected.

which is also a form of Weber's equation and permits solutions as a function of $n = \frac{E}{2\beta} - \frac{1}{2}$. In order for the solutions to be quadratically integrable, it is necessary that n take on integer values: $n = 0, 1, 2, \ldots$ (Morse & Feshbach, 1953, p. 1641). With normalization factors, the solutions are the Weber–Hermite or parabolic cylinder functions:

$$\psi_n(t) = \frac{1}{\sqrt{2^n n!}} \left(\frac{\alpha}{\pi} \right)^{1/4} \exp\left(-\frac{\alpha t^2}{2} \right) H_n(t\sqrt{\alpha}),$$

where $\alpha = m\omega$ and the H_n are Hermite polynomials, and α is a time-frequency trade parameter/variable.

If in the case of a function, $f(x)$, an expansion of the form:

$$f(x) = \alpha_0 \psi_0(x) + \alpha_1 \psi_1(x) + \cdots + \alpha_n \psi_n(x) + \cdots$$

exists, and if it is legitimate to integrate term-by-term between the limits $-\infty$ and $+\infty$, then:

$$a_n = \frac{1}{\sqrt{(2\pi)^{1/2} n!}} \int_{-\infty}^{+\infty} \psi_n(t) f(t) dt.$$

The first 6 WHWFs (WHWF 0-5) are shown in the time and frequency domains in Fig. 17. It is apparent that the WHWFS increase in (temporal

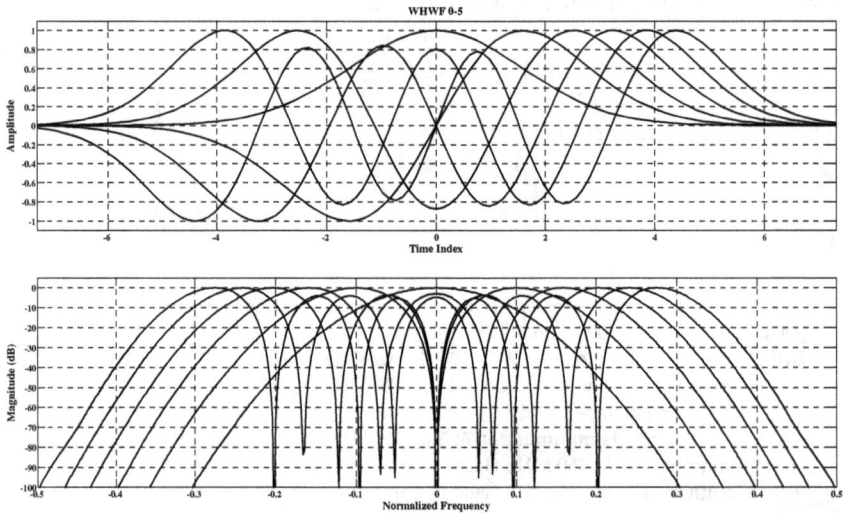

Fig. 17   The first 6 WHWFs (upper) and their Log Magnitude Spectra (lower) labeled according to an increasing n ($n = 0, 1, 2, 3, 4, 5$). As n increases, the temporal length and bandwidth increase, i.e., time-bandwidth product increases.

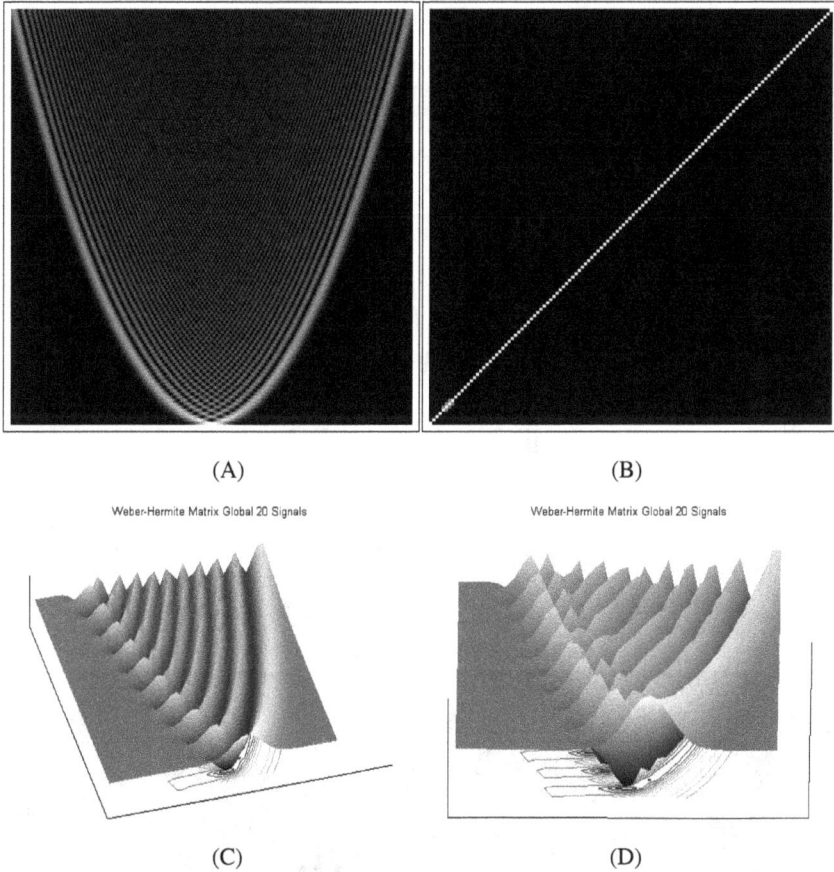

(A)   (B)

Weber-Hermite Matrix Global 20 Signals   Weber-Hermite Matrix Global 20 Signals

(C)   (D)

Fig. 18   A: a 128×256 WH (Global) matrix (magnitude). If the complex WH matrix is designated, W, then, as W is a unitary matrix, $WW\dagger = I$, where $W\dagger$ is the conjugate transpose (Hermitian adjoint) of W, and I is the identity matrix shown in B. The inverse of W, or W^{-1}, is equal to the conjugate transpose: $W^{-1} = W\dagger$. C and D are exploded views of the first 20 signals in A.

length $\times$ bandwidth), i.e., in time-bandwidth product, as n increases (Fig. 18).

In a manner similar to that of the Fourier transform, a WH transform can be constructed by matrix methods. Figure 18A shows a 128×256 WH (magnitude) matrix. If the complex WH matrix is designated, W, then, as W is a unitary matrix, $WW\dagger= I$, where $W\dagger$ is the conjugate transpose (Hermitian adjoint) of W, and I is the identity matrix shown in 18B. The inverse of W, or W^{-1}, is equal to the conjugate transpose: $W^{-1}=W\dagger$.

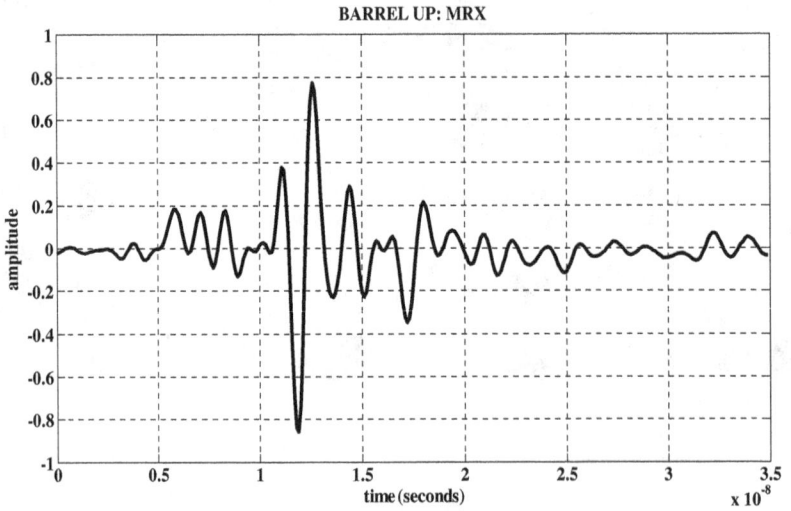

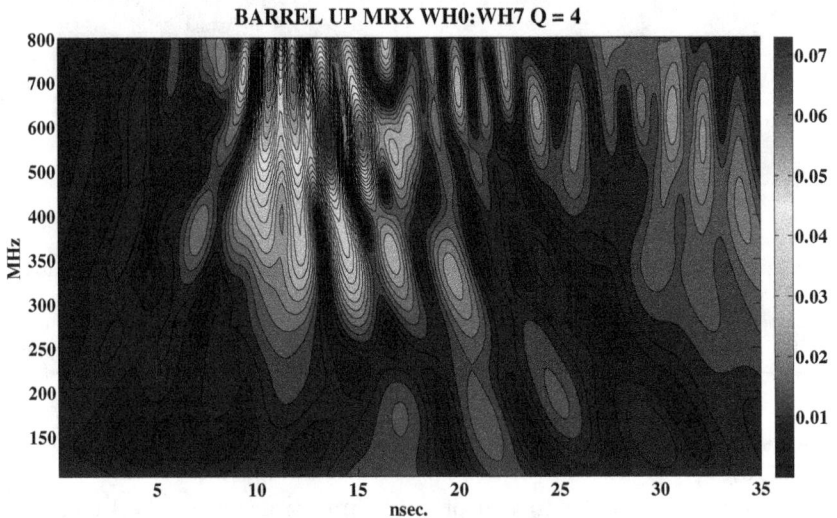

Fig. 19   A: a time domain averaged MRX from a target barrel in upright position.
B: multiple window time-frequency analysis (MWTFA) spectrum of the MRX using 8
WHWFs: WH0:WH7 to obtain 8 time-frequency spectra that were then averaged.

Using WHWFs it is possible to take a time domain signal, e.g., an MRX signal elicited from a barrel target (Fig. 19A), and calculate a multiple-window time-frequency analysis (MWTFA) spectrum (Fig. 19B). The MWTFA is a WHWF expression of Thomson's multiple window method (Thomson, 1982), that is a time-frequency distribution estimator for a random process and is a substitute for a Fourier transform periodogram — an inappropriate estimator for short, time-limited signals. Thomson's approach to spectral estimation of such signals is to compute several periodograms using a set of orthogonal windows that are locally concentrated in frequency and then averaged (Xu *et al.*, 1999). Optimal windows are the PSWFs, but WHWFs are used in the following analyses, being a good approximation, and whereas the WHWFs are analytic, the PSWFs are not.

6. Treatment of Nonstationary Signals

The term "non-stationary" is used in multiple senses. Here we use it in the sense shown in Fig. 20. In the case of short, time-limited signals, accurate averaging requires that those signals be aligned in time as shown on the left of this figure. Jitter destroys alignment. If similar signals are not aligned, as shown on the right, their "average" will not be a good estimator.

As observed, the terms "stationary" and "non-stationary" are used in multiple senses. In the statistics literature, all signals shown here are

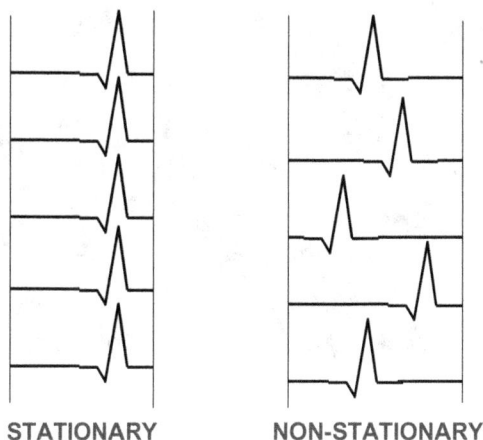

STATIONARY   **NON-STATIONARY**

Fig. 20   Collections of "stationary" and "non-stationary" signals that require averaging. The way these terms are used is described below. The stationary signals on the left can be summed and averaged as they are aligned in time. The jittered non-stationary signals on the right are not aligned and cannot be accurately averaged.

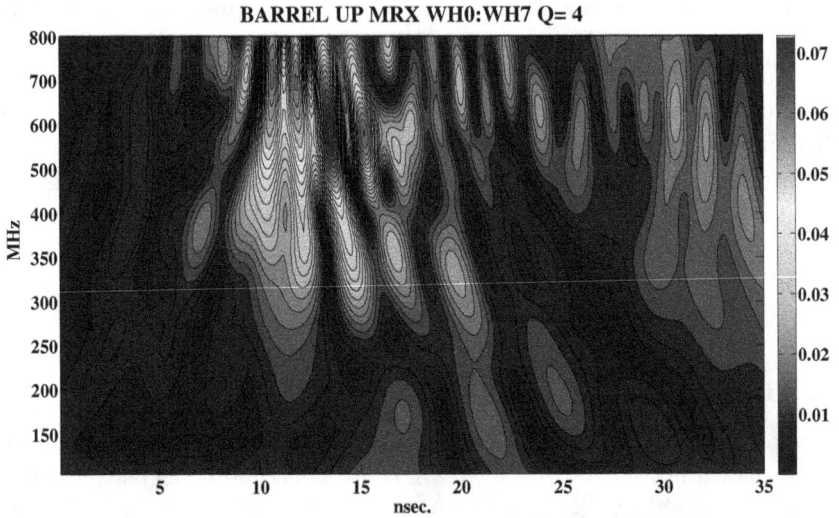

(A)

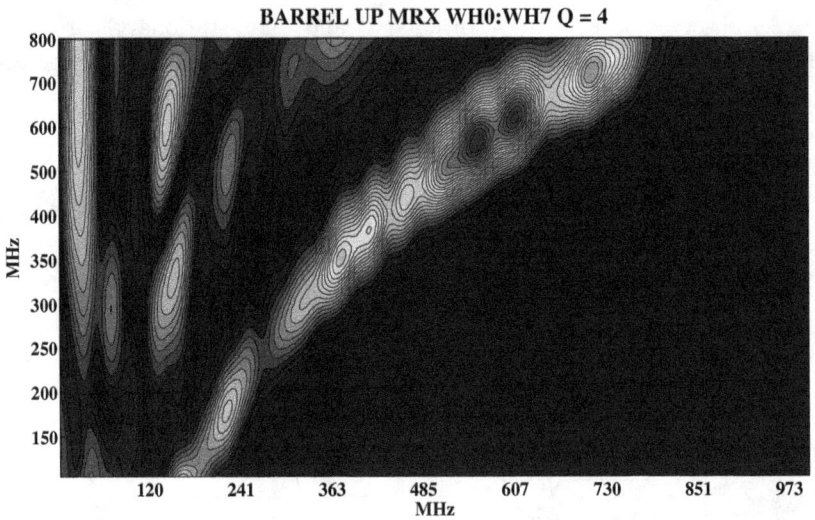

(B)

Fig. 21　(Continued)

considered non-stationary because they are not periodic. The statistics literature usually assumes that the signal commencement, e.g., at time t_0, is the same for all signals. However, in the engineering literature the assumption is that the signal commencement may not be the same for all signals, and many times is not. As it is mostly the case that the distance between the transceiver and a target is not constant, radar signals should be considered non-stationary. In this engineering sense, the left column of signals is stationary, and the right column is non-stationary, and this is the case also from a receiver's point-of-view. We shall refer to "signals subjected to jitter", when referring to non-stationary signals in the engineering sense, and as exemplified in the right column. Time averaging of the first column of signals results in a valid temporal average. Attempts at time "averaging" of the second column of signals would result in a spurious average.

If the reception of a series of short, time-limited signals are subject to jitter and are non-stationary in the sense we have declared the use of the term, taking their time average is difficult. To circumvent this problem, and also to provide insight into the spectral characteristics of both an RX's carrier and envelope, we introduced a new double frequency spectrum described in the next section.

7.   Carrier Frequency-Envelope Frequency (CFEF) Spectra

If return signal (RX) information in the time domain can be neglected, then the following frequency-frequency or Carrier Frequency-Envelope Frequency (CFEF) spectrum can be used (Fig. 21). The CFEF is obtained

---

Fig. 21   A: multiple window time-frequency analysis (MWTFA) spectrum of the Barrel UP target MRX using 8 WHWFs: WH0:WH7 to obtain 8 time-frequency spectra that were then averaged. This is Fig. 19B. B: Carrier Frequency-Envelope Frequency (CFEF) spectrum. This spectrum is composed of the magnitudes of the Fourier transforms of the cuts at each frequency through the time-frequency spectrum shown in Figure A. The spectra shown indicate both the spectra of the MRX carrier frequencies as well as of the modulating envelopes of those carriers. It can be seen that although there are peaks in these spectra at approximately the carrier frequencies — by reading across to the y-axis — that indicate short, almost monocycle, bursts at the appropriate frequency, by reading down to the x-axis it can be seen that the frequencies of the envelopes modulating the carriers are also indicated. On the far left a broad band carrier signal is indicated that is modulated by a low frequency envelope. Any long vertical band indicates a broad band spike burst. In the middle, there are "islands" indicating short duration envelopes modulating "carrier frequencies", i.e., wave packets. There are also wave packets located at approximately the same frequency on the y- and x-axes indicating wave packet envelope modulations of the "carrier frequency" components of the RX signal at the same frequency as those "carrier frequency" components — a *double modulation involving both a "carrier" and envelope at the same frequency*. See also Section 2.2.9, below.

by the Fourier transformation of each line in a MWTFA (Fig. 19B). The CFEF y-axis is identical to the MWFTA y-axis, but the CFEF x-axis is also a frequency axis and the influence of temporal jitter has been removed. To obtain an averaged signal, multiple CFEFs are averaged and the jitter is not an influence on the average. The average is obtained for a penalty of loss time of RX arrival information.

The CFEF method provides detailed spectral information concerning a target return. As shown in Fig. 21B, information concerning a signal's "envelope" frequencies, as well as "carrier" frequencies is revealed.

8.   Polarization

The polarization and polarization changes of RX signals elicited by differently polarized TX signals constitute important information that can be exploited in a number of ways, but will not be addressed in this book. This subject is complex and requires a treatment that itself would be book-length.

In anechoic chamber tests (Section 2.2.0, below) the RXs elicited by two different orthogonally polarized TX signals were collected, but no difference in these RX signals was detected. For the targets tested, TX-RX polarization was not a discriminating variable.

PART 1 — Ka-BAND MAP PROTOTYPE

1.1. Ka-Band MAP System

The MAP Ka-Band prototype system high-level view is shown in Fig. 1.1.1. The carrier frequency for all transmitted signals was 35.3 GHz. However, it is important to note that it is the envelope modulations in both the TX and RX signals that are of interest; and it is the TX envelope that was amplitude modulated to match target resonances — not the carrier. The chosen carrier frequency of 35.3 GHz was arbitrary, as there were no significant absorbing/scattering media between the transceiver and the targets.

It is noteworthy that a high sample rate scope was used both as an essential part of the receiver, and also to view the RX signals. Sampling rate was 10 GSs. Systems that transmit an impulse and receive and process the RX signal using a sampling receiver are considered to operate in the time domain. Systems that transmit individual frequencies in a sequential manner or as a swept frequency, i.e., as in LFM, and receive and process the RX signal using a frequency conversion receiver, are considered to operate in the frequency domain. Therefore, the MAP receiver is a time-domain receiver.

1.2. Targets Addressed by the Ka-Band System

The targets addressed in the Ka-Band system tests were both large scale (Fig. 1.2.1) and dimensionally-corrected, small scale (Fig. 1.2.2).

1.3. Mie (Resonance), Optical & Rayleigh Scattering

The tests used MTX and DTX signals that were optimized for maximum target RX echos and that are the result of Mie or resonance scattering. As mentioned in the INTRODUCTION, unlike conventional radars which usually operate in the optical or Rayleigh scattering regions, RAMAR MAP

Ka-Band Prototype System

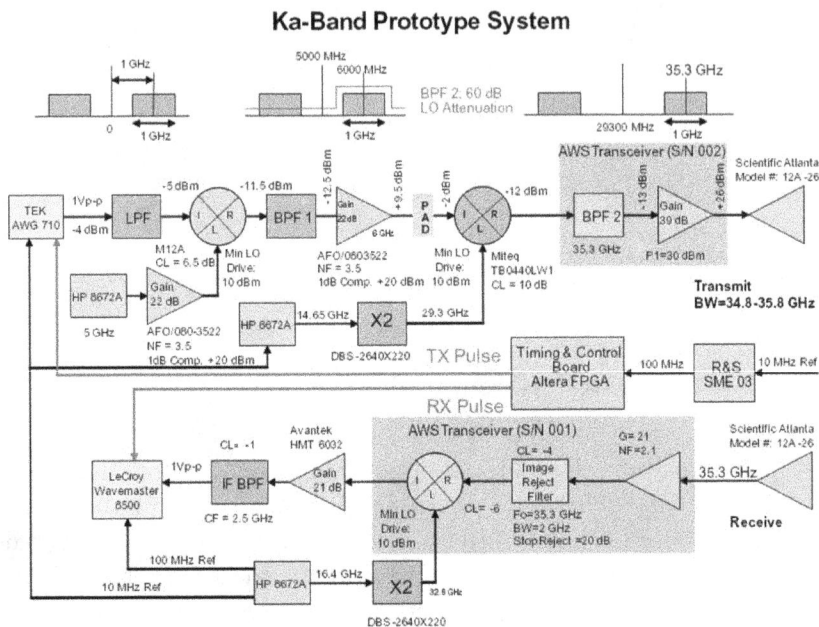

Fig. 1.1.1   Ka-Band Prototype System. MAP reception requires the preservation of all frequency components in the RX signal and thus a high sample rate. A LeCroy high sample rate scope was used both as an essential part of the receiver, and also to view the RX signals. Systems that transmit an impulse and receive and process the RX signal using a sampling receiver are considered to operate in the time domain. Systems that transmit individual frequencies in a sequential manner or as a swept frequency, i.e., as in LFM, and receive and process the RX signal using a frequency conversion receiver, are considered to operate in the frequency domain. Therefore, the MAP receiver is a time-domain receiver. The transmit (TX) path consists of the output from an arbitrary waveform generator (AWG) with $f_c = 1$ GHZ and bandwidth (BW) = 1 GHz (max $f_m = 1$ GHz). The first carrier up-conversion is from $f_c = 1$ GHz to $f_c = 6$ GHz; and the second to $f_c = 35.3$ GHz. The receive (RX) path consists of a down-conversion to $f_c = 2.5$ GHz and the modulating envelope of that signal is sampled and displayed. It is important to note that it is the envelope modulation of the TX and RX signals that is of interest and it is the TX envelope that was matched to target resonances — not the carrier. The carrier chosen is arbitrary — here it is 35.3 GHz. The oscilloscope (LeCroy Wavemaster 8500) served as both a fast sampling time domain receiver and display. $F_s = 10e9$ samples/second.

radar/sensors operate in the Mie scattering region. Tests were conducted to verify that resonance scattering was addressed. Some definitions referring to a sphere target (Fig. 1.3.1) are:

In the *Rayleigh region* ($2\pi a/\lambda \ll 1$, where $a = $ radius of a sphere and λ is the radar TX wavelength), the radar cross-section is proportional to f^4 where $f = c/\lambda$.

Fig. 1.2.1   Targets: Humvee (top) and Ford Truck (bottom).

In the *optical region* $(2\pi a/\lambda \gg 1)$, the radar cross-section approaches the physical area of the sphere as TX frequency is increased. Moreover, scattering takes place from small bright spots at the tip of the sphere.

In the *Mie or resonance region* $(2\pi a/\lambda \sim 1)$, the target oscillates as a function of the frequency $f = 2\pi a/\lambda$ with a maximum at $f = 2\pi a/\lambda = 1$, and is roughly 5.6 dB or more greater than in the optical region (Skolnik, 2001).

In the tests conducted, the MTX carrier, f_c, was at 35.3 GHz, and far into the optical region for the target sphere. However, it was the frequency of the modulation of the carrier, f_m, that was matched to the sphere's resonance scattering frequency and gave the results presented.

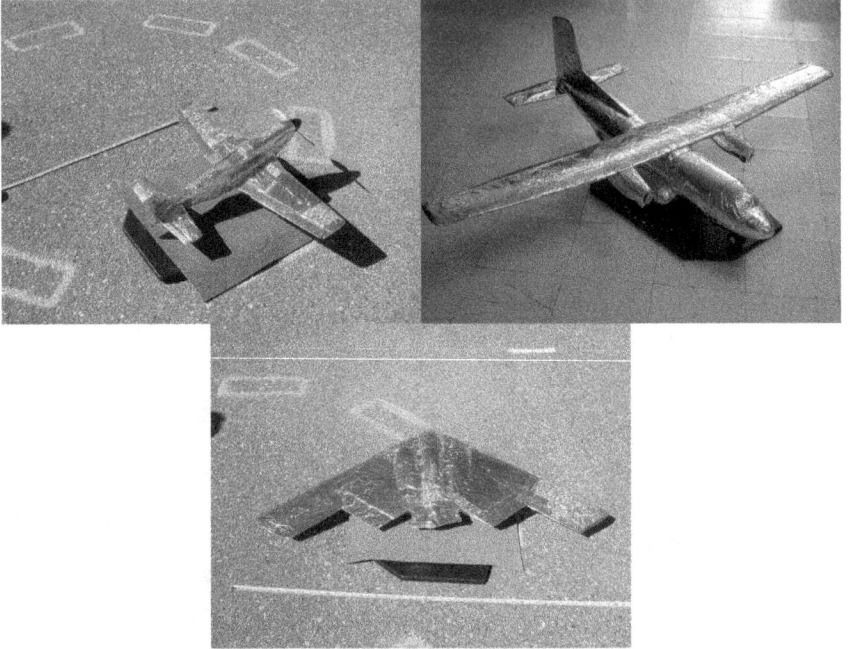

Fig. 1.2.2   Model Targets: P-51 (38 inch wingspan); C-160 (72 inch wingspan); and B-2 (40 inch wingspan). All are covered with aluminum foil.

For a sphere of 6 inches, the radius is $0.0762\,\text{m}$. For a TX to be in resonance with this sphere, i.e., for

$$\frac{2 \times \pi \times r}{\lambda} = 1 = 10^0$$

then

$$\frac{2 \times \pi \times r}{\lambda} = \frac{2 \times \pi \times 0.0762}{0.4788} = 1$$

and $\lambda = 0.4788\,\text{m} \to f = 626.59\,\text{Hz}$.

The following Figs. 1.3.2–1.3.3 provide supporting evidence that the MRX elicited by an MTX is Mie (resonance), rather than Rayleigh or optical scattering.

1.4.   Return Signal SNR Enhancement

Examples of enhanced target return, RTX, SNR achieved by MTX over energy-matched PTX (or UWB) signals are shown in Figs. 1.4.1–1.4.3.

SPHERE 6" DIAMETER (626.59 MHz)

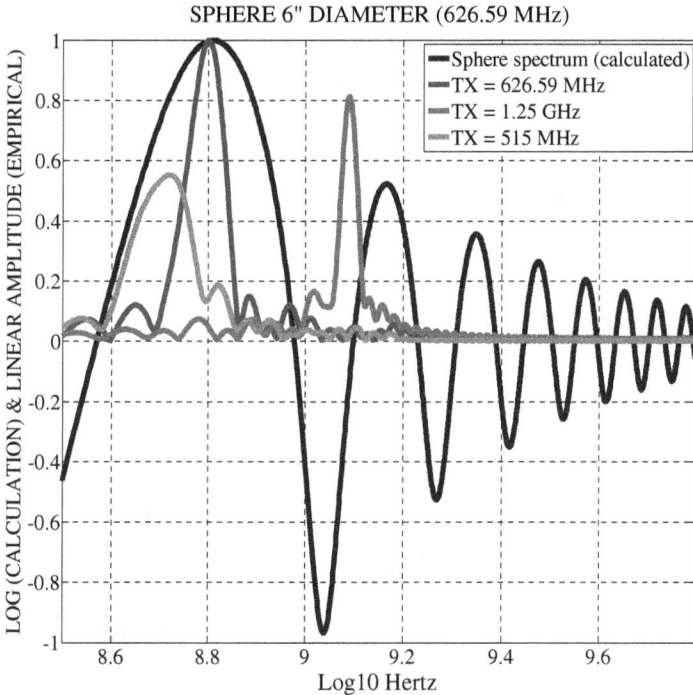

Fig. 1.3.1A   Sphere Target, 6″ Diameter. Log-Log spectra: Calculated power spectrum, Empirical DRX spectrum (DTX = 626.59 MHz), Empirical DRX (DTX = 1.25 GHz), and Empirical DRX (DTX = 515 MHz). Log10 (626.59e6) = 8.797.

The degree of SNR RX enhancement is target-dependent and these examples are representative of their class of targets.

1.5.   Corner Reflector Tests

Control tests were conducted using a target corner reflector. As an ideal corner reflector acts like a reflecting mirror, returns "all TX energy" (in some direction), and there are no target resonances to exploit, the expectancy was that there would be no significant enhancement in the return signal energy by using MTX signals, versus PTX signals. This expectancy was confirmed (Fig. 1.5.1). There is little difference between the RXs elicited by PTX and MTX in the case of the corner reflector. The MAP technique exploits spectral diversity offered by the target. However, a corner reflector, being "mirror-like", returns "all" spectral components in

SPHERE 6" DIAMETER (626.59 MHz)

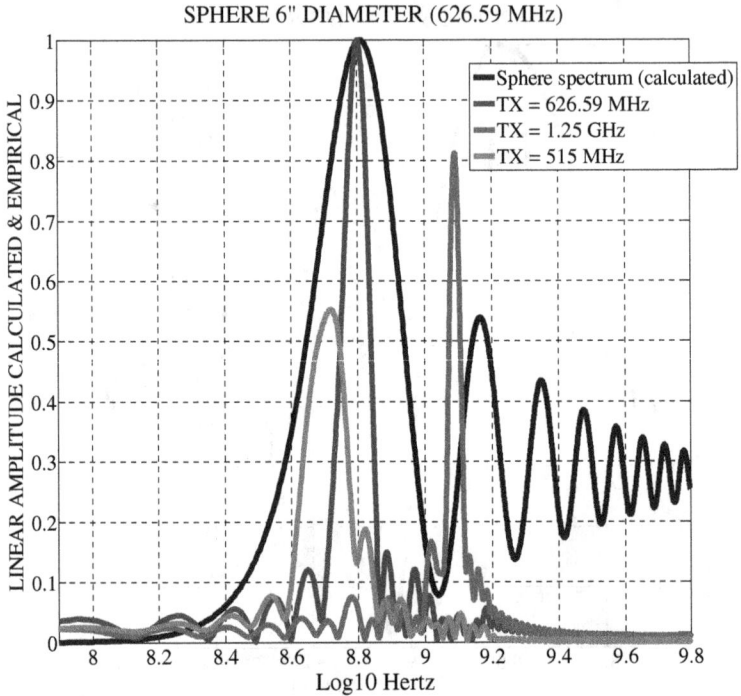

Fig. 1.3.1B Sphere Target, 6″ Diameter. Linear-Log spectra: Calculated power spectrum, Empirical DRX spectrum (DTX = 626.59 MHz), Empirical DRX (DTX = 1.25 GHz), and Empirical DRX (DTX = 515 MHz).

the TX signal. Therefore there is no substantial difference in the PRX and MRX signals elicited by PTX and MTX signals.

1.6. Exclusive & Inclusive Optimum Transmit Signal Design for Target Aspect Independent Recognition

The following approaches to designing a transmit signal (TX) for target aspect independent recognition rely on two different philosophies: (1) the *exclusive*, in which resonating frequencies are retained in the TX, only if they are present in the PRX at *all* tested aspect angles; and (2) the *inclusive*, in which all resonating frequencies are retained in the TX, if they are present in the PRX at *any* tested aspect angle. Thus:

An ***exclusive*** MTX is constructed by the *multiplication* of the spectrum of the short duration pulse target time reversed PRX returns

NORMALIZED RADAR CROSS-SECTION OF A SPHERE
CALCULATED USING MIE SERIES
After: Skolnik, M.I., "Introduction to Radar Systems", 3rd Edition, 2001, p. 51

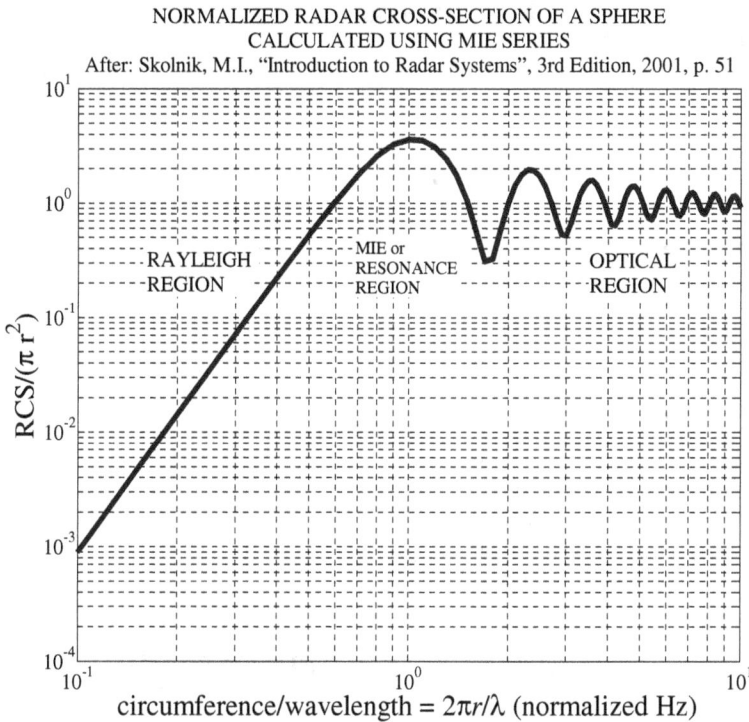

circumference/wavelength = $2\pi r/\lambda$ (normalized Hz)

Fig. 1.3.1C   Calculated normalized radar cross section of a sphere as a function of its circumference $(2\pi a)$ measured in wavelengths. $a =$ radius, $k = 2\pi/\lambda$. After Skolnik, M.I. *Introduction to Radar Systems*, 3rd Edition, McGraw-Hill, New York, 2001, p. 51.

at target aspect angles $0, 10, 20, \ldots, 180°$. The result is then inverse Fourier transformed to obtain the exclusive aspect independent MTX. The objective is to obtain an MTX matched to resonances excited at *all* target aspect angles, e.g., large structure components and whole or large body resonances.

An *inclusive* MTX is constructed by the *addition* of the spectrum of the time reversed PTX at aspect angles $0, 10, 20, \ldots, 180°$. The result, again, is then inverse Fourier transformed to obtain the inclusive aspect independent MTX. The objective is to obtain an MTX matched to resonances excited at *any* target aspect angles, e.g., large structure components, whole or large body resonances, and also subcomponent resonances, i.e., the resonances due to small structures only visible at particular angles of a complex target.

SPHERE 6 inch D

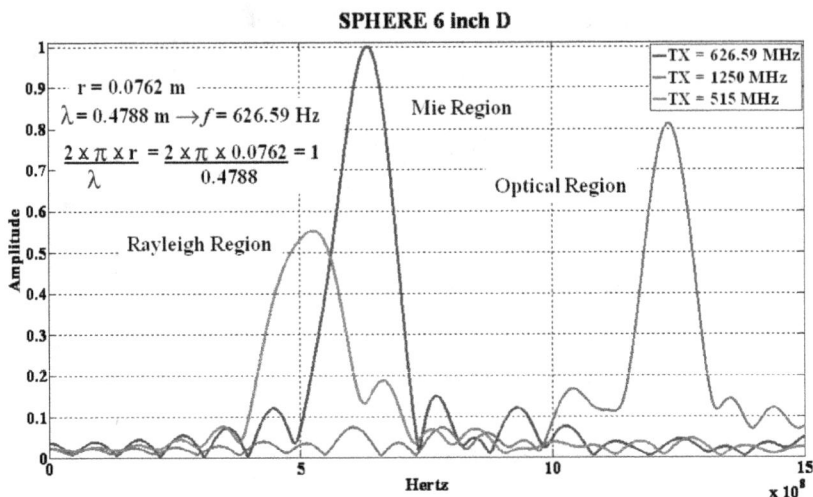

Fig. 1.3.2   DRX signal spectra, target: a 6 inch diameter aluminum sphere, elicited by three different DTX pulses of different f_m frequencies: 515, 626.59 and 1250 MHz — all modulating the same carrier frequency: $f_c = 35.3$ GHz. These spectra are also shown in Figs. 1.3.1a & b, above.

SPHERE 6" DIAMETER (626.59 MHz)

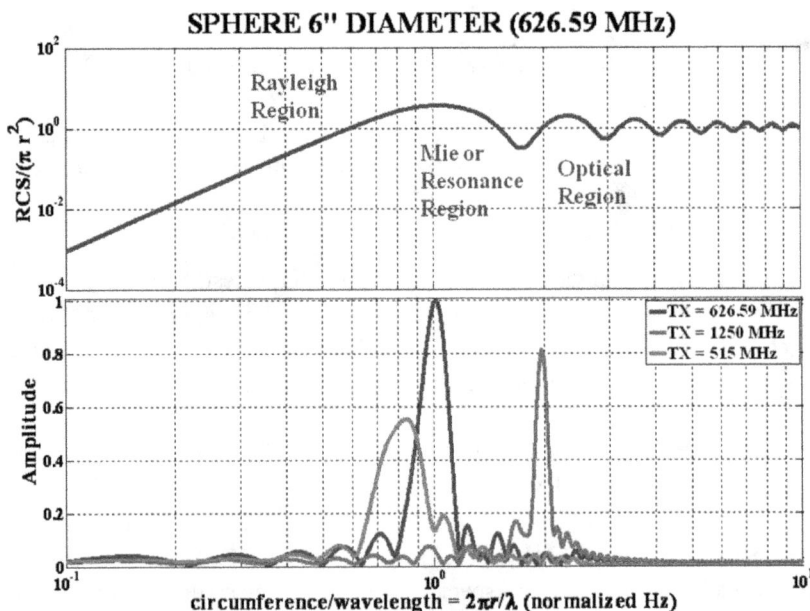

Fig. 1.3.3   A composite of Figs. 1.3.1c and 1.3.2 indicating that the optimum DTX envelope modulation is at 626.59 MHz. (For 6 inch diameter sphere, $a = 0.07620$; and for TX = 626.59 MHz, $\lambda = 0.4788$ m. Therefore, $2 \times \pi \times 0.0762/0.488 = 1$, and as $\log_{10}(1) = 0$, the x-axis shows the maximum at 0 normalized Hz).

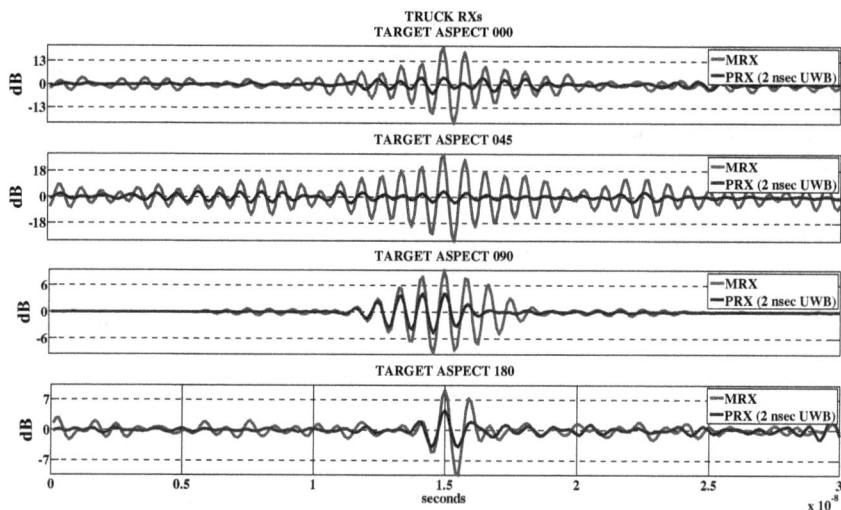

Fig. 1.4.1 Time domain records of MRX showing enhancement (in dB) with respect to PRX (or UWB RXs); MTX and PTX are energy matched. Truck target at aspect angles 0°/45°/90° and 180°.

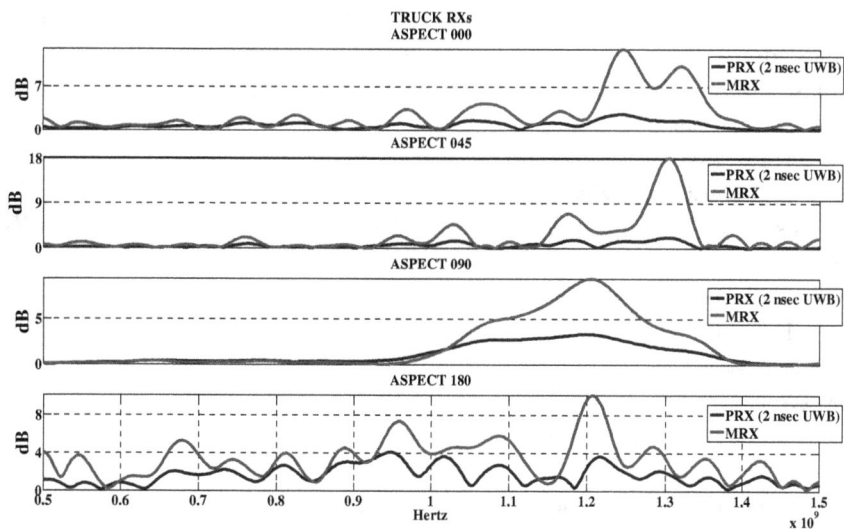

Fig. 1.4.2 Spectra of MRX enhancement (in dB) with respect to PRX (or UWB RXs); MTX and PTX are energy matched. Truck target at aspect angles 0°/45°/90° and 180°.

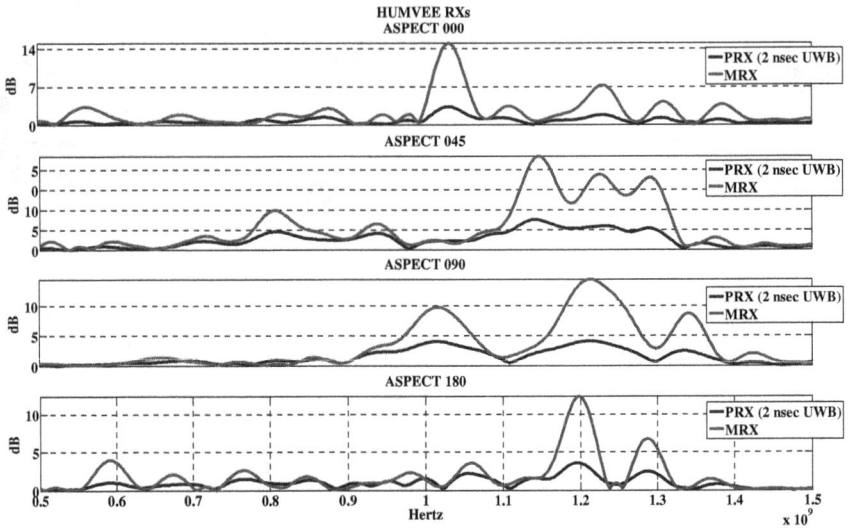

Fig. 1.4.3 Spectra of MRX enhancement (in dB) with respect to PRX (or UWB RXs); MTX and PTX are energy matched. Humvee target at aspect angles 0°/45°/90° and 180°.

The exclusive and the inclusive MTX signals for the C-160 and P-51 targets are shown in Figs. 1.6.1. and 1.6.2 provides the MTX individual components from which the exclusive MTX was constructed, and Figs. 1.6.3–1.6.5 provide the exclusive MRXs in three dimensions (aspect angle, frequency and amplitude). These results indicate that the MRX spectra of the two targets are distinct.

Figure 1.6.6 provides the MTX individual components from which the inclusive MTX was constructed, and Figs. 1.6.7–1.6.9 the inclusive MRXs in three dimensions (aspect angle, frequency and amplitude). These results also indicate that the MRX spectra of the two targets are distinct.

Both targets can be detected using *exclusive and inclusive* MTX signals, and the target orientation can also be detected using *inclusive* MTX signals, as there is aspect-dependence in the amplitude of certain frequency bands at certain aspect angles. Target ID can also be achieved with a combination of the two forms of MTX, and using an MTX:MRX question-and-answer protocol composed of a sequence of MTXs and MRX yes/no amplitude confirmation "replies". With this aim in mind, two new methods of demonstrating the aspect-dependence/independence are shown in Figs. 1.6.10 and 1.6.11.

CORNER REFLECTOR

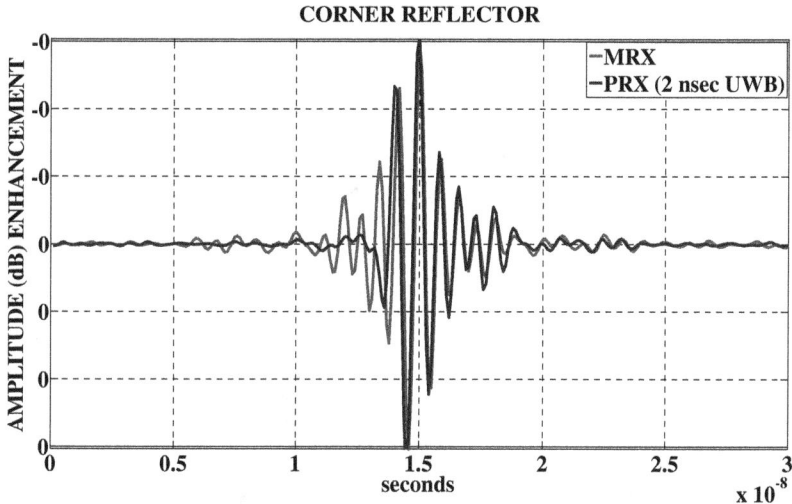

(a)

CORNER REFLECTOR

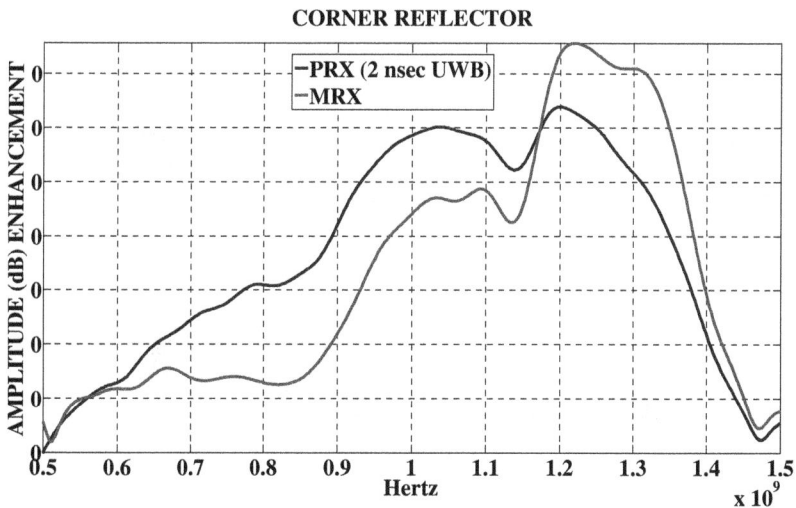

(b)

Fig. 1.5.1   Corner Reflector Target. a: PRX and MRX signals in the time domain. b: the same PRX and MRX signals but in the frequency domain. There is little difference between the echos elicited by PTX and MTX in the case of the corner reflector. The corner reflector acts as a control test target. The MAP technique takes advantage of spectral diversity offered by the target. However, a corner reflector, being "mirror-like", returns "all" spectral components in the TX signal. Therefore there is no substantial difference in the PRX and MRX target echos elicited by PTX and MTX signals.

Table 1.6.1   Target: C-160 Model.

MRX Band (GHz)	MTX Form	Probability of belonging to a C-160 MRX Spectrum
0.87–0.91	Exclusive	High
0.91–0.92	Exclusive	Low
0.92–0.93	Exclusive	Very Low
0.70–0.80	Inclusive	Medium
0.90–0.98	Inclusive	Medium
0.91–1.10	Inclusive	High
1.10–1.20	Inclusive	Low

Table 1.6.2   Target: P-51 Model.

MRX Band (GHz)	MTX Form	Probability of belonging to a C-160 MRX Spectrum
0.93–0.94	Exclusive	High
0.78–0.82	Inclusive	Low
1.00–1.05	Inclusive	High
1.324–1.375	Inclusive	Low

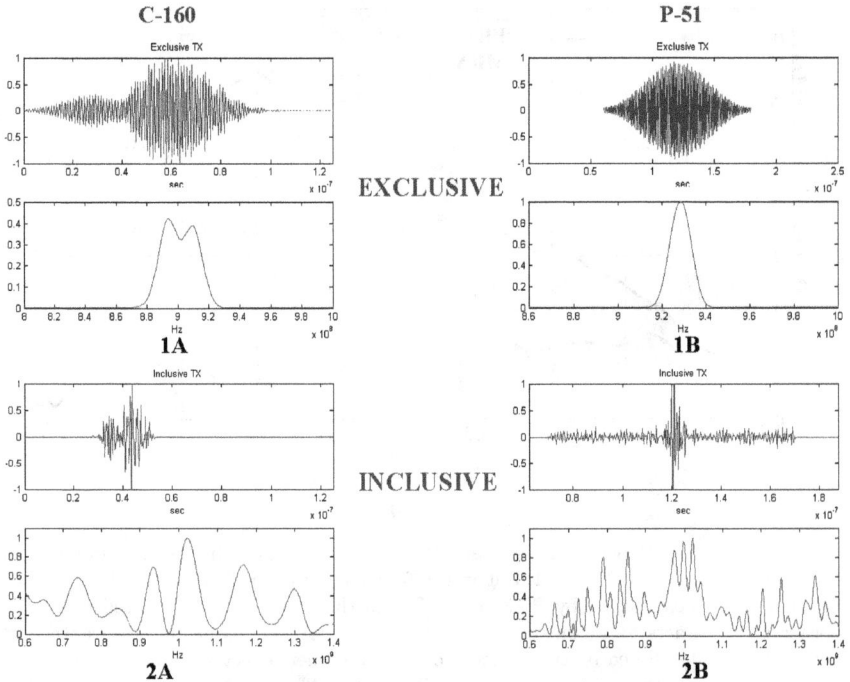

Fig. 1.6.1 A: *Exclusive* MTX signals, time and frequency domains. A: The C-160 exclusive MTX. B: The P-51 exclusive MTX. B: *Inclusive* MTX signals.

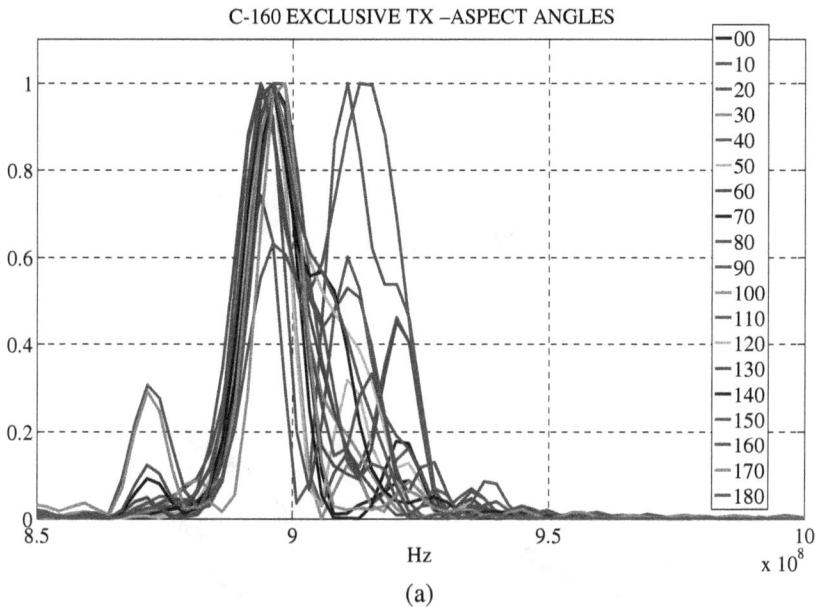

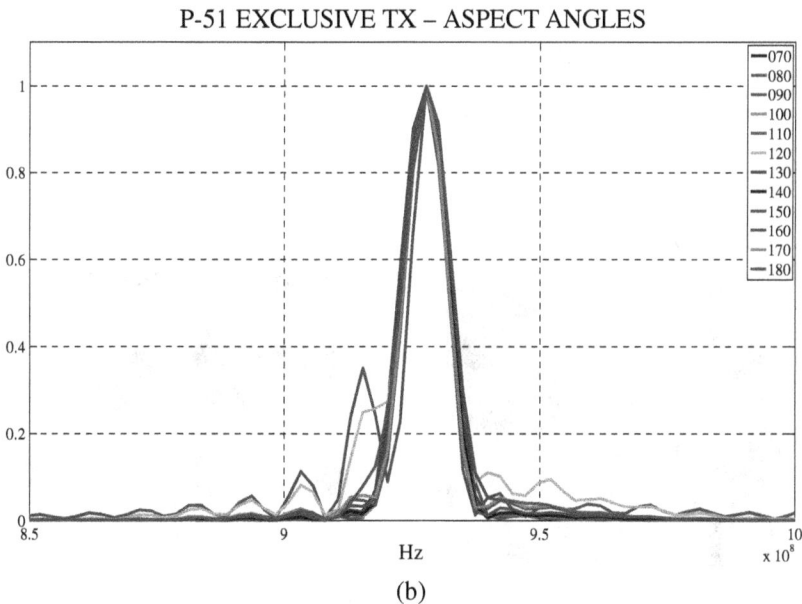

Fig. 1.6.2 *Exclusive* MTX signal components. a: the selected C-160 MTX signal components at the aspect angles 00–180° (at 10 degree intervals). b: the P-51 MTX signal components at the aspect angles 70–180° (at 10 degree intervals).

C-160

EXCL. MRX

P-51

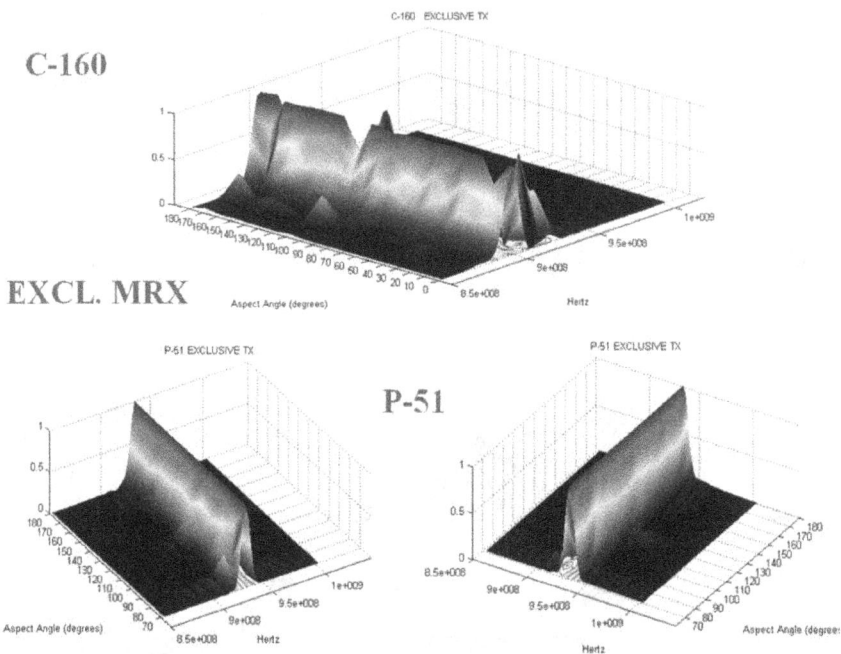

Fig. 1.6.3   Aspect-Frequency Spectra of MRX signals elicited by *exclusive* MTX signals at the aspect angles shown. a: C-160 MRX exclusive spectrum. b & c: Two views of a P-51 MRX exclusive spectrum.

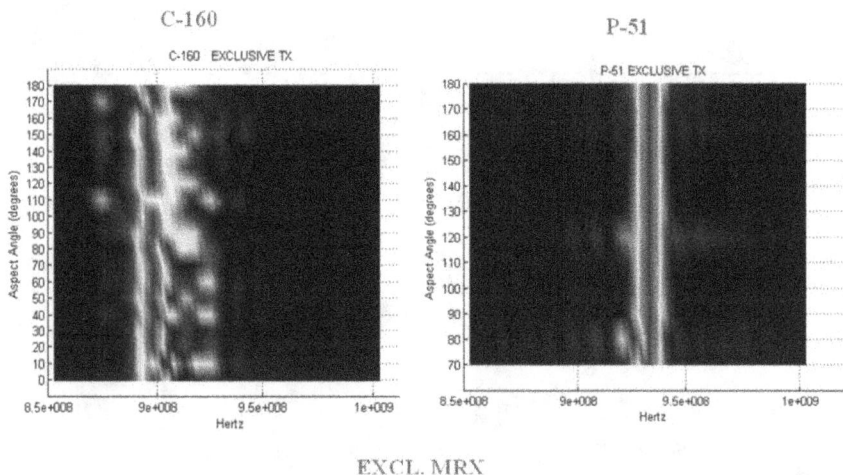

C-160   P-51

EXCL. MRX

Fig. 1.6.4   MRX Aspect-Frequency *exclusive* spectra — top view.

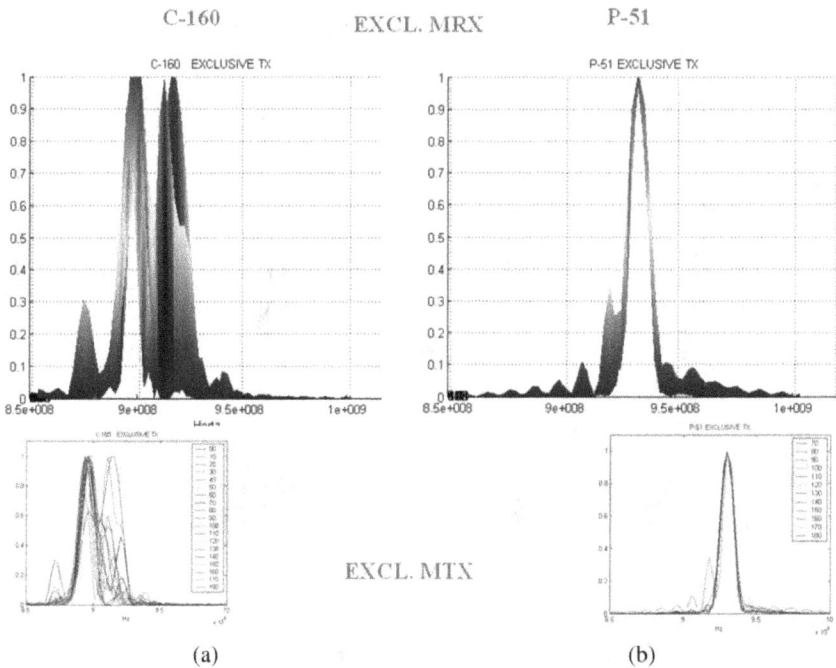

(a)   (b)

Fig. 1.6.5   Spectra of MRX signals elicited by *exclusive* MTX signals together with the components on which the exclusive MTX were based. a: Target, C-160. b: Target, P-51.

These spectra provide the relative amplitudes of individual frequency bands in the target RX spectra at the indicated aspect angles. The relative individual band amplitudes can be used to provide probability weightings that a detected specific RX spectral resonance band originated from a known target. Table 1.6.10 shows representative spectral bands for the C-160 model taken from Figs. 1.6.2 and 1.6.6 and weighted according to the probability that these RX bands belong to a C-160 RX spectrum; and Table 1.6.11 shows representative spectral bands for the P-51 model also taken from Figs. 1.6.2 and 1.6.6 and weighted according to the probability that these bands belong to a P-51 RX spectrum.

1.7.   Vehicle Targets

Examples of time domain MRXs and PRXs are shown in Fig. 1.7.1 for the Truck and Humvee targets at orientations of 000°, 045°, 090° and 180°. The maximum band amplitude enhancement for these two targets — MRX

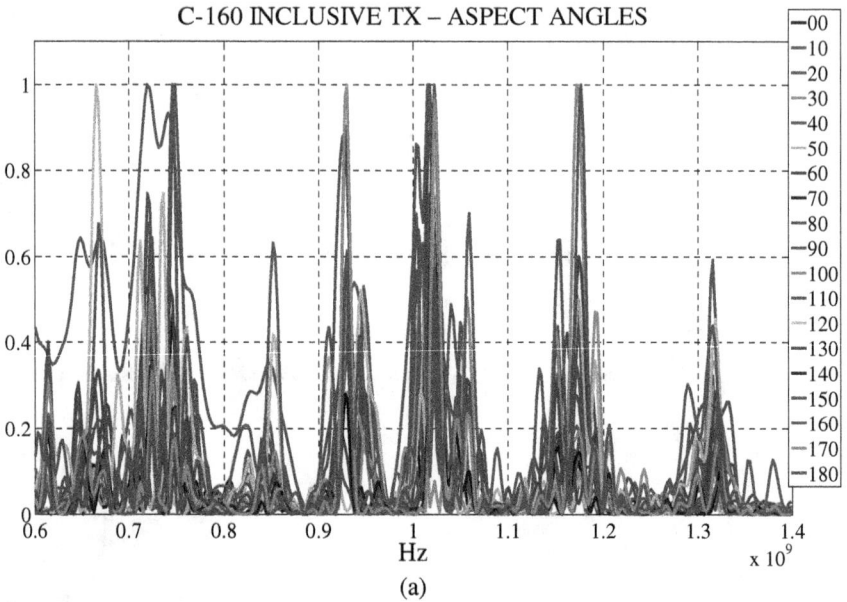

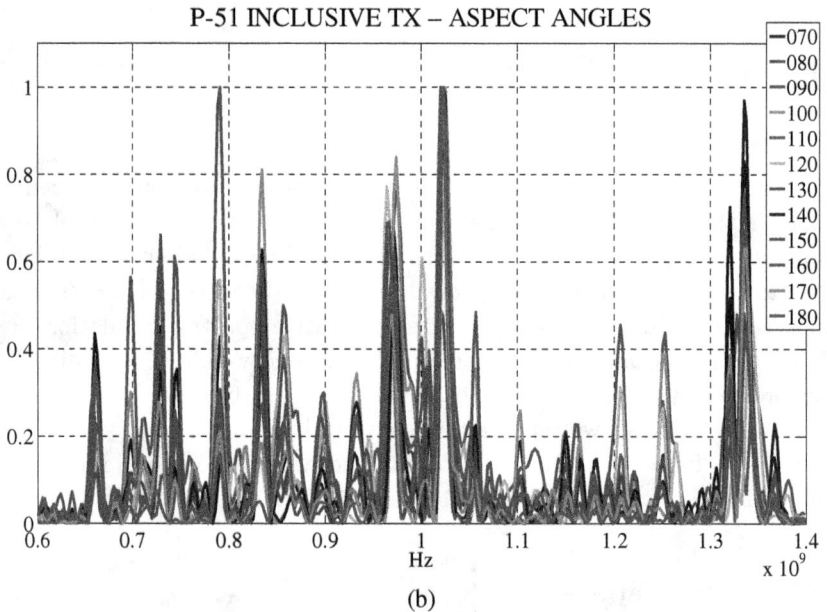

Fig. 1.6.6 *Inclusive* MTX signal components. a: the selected C-160 MTX signal components at the aspect angles 00–180° (at 10 degree intervals). b: the P-51 MTX signal components at the aspect angles 70–180° (at 10 degree intervals).

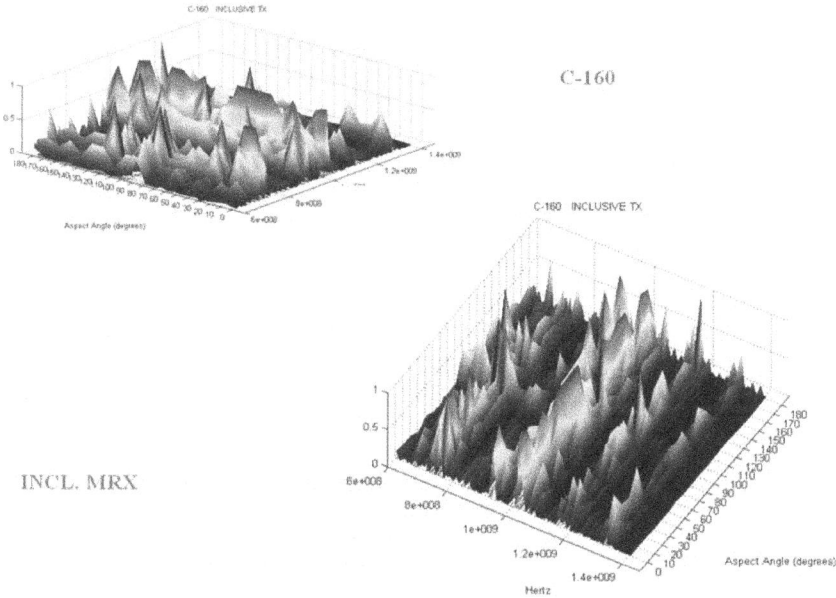

Fig. 1.6.7   Aspect-Frequency Spectra of MRX signals elicited by *inclusive* MTX signals. Two views of the spectra for the C-160 target. These MRX spectra show the signature of both aspect-independence (clear amplitude troughs and valleys at set frequencies) and also aspect-dependence (higher amplitude peaks are certain aspect angles).

versus PRX or MRX/PRX — is summarized in Table 1.7.1. The spectra for these examples are shown in Fig. 1.7.2.

The time-frequency spectra (Fig. 1.7.3) and the time-frequency-amplitude spectra (Fig. 1.7.4) indicate that, in contrast with the PRX spectra, the MRX spectra exhibit the expected time domain convolutional symmetry as indicated in Figs. 7 and 8 of the Introduction.

In the time domain the RXs are target aspect dependent. In contrast, in the frequency domain the frequency locations of the bands are aspect independent, but the amplitudes of those bands can vary with aspect.

1.8.   Model Targets

The results of tests on the different aluminum-covered model aircraft: P-51, C-160 and B-2, are similar for the three models tested:

- The amplitude of the spectral MRX in comparison with PRX (UWB) spectral bands were increased by up to 7 times (8.5 dB) — Figs. 1.8.1, 2, 4, 6, 7 and 8.

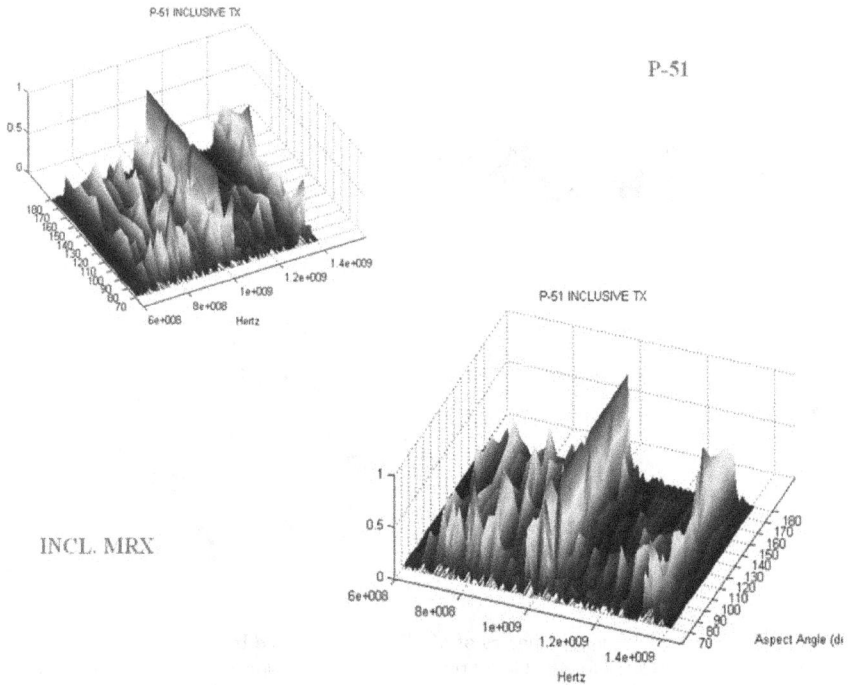

Fig. 1.6.8   Aspect-Frequency Spectra of MRX signals elicited by *inclusive* MTX signals. Two aspects of the spectra for the P-51 target. These MRX spectra again show the signature of both aspect-independence (clear amplitude troughs and valleys at set frequencies) and also aspect-dependence (higher amplitude peaks are certain aspect angles).

- While the frequency location of the RX spectral bands are aspect-independent, the amplitudes of those spectral bands are aspect-dependent — Figs. 1.8.2, 6 and 8.
- Time-frequency spectra for the MRX show the expected time domain convolutional symmetry as indicated in Figs. 7 and 8 of the Introduction.

1.9.   Nonlinear Combination of Separate Subcomponent Target Minor Resonances

The question of whether a complex target's RX signal is merely composed of the superposition of the subcomponent minor resonances composing the target is relevant to the aim of simulating, and perhaps predicting, the complex target's RX spectrum. The tests conducted indicate that

C-160   INCL. MRX   P-51

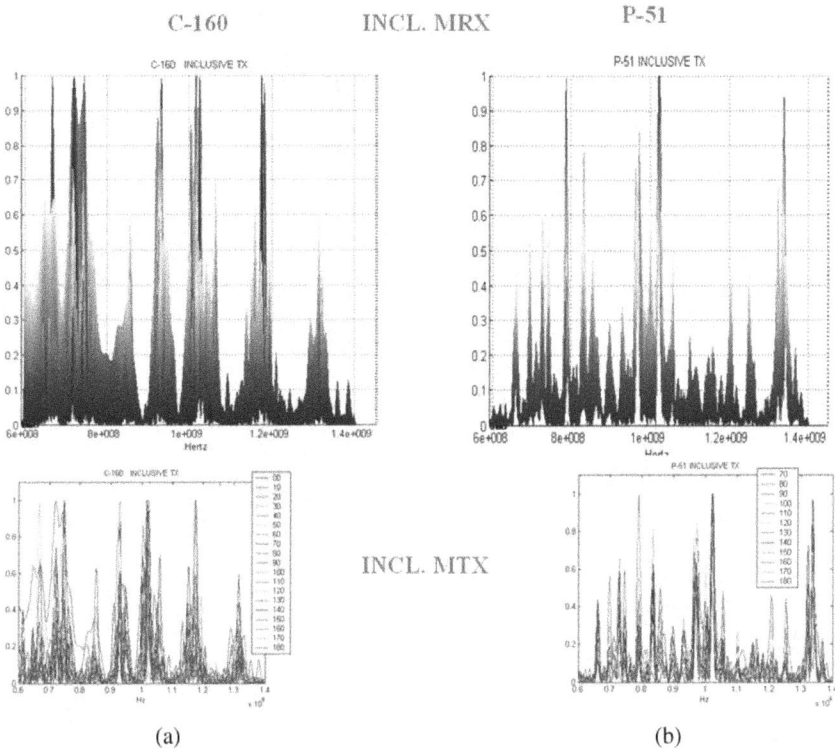

(a)   (b)

Fig. 1.6.9   Spectra of MRX signals elicited by *inclusive* MTX signals together with the components on which the inclusive MTX were based. a: Target, C-160. b: Target, P-51.

superposition of separate subcomponent resonances does not apply, and simulation of an empirically untested complex target's RX spectrum is difficult due to nonlinearities. As discussed in the INTRODUCTION, this nonlinearity of the addition of resonant target cross-section parts, does not necessarily mean that, once added, the whole provides a nonlinear impulse response. It does mean that it is difficult to predict the impulse response of the resulting cross-section. Therefore these tests address a nonlinearity of changes in the impulse response on addition of subcomponents, rather than a nonlinearity in the impulse response on the excitation by test signal frequencies.

Figures 1.9.1–1.9.3 show simple minor target arrangements in which the distance separating the components was varied. The test results indicate that the formation of complex target resonances is initially by near-field coupling of the minor component resonances. The conclusion drawn from

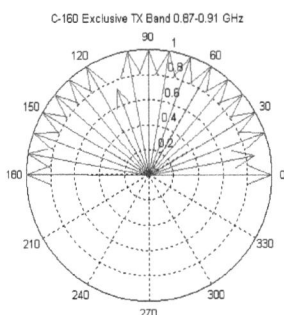

A1

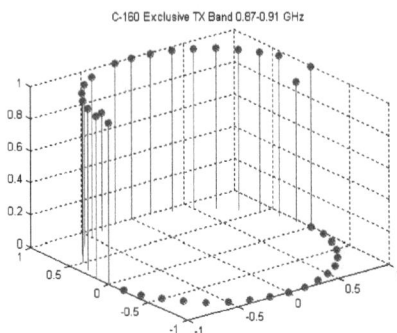

A2

C-160 Exclusive TX: **Band 0.87-0.91 GHz.**

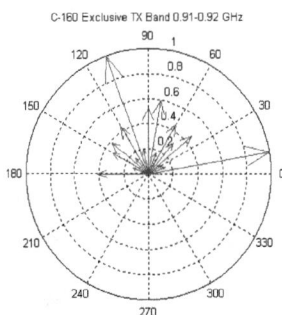

B1

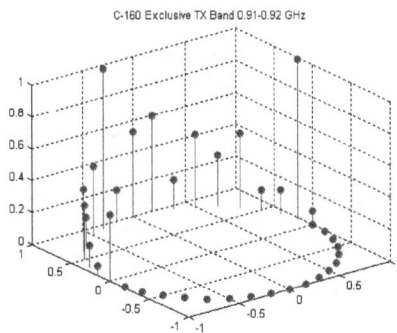

B2

C-160 Exclusive TX: **Band 0.91-0.92 GHz.**
Notice that the RXs at two aspect angles that are <1 in A, above, are = 1, here.

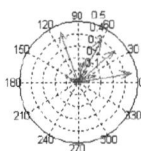

C1

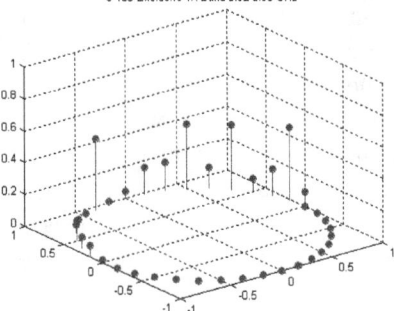

C2

C-160 Exclusive TX: **Band 0.92-0.93 GHz.**

Fig. 1.6.10   Target: C-160 Model. Two representations of MRX spectral band aspect angle dependence. The MRX spectra elicited by exclusive (A, B, C) and inclusive (D, E, F, G) MTX signals.

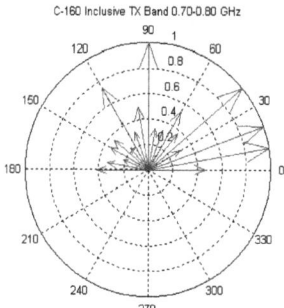

D1

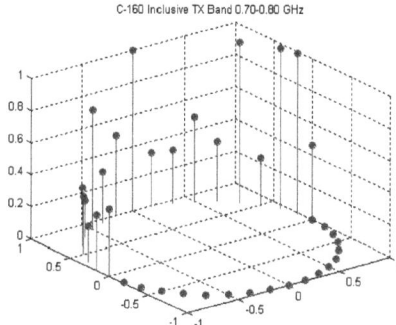

D2

C-160 Inclusive TX: **Band 0.70-0.80 GHz.**

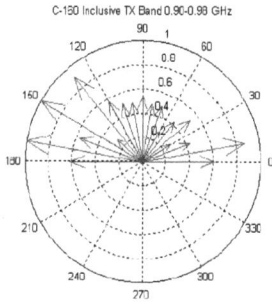

E1

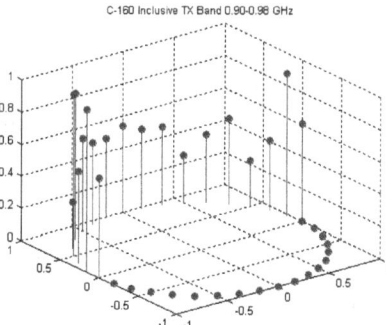

E2

C-160 Inclusive TX: **Band 0.90-0.98 GHz.**

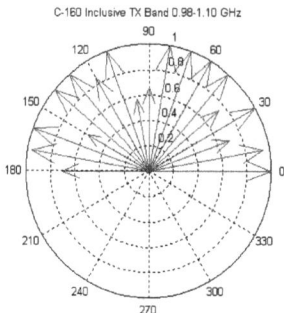

F1

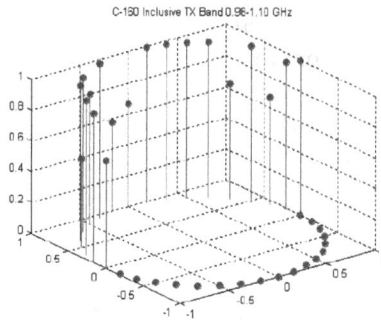

F2

C-160 Inclusive TX: **Band 0.91-1.10 GHz.**

Fig. 1.6.10 (*Continued*) The exclusive MRX spectral bands facilitate target identification at all angles.

The inclusive MRX spectral bands facilitate both target identification and orientation detection.

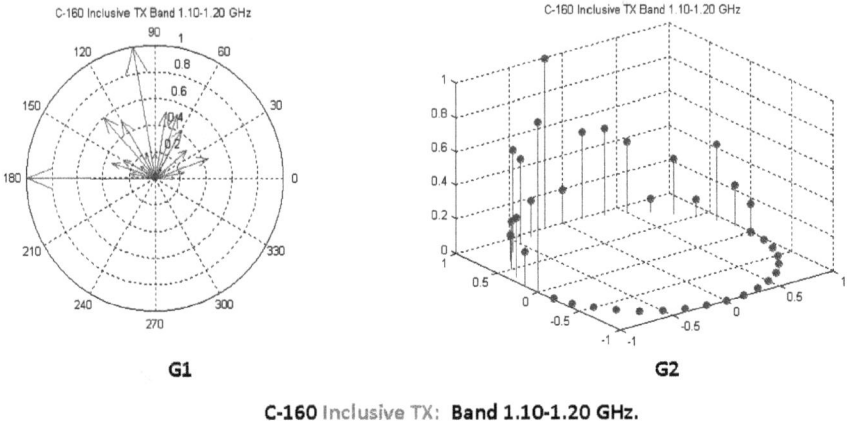

C-160 Inclusive TX Band 1.10-1.20 GHz

G1

C-160 Inclusive TX Band 1.10-1.20 GHz

G2

C-160 Inclusive TX: **Band 1.10-1.20 GHz.**

Fig. 1.6.10   (*Continued*)

these tests is that the closer the near-field coupling the more the amplitudes of the RX resonances of the minor targets diminish. At close coupling other resonances develop indicative of the composite target and which were perhaps generated at frequencies outside the 1 GHz of the Ka-band system's bandwidth. Thus simulation, modeling and prediction of a complex target's spectrum is not a simple matter of separate subcomponent minor resonance addition. These results agree with the conclusion of Jofre *et al.* (2009), made in the context of a study of a UWB tomographic radar, that when highly inhomogeneous objects are observed, multiple and high-contrast scattering occurs. However, once the response of the inhomogeneous object is composed, whether that object's impulse response is nonlinear or not can only be answered on a case by case basis. In the case of second harmonic generation: recently, Crowne & Fazi (2009) calculated that metallic targets are poor emitters of second-harmonic radiation elicited by RF radiation — although the possibility of detection remains in the case of high power radiation and privileged grazing incidence.

1.10.   Target Identification

With an archive of PRX *a priori* information about a target, or class of targets, identification of a target on the basis of a specific RX signal can proceed in a number of ways, and here we describe one possible method. The method is based on a model that exploits a set or matrix of unknown aspect independent TX signals, UTX. The UTX enter a channel, and a known set of signals, KRX, is received (Fig. 1.10.1). The role of KRX can

A1   A2

P-51 Exclusive TX: Band 0.93-0.94 GHz

B1   B2

P-51 Inclusive TX: Band 0.78-0.82 GHz

C1   C2

P-51 Inclusive TX: Band 1.00-1.05 GHz

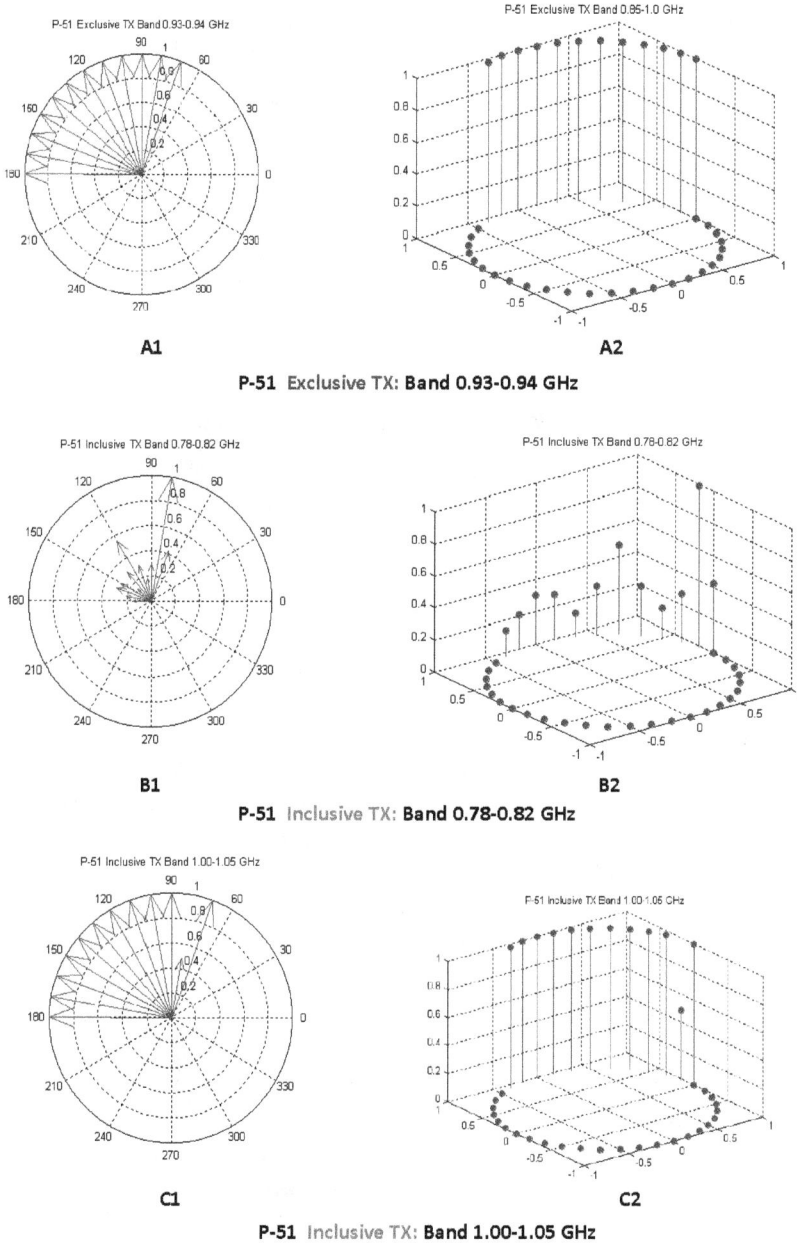

Notice that the RX at one aspect angle that is <1 here, in B, above, is = 1.

Fig. 1.6.11   Target: P-51 Model. Two representations of MRX spectral band aspect angle dependence. The MRX spectra elicited by exclusive (A, B, C) and inclusive (D, E, F, G) MTX signals.

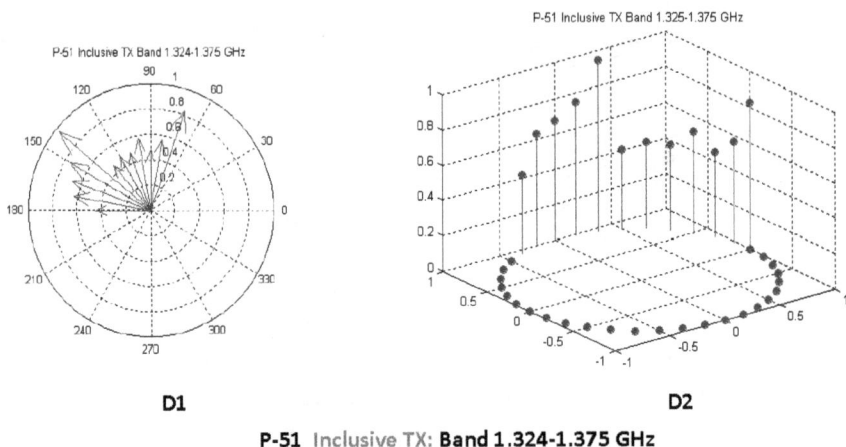

P-51 Inclusive TX: **Band 1.324-1.375 GHz**

Fig. 1.6.11 (*Continued*)   The exclusive MRX spectral bands facilitate target identification at all angles.
The inclusive MRX spectral bands facilitate target identification and orientation detection.

be played either by a set of PRXs or MRXs obtained with the target at a number of aspect angles. By assuming some statistical properties of the channel and the UTX, the KRX can be decomposed to obtain an estimate of the common aspects of the RX signals under the explicit assumptions adopted for the UTX, i.e., a set of signals designated: ERX. Therefore the ERX are an estimate of the UTX, or [UTX]. Any RX signal received later can then be correlated with any of the ERX/[UTX] set for target identification purposes, and the correlations provide a measure of identification confidence. The role of the target in this model is a mixing matrix role. That is the basic model, but there are, however, several strategies for extracting the source signals ERX/[UTX], e.g., PCA, SVD, ICA, GSO, and any method used is only as valid as the assumptions implicit in that method. We turn now to quickly introduce some of these methods and their assumptions.

Principal Component Analysis (PCA) is a general multivariate technique whereby signals are decorrelated and components are extracted according to the decreasing order of their variances, and is a technique for computation of the eigenvectors and eigenvalues of the, e.g., RX matrix of signals (Joffre, 2002). The technique involves the linear transformation of a group of correlated variables to achieve certain optimum conditions, the most important of which is that the transformed variables are uncorrelated

TRUCK

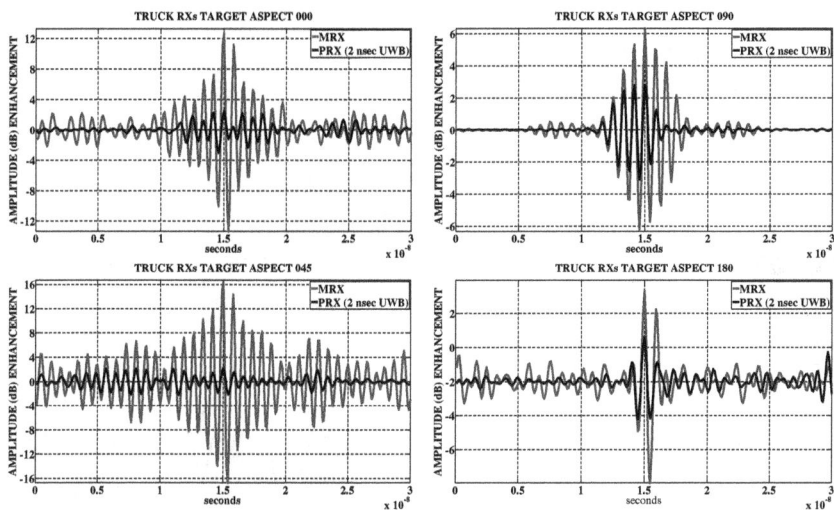

(a)

HUMVEE

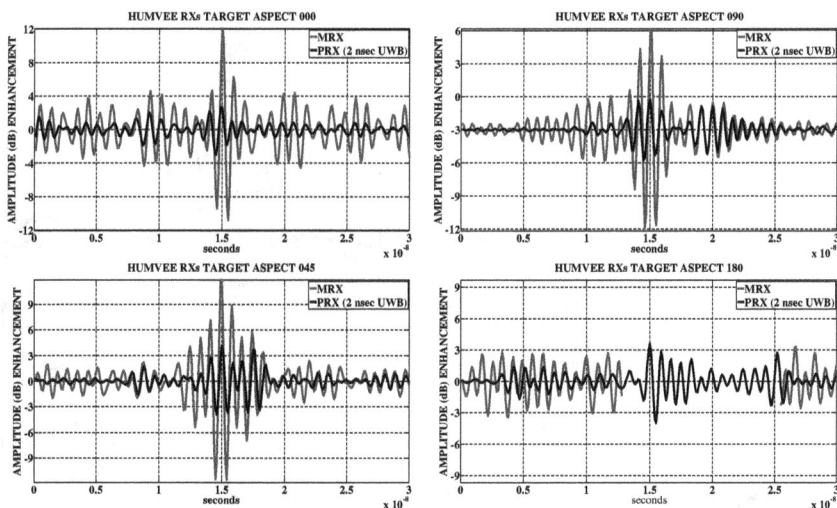

(b)

Fig. 1.7.1   Time domain examples of PRXs (or UWB) and MRXs for a: the Truck target; and b: the Humvee target. Rows in each case are for the targets at aspect angles 00° & 45° and 90° & 180°. The maximum band amplitude enhancements — MRX versus PRX or MRX/PRX — are summarized in Table 1.7.1.

TRUCK PRXs & MRXs

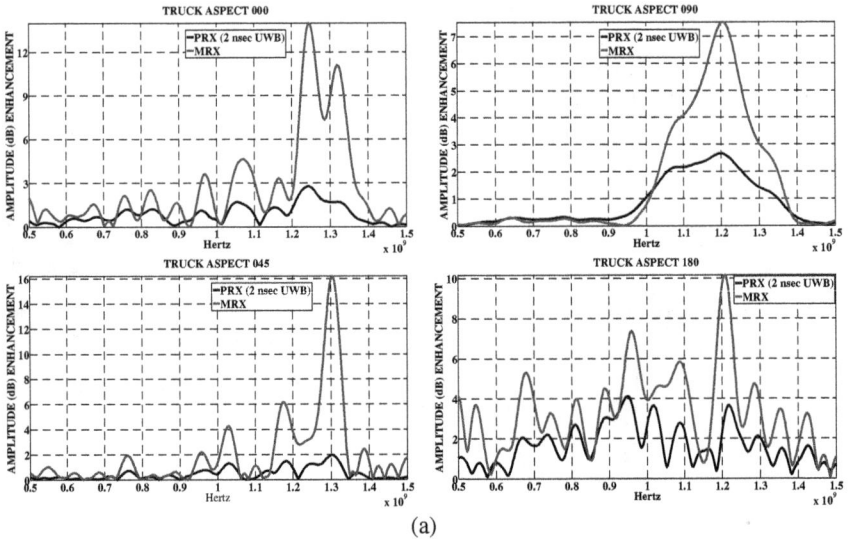

(a)

HUMVEE PRXs & MRXs

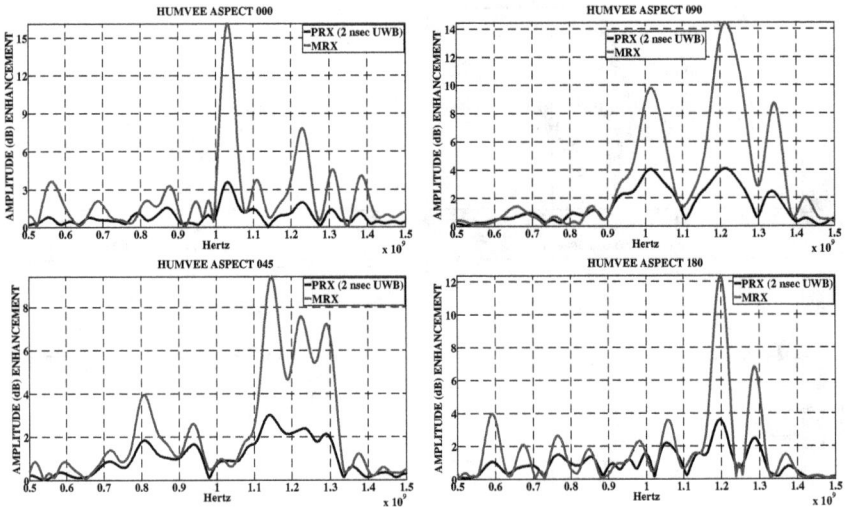

(b)

Fig. 1.7.2   Spectra of PRXs (or UWB) and MRXs for a: the Truck target; and b: the Humvee target. Rows in each case are for the targets at aspect angles 00° & 45° and 90° & 180°. The maximum band amplitude enhancements — MRX versus PRX or MRX/PRX — are summarized in Table 1.7.1.

Table 1.7.1 MRX/PRX Maximum Band
Amplitude Enhancements.

Target Aspect Angle (deg)	Ford Truck (dB)	Humvee (dB)
000	12.7	11.8
045	17.1	7.9
090	5.8	9.6
180	6.9	7.5

(Jackson, 2003). PCA assumes that the UTX signals have Gaussian probability distributions.

Singular Value Decomposition (SVD) is a PCA matrix-based method for obtaining principal components without having to obtain the covariance matrix. SVD requires the assumption that the "important" source signals in the matrix of signals, UTX, required for target identification, are not among the smaller eigenvectors. Another method to decorrelate a set of signals is Gram–Schmidt orthogonalization (GSO). GSO depends on the initial choice among the ERX/[UTX] signals, removing that chosen signal, and projecting the remaining reduced set of signals onto a lower dimensional plane. The correct initial choice is thus critical.

Independent Component Analysis (ICA) is a generalization of PCA, of factor analysis (FA), and of a form of blind source separation (BSS). ICA assumes that the UTX and any noise components, are non-Gaussian and statistically independent, and uses different optimization criteria. The requirement of statistical independence sets ICA apart from the other procedures. Whereas PCA and FA find a set of signals, ERX/[UTX], that are uncorrelated with each other, ICA finds a set of signals, ERX/[UTX], that are statistically independent from each other. It should be noted that a lack of correlation is a weaker property than independence: whereas independence implies a lack of correlation, a lack of correlation does not imply independence (Stone, 2004). ICA can use a PCA pre-processing step that decorrelates (i.e., whitens) the KRX. There are a number of different forms of ICA, e.g., temporal, spatio-temporal, local (Hyvärinen *et al.*, 2001; Cichocki & Amari, 2002). ICA is more noise-sensitive than PCA.

Another method, Complexity Pursuit, assumes that the source signals, UTX, have informational temporal or spatial structure. Complexity Pursuit uses a measure of complexity, e.g., Kolmogorov complexity (Cover & Thomas, 1991), and the signal with the lowest complexity is extracted

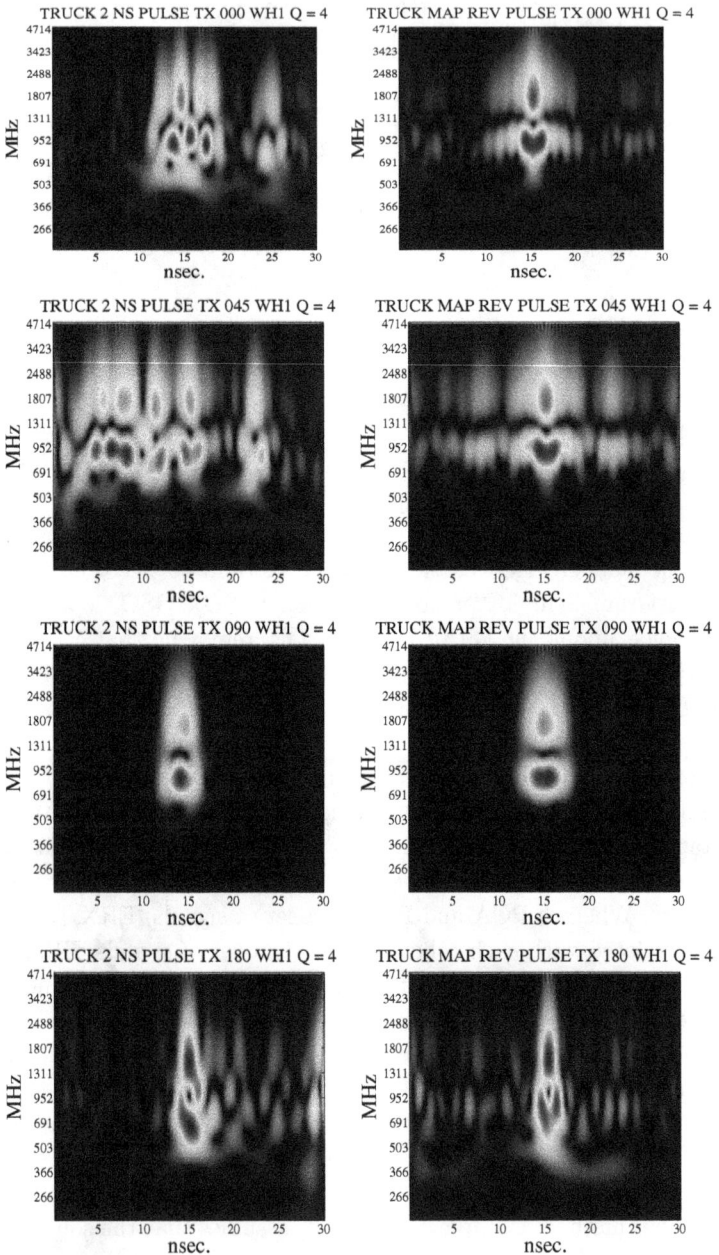

Fig. 1.7.3   Time-frequency spectra for PRX (or UWB) — left — and MRX — right — for a: Truck target; and b: Humvee target. Rows top to bottom in each case are for the targets at aspect angles 00°, 45°, 90° and 180°. All spectra were obtained from the raw RX data using WH-1, a differentiating wavelet filter. In contrast with the PRX spectra,

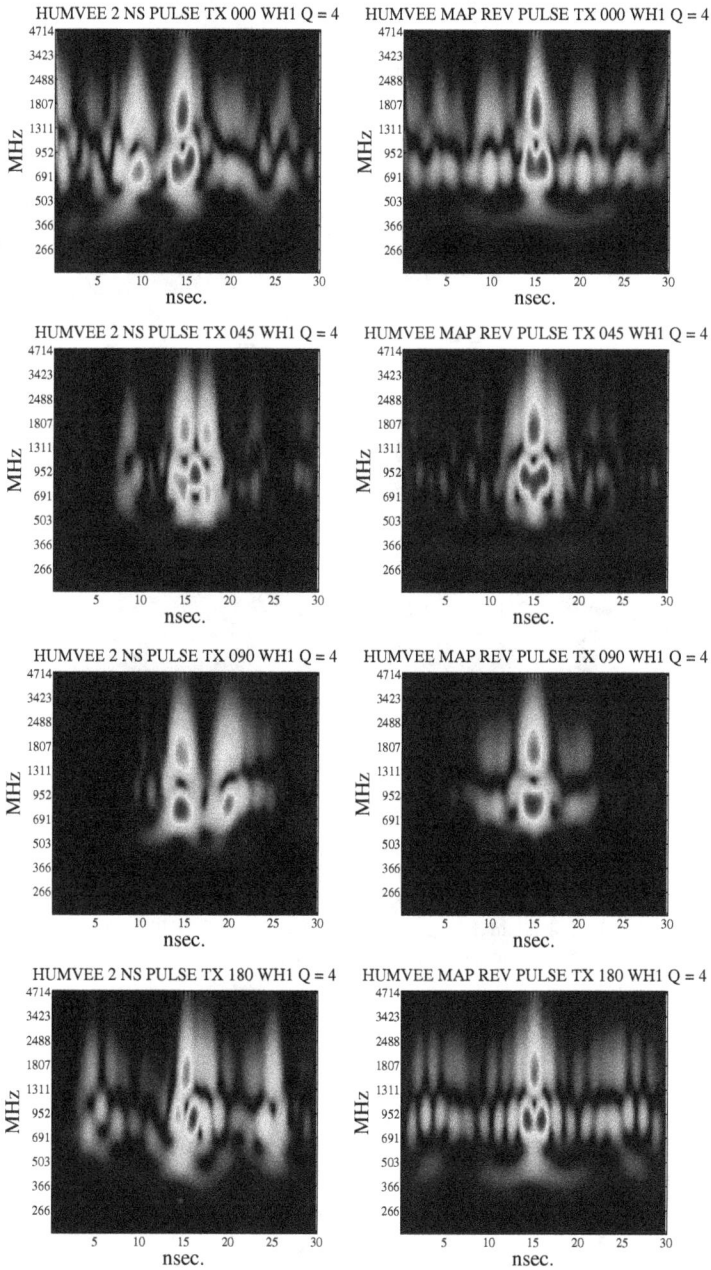

HUMVEE 2 NS PULSE TX 000 WH1 Q = 4

HUMVEE MAP REV PULSE TX 000 WH1 Q = 4

HUMVEE 2 NS PULSE TX 045 WH1 Q = 4

HUMVEE MAP REV PULSE TX 045 WH1 Q = 4

HUMVEE 2 NS PULSE TX 090 WH1 Q = 4

HUMVEE MAP REV PULSE TX 090 WH1 Q = 4

HUMVEE 2 NS PULSE TX 180 WH1 Q = 4

HUMVEE MAP REV PULSE TX 180 WH1 Q = 4

Fig. 1.7.3 (*Continued*) the MRX spectra exhibit the expected time domain symmetry due to the convolution of the target transfer function with its complex conjugate (the MTX), as indicated in Figs. 7 and 8 of the INTRODUCTION.

TRUCK 2 NS PULSE TX 000 WH1 Q = 4

TRUCK MAP REV PULSE TX 000 WH1 Q = 4

TRUCK

TRUCK 2 NS PULSE TX 045 WH1 Q = 4

TRUCK MAP REV PULSE TX 045 WH1 Q = 4

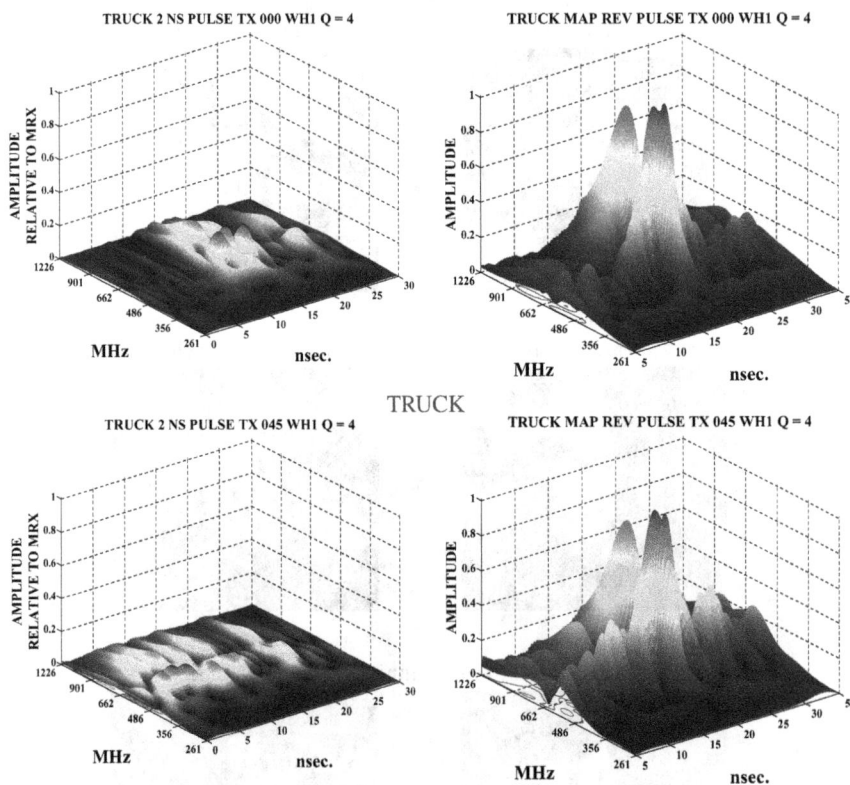

Fig. 1.7.4   Time-frequency-amplitude spectra for PRX (or UWB) — left — and MRX — right — for a: Truck target; and b: Humvee target. Rows in each case are for the targets at aspect angles 00° & 45° and 90° & 180°. All spectra were obtained from the raw RX signals using WH-1, a differentiating wavelet filter. In contrast with the PRX spectra, the MRX spectra exhibit the expected convolutional symmetry in the time domain as indicated in Figs. 7 and 8 of the INTRODUCTION. The maximum band amplitude enhancements — MRX versus PRX or MRX/PRX — are summarized in Table 1.7.1.

(Hyvärinen, 2001). The method can be used on mixtures, KRX, that are super-Gaussian, Gaussian, and sub-Gaussian.

These methods are also related to blind equalization, as system identification is similar to the problem of equalizing a linear channel (Ding & Li, 2001). The objective of blind equalization in this case is to recover the underlying target response characteristics based solely on the probabilistic and statistical properties of the *a priori* data.

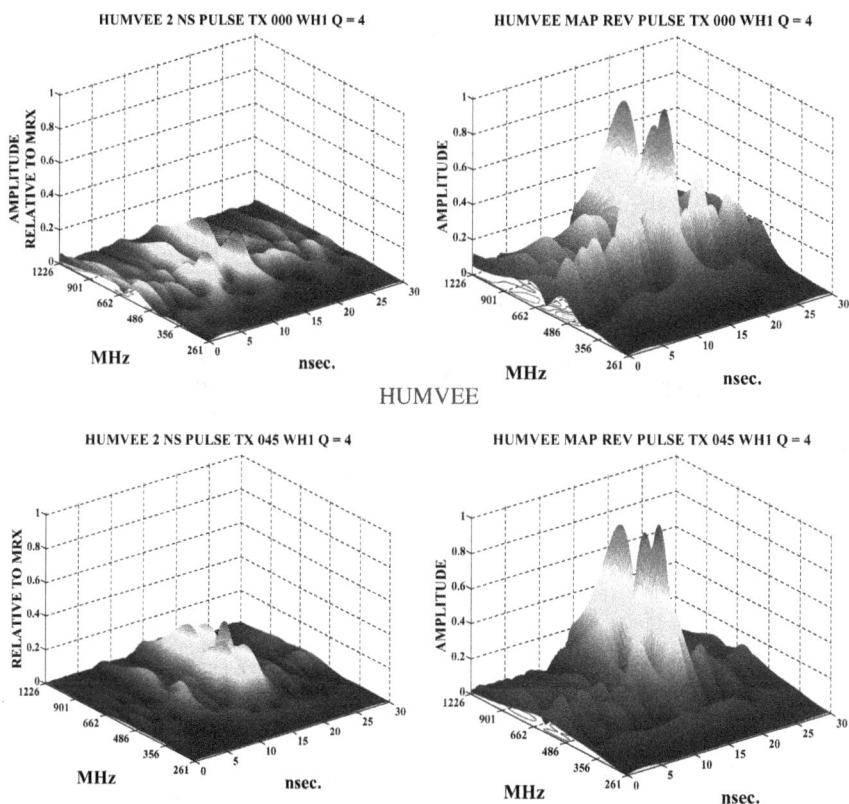

HUMVEE 2 NS PULSE TX 000 WH1 Q = 4

HUMVEE MAP REV PULSE TX 000 WH1 Q = 4

HUMVEE

HUMVEE 2 NS PULSE TX 045 WH1 Q = 4

HUMVEE MAP REV PULSE TX 045 WH1 Q = 4

Fig. 1.7.4   (*Continued*)

1.10.1. *Singular value decomposition*

The following Fig. 1.10.1.1 indicates that under the assumption required by the SVD method — that the UTX signals are Gaussian — P-51 and C-160 model targets can be identified on the basis of their MRXs and SVD-originated ERX/[UTX]s.

1.10.2. *Independent component analysis*

The following Fig. 1.10.2.1 shows that under the assumption required by ICA — that the UTX signals are non-Gaussian and statistically independent — P-51 and C-160 model targets can also be identified

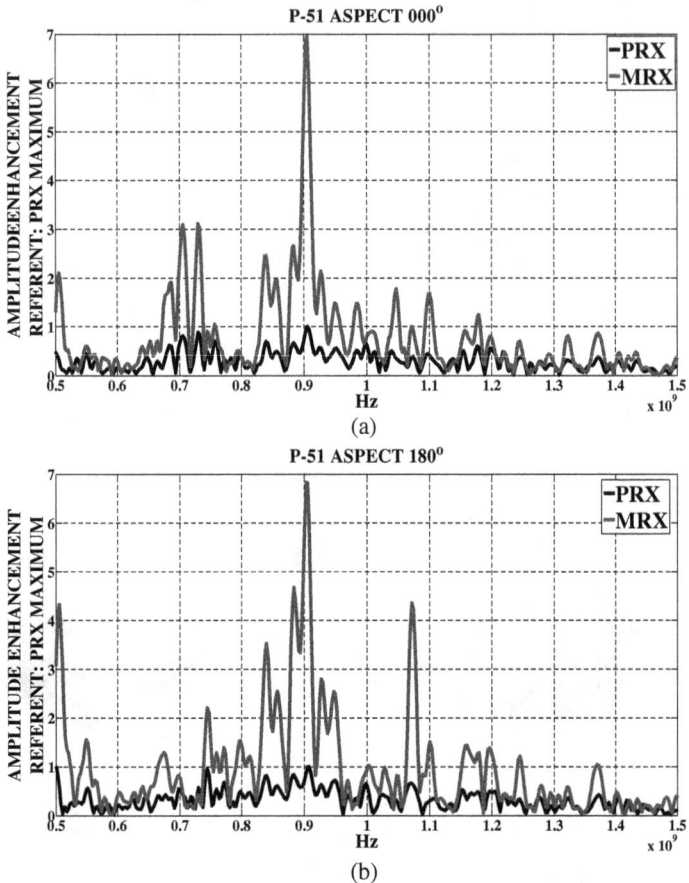

Fig. 1.8.1 MRX and PRX spectra for the P-51 Model at 0° and 180° aspect angles. The amplitude enhancement afforded by the MTX signal, in comparison with the PTX (or UWB) signal, is indicated by referencing the amplitude of the MRX to the maximum of the PRX. With changes in aspect there are changes in the amplitude of spectral bands, while the frequency location of those bands remain constant.

on the basis of their MRXs and ICA-originated [UTX]s. Figure 1.10.2.2 demonstrates similarly that Truck and Humvee targets can be identified.

1.10.3. *Aspect independence*

The success of both SVD and ICA methods, requiring different UTX assumptions, may be due to the quite different PRX and MRX spectra (Fig. 1.10.3.1).

AVERAGE P-51 MRXs; ASPECTS 000:180; REFERENTS:PRX MAXIMA

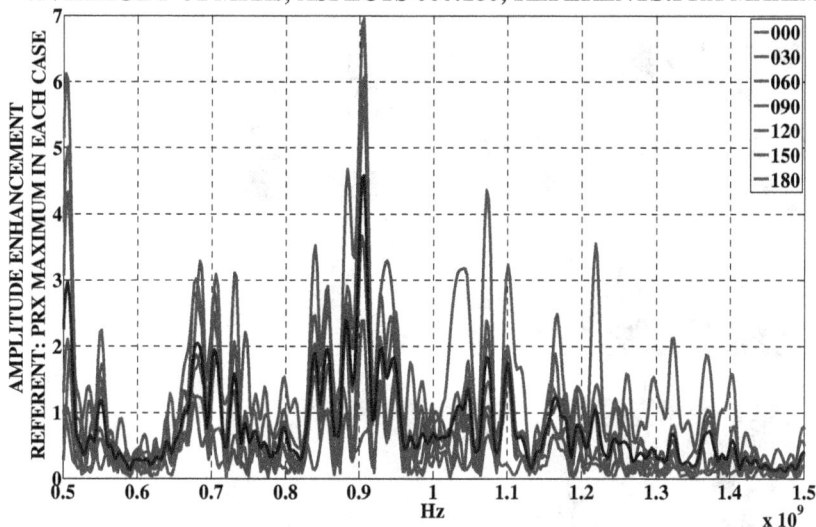

Fig. 1.8.2   The average of MRX spectra for the P-51 Model at 7 aspect angles overlaid on the unaveraged 7 spectra. The main spectral bands are indicated.

1.11.   Selective Enhancement of Target Major & Minor Resonances

If the PRX/MRX spectra for different targets have minimal overlapping frequency bands, it is possible to design MTX signals that selectively enhance one or the other's spectral return. For example, Fig. 1.11.1 shows the MRX spectra obtained for the two targets: the extended Sea Sparrow missile mockup (ESSM) and its container. Using the MTX for the ESSM on the container target, or using the MTX for the container on the ESSM target, results in reduced amplitude MRXs.

Minor target resonances can also be selectively addressed. Figure 1.11.2 shows the MRX spectra for two flat aluminum strips, one of length 0.30 m, and the other of length 0.55 m. The resonant frequency of the smaller strip is 1 GHz, and of the longer strip, 0.5455 GHz. Changing the frequency of MTX from mismatching to matching the relevant strip's resonant frequency enhanced the MRX by 5.1 dB in the case of the 0.55 m strip, and 21.9 dB in the case of the 0.30 m strip.

In other instances, the spectral bands in different target spectra may overlap. Many targets have shared frequency bands in their PRX/MRX

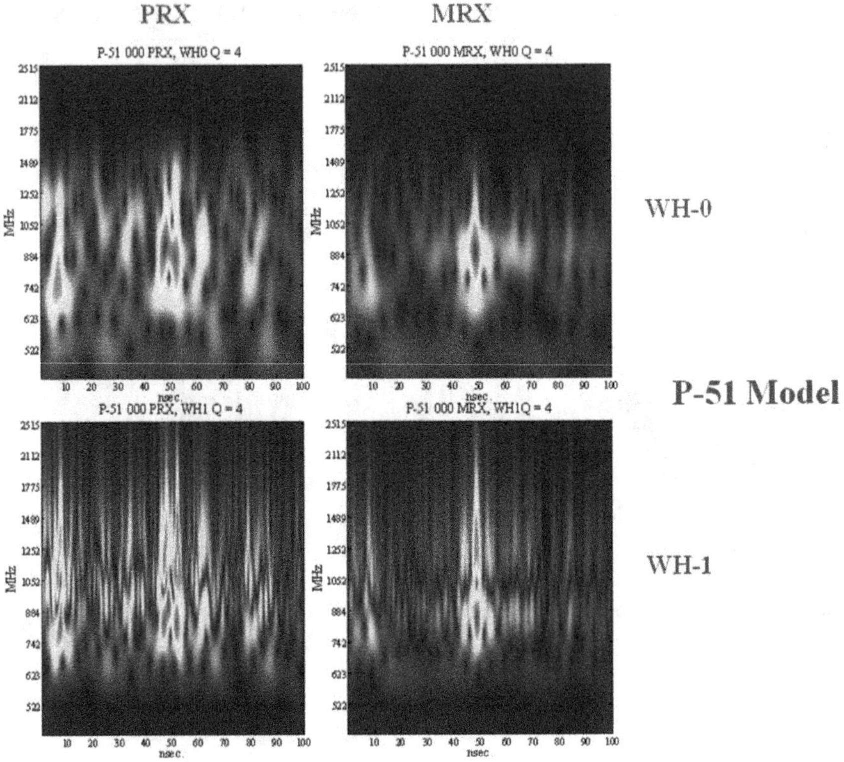

Fig. 1.8.3   Time-frequency spectra for PRX (or UWB) — left — and MRX — right — for the P-51 Model at 0° aspect angle. Upper spectra obtained from raw RX signals using WH-0, an averaging scaling function filter. Lower spectra obtained using WH-1, a differentiating wavelet filter. In contrast with the PRX spectra, the MRX show the expected time domain convolutional symmetry as indicated in Figs. 7 and 8 of the INTRODUCTION.

spectra. To selectively enhance the uniquely discriminating bands from one target with respect to another, MTX signals can be designed for which the shared frequency components are masked, or set to zero. Figures 1.11.3–1.11.5, demonstrate that it is possible to design MTX signals that enhances minor resonances present in the target PRXs/MRXs, while eliciting minimum signal return from the major target resonances. Thus it is possible to "zoom-in" on designated target minor resonances, and conduct "questions-and-answers" in a series of TX-RX exchanges with the target, in which a TX "question" is, or is not, confirmed by the amplitude of the RX "answer".

C160 ASPECT 000⁰

(a)

C160 ASPECT 050⁰

(b)

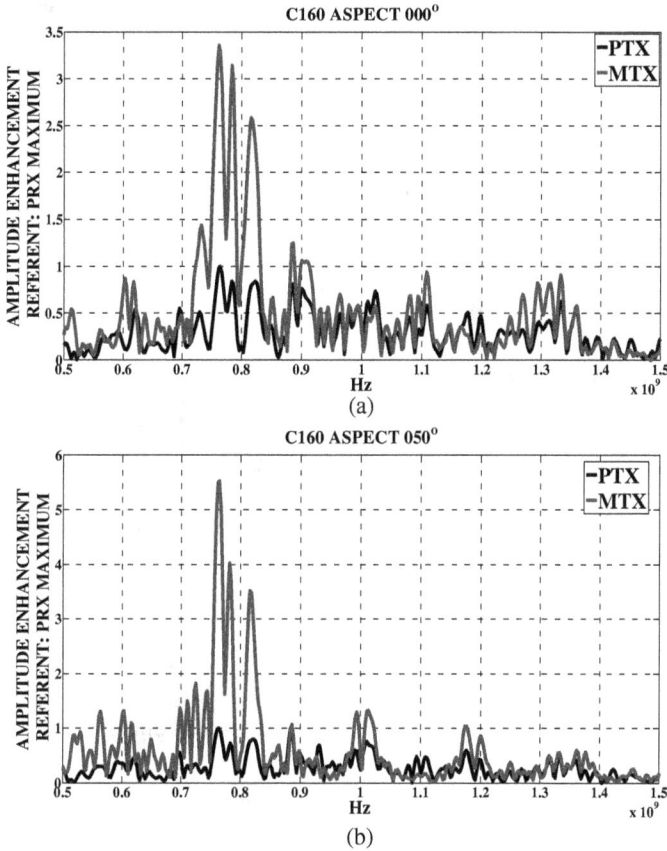

Fig. 1.8.4   MRX and PRX spectra for the C-160 Model at 0° and 50° aspect angles. The amplitude enhancement afforded by the MTX signal, in compoarison with the PTX (or UWB) signal, is indicated by referencing the amplitude of the MRX to the maximum of the PRX. With changes in aspect there are changes in the amplitude of spectral bands, while the frequency location of those bands remain constant.

1.12.   Target Surface Detection

In some cases, the quadratic Fractional Fourier transform (FRFT) — addressed more fully in Section 1.15, below — can provide information of differences in the spread of current on target surfaces of different material composition. Whereas the conventional Fourier transform is based on cisoidal, and steady state, basis functions, the FRFT is based on frequency modulated or chirp basis functions that convolve maximally with signals changing in frequency or dispersing; and such dispersion, and differences

AVERAGE C-160 MRXs ASPECTS 000:90, ADJUSTED TO PRXs, TX 1 ns

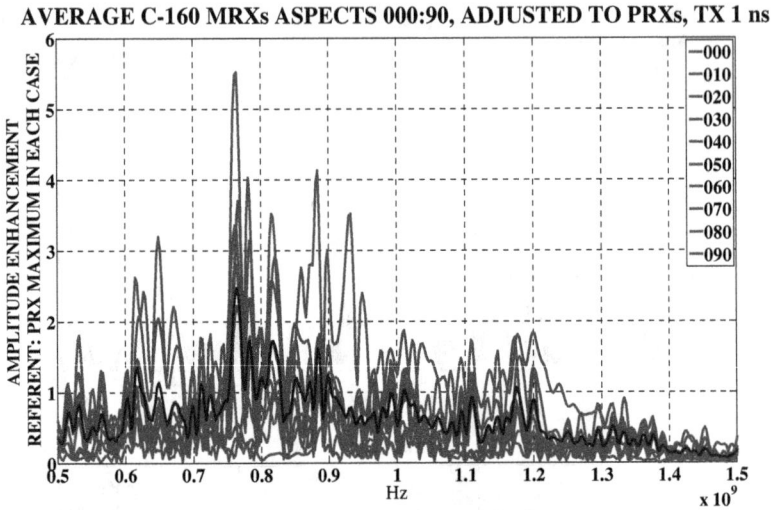

Fig. 1.8.5 The average of MRX spectra for the C-160 Model at 10 aspect angles overlaid on those unaveraged 10 spectra. The main spectral bands are indicated.

in dispersion, can occur on the surface of targets. In the following set of equations, the variable, a, ranges from -2 to $+2$. When $a = +1$, the transformation is the conventional forward Fourier transformation; when $a = -1$, the transformation is the conventional inverse Fourier transformation. When $a = 0$, the untransformed time-domain signal is obtained. At other values of a, the a'th-order FRFT is obtained. The relations are described in the following equations:

$$FRFT_a = \int_{-\infty}^{+\infty} K_a(f,t)f(t)dt,$$

where

$$K_a(f,t) = A_a \exp[i\pi(\cot(\alpha f^2) - 2csc(\alpha ft) + \cot(\alpha t^2))],$$
$$A_a = \sqrt{1 - i\cot(\alpha)},$$
$$\alpha = a\pi/2.$$

In the case $a = 1$, then $\alpha = \pi/2$, and

$$FRFT_1(f) = FT(f) = \int_{-\infty}^{+\infty} \exp[-i2\pi ft]f(t)dt.$$

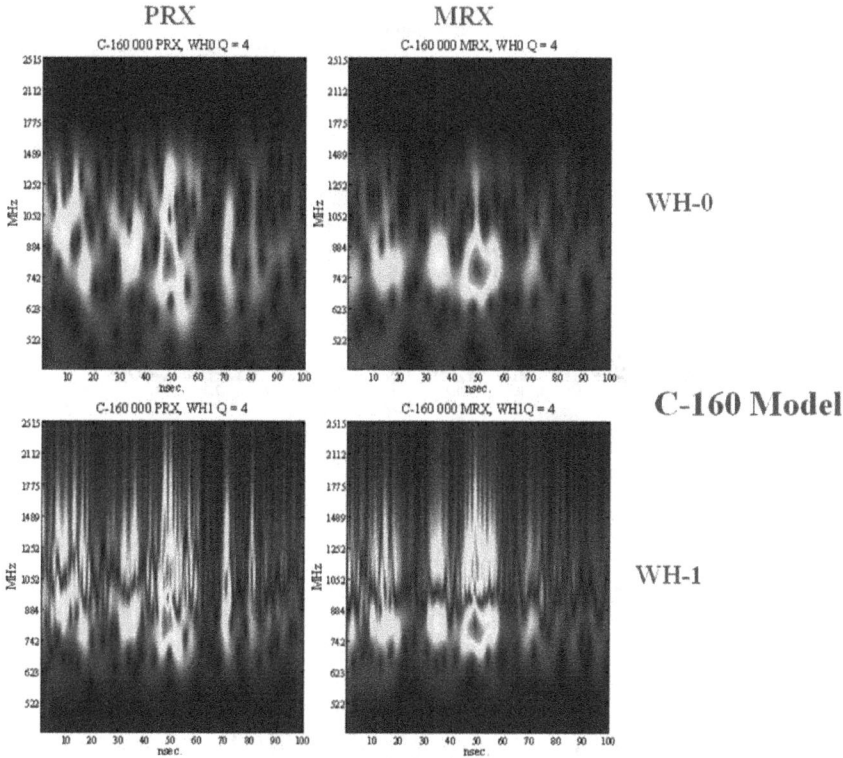

Fig. 1.8.6 Time-frequency spectra for PRX (or UWB) — left — and MRX — right — for the C-160 Model at 0° aspect angle. Upper spectra obtained from the raw RX signals using WH-0, an averaging scaling function filter. Lower spectra obtained using WH-1, a differentiating wavelet filter. In comparison with the PRX spectra, the MRX show the expected time domain convolutional symmetry as indicated in Figs. 7 and 8 of the INTRODUCTION.

Figure 1.12.1 compares the FRFT spectra for both PRXs (left column) and MRXs (right column), and for the truck (a) and humvee (b) as targets, at various aspect angles. Apart from the higher symmetry and definition in the MRX FRFT spectra (right column) versus the PRX FRFT spectra (left column), there are major differences between the truck MRX FRFT spectra (a) and the humvee FRFT spectra (b). Figures 1.12.2–1.12.4 further highlights the differences in the truck versus humvee FRFT spectra.

Figure 1.12.5 compares the MRX FRFT Spectra for aluminum and rubber surfaced cone targets at various aspect angles. The differences in these spectra indicate that surface composition differences in targets of identical geometry can be detected.

B-2 ASPECT 000⁰

(a)

B-2 ASPECT 120⁰

(b)

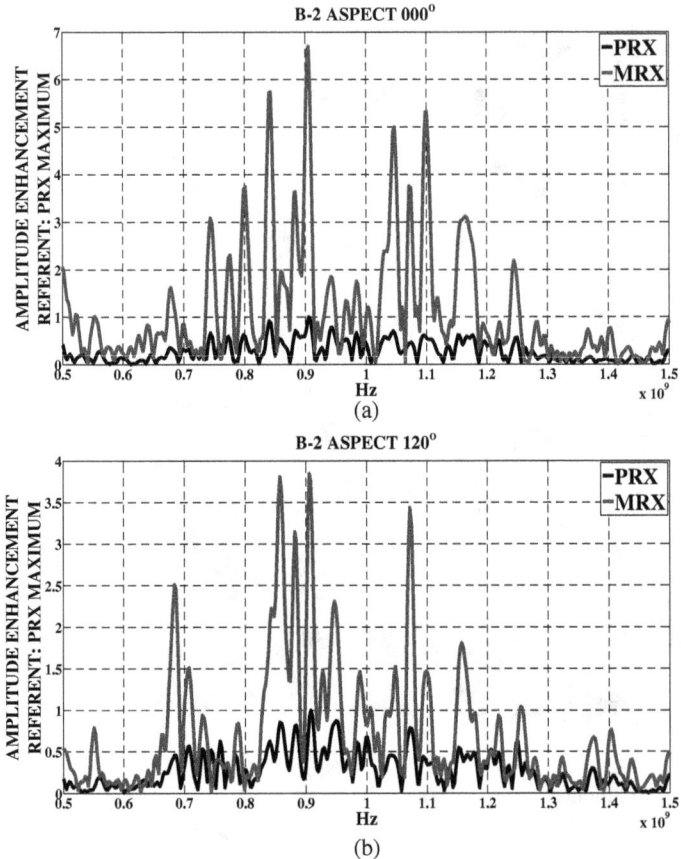

Fig. 1.8.7   MRX and PRX spectra for the B-2 Model at 0° and 120° aspect angles. The amplitude enhancement afforded by the MTX signal, in comparison with the PTX (or UWB) signal, is indicated by referencing the amplitude of the MRX to the maximum of the PRX. With changes in aspect there are changes in the amplitude of spectral bands, while the frequency location of those bands remain constant.

1.13.   Nonlocal Transformations: Wigner-Ville Distribution & Ambiguity Function

Time-frequency methods, including the spectrogram, are designed for linear, but non-stationary, data, i.e., for LTV system signals. Time-frequency distributions are obtained from the product of a signal at a past time and the signal at a future time, or, equivalently, from the product of a signal's higher frequency components and a signal's lower frequency components. Therefore, time-frequency analyses are *nonlocal or global transformations*.

AVERAGE B-2 MRXs ASPECTS 000:150, ADJUSTED TO PRXs, TX 1 ns

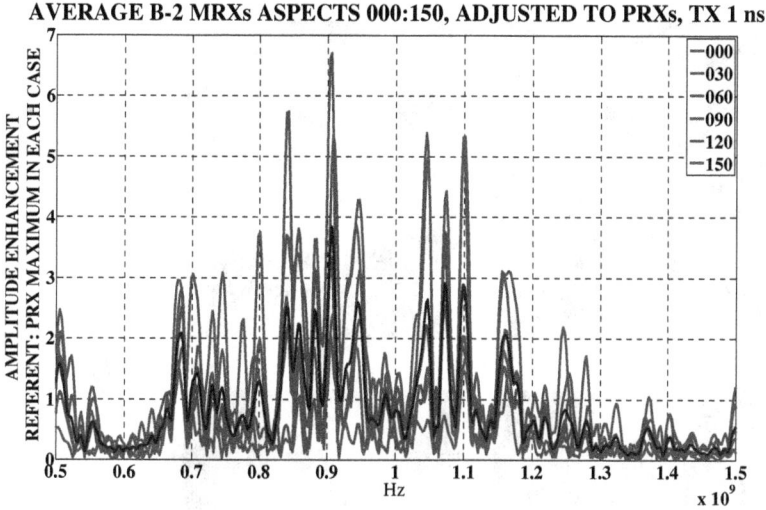

Fig. 1.8.8   The average of MRX spectra for the B-2 Model at 6 aspect angles overlaid on those unaveraged 10 spectra. The main spectral bands are indicated.

Because the analyzed signal is used twice in these definitions, the time-frequency distribution is said to be *bilinear* or *quadratic*. Time-frequency analysis is effective in analyzing non-stationary signals, whose frequency distribution and magnitude vary with time, and is phase-shift invariant. A drawback to these methods is potential cross-term "contamination", which can occur when analyzing multi-component signals. However, by using window functions, the cross-terms can be mitigated with the penalty of some loss of resolution. Time-frequency methods are suitable for signals with narrow instantaneous bandwidth and specifically, signals from LTV systems.

The *general class, or Cohen's class*, C, for all time-frequency representations is:

$$C(t,\omega) = \frac{1}{4\pi^2} \iiint_{-\infty}^{+\infty} s^*(u - \tau/2)s(u + \tau/2)\varphi(\theta,\tau)$$
$$\times \exp[-i\theta t - i\tau\omega + i\theta u]dud\tau d\theta,$$

or

$$C(t,\omega) = \frac{1}{4\pi^2} \iiint_{-\infty}^{+\infty} S^*(u + \theta/2)S(u - \theta/2)\varphi(\theta,\tau)$$
$$\times \exp[-i\theta t - i\tau\omega + i\tau u]d\theta d\tau du.$$

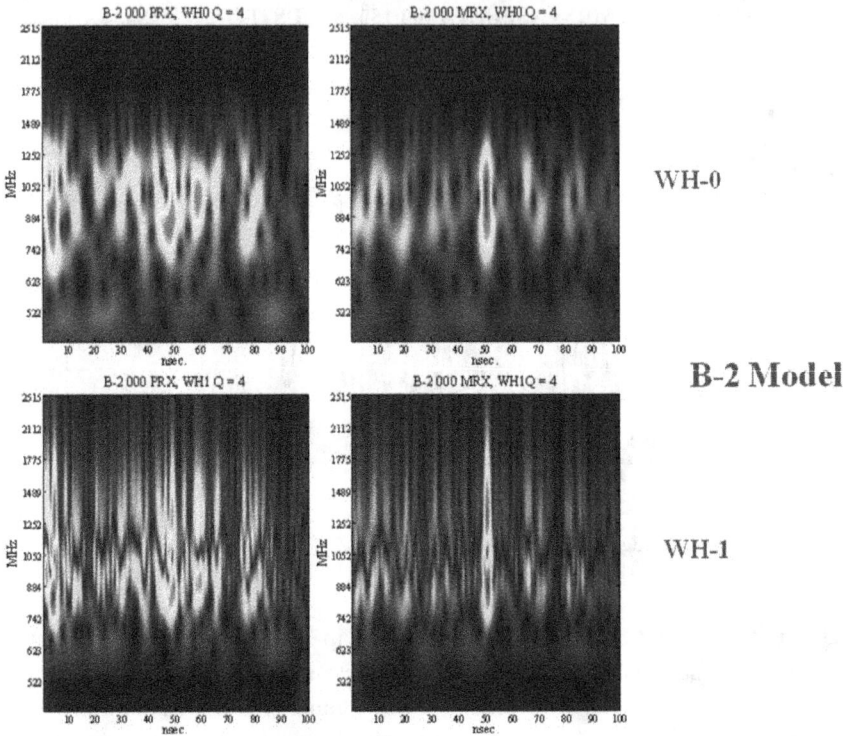

Fig. 1.8.9   Time-frequency spectra for PRX (or UWB) — left — and MRX — right — for the B-2 Model at 0° aspect angle. Upper spectra obtained from the raw RX signals using WH-0, an averaging scaling function filter. Lower spectra obtained using WH-1, a differentiating wavelet filter. In comparison with the PRX spectra, the MRX show the expected convolutional symmetry as indicated in Figs. 7 and 8 of the INTRODUCTION.

where $\varphi(\theta, \tau)$ is a function called the kernel (Claasen & Mecklenbräuker, 1980a–c; Janssen, 1981, 1982, 1984), and s is a time domain signal.

The general class can be defined in terms of the *characteristic function, M*:

$$C(t, \omega) = \frac{1}{4\pi^2} \iint_{-\infty}^{+\infty} M(\theta, \tau) \exp[-i\theta t - i\tau\omega] d\theta d\tau,$$

where

$$M(\theta, \tau) = \varphi(\theta, \tau) \int_{-\infty}^{+\infty} s^*(u - \tau/2) s(u + \tau/2) \exp[i\theta u] du$$

$$= \varphi(\theta, \tau) A(\theta, \tau),$$

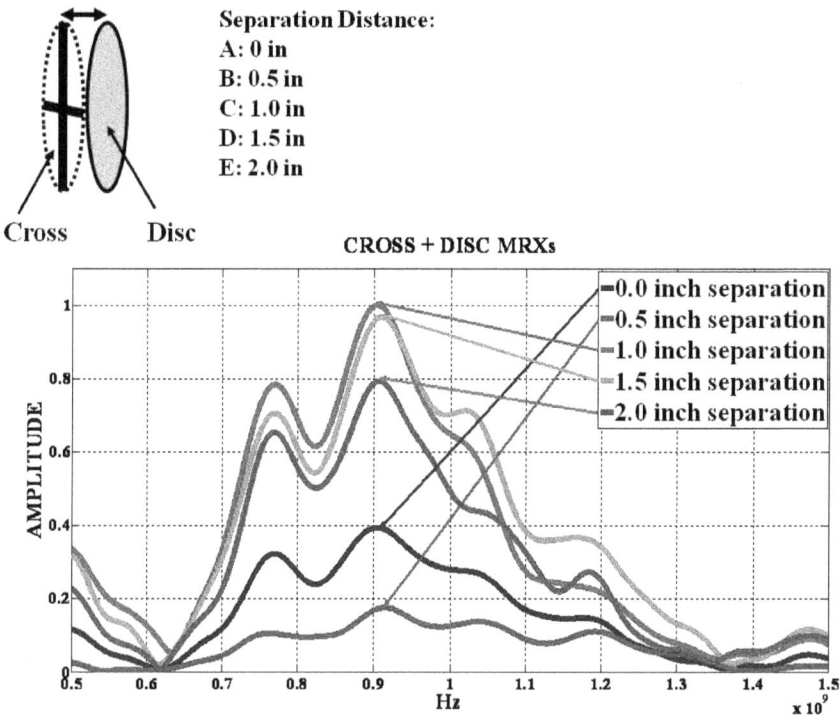

Separation Distance:
A: 0 in
B: 0.5 in
C: 1.0 in
D: 1.5 in
E: 2.0 in

Fig. 1.9.1 PRX spectra of a cross and a disk separated by determined distances. At separations of 1.0 inch and greater, the two targets respond independently and additively. At a separation of 0.5 inches or less the two minor targets are near-field coupled, the resonances within the RX bandwidth are drastically lowered in amplitude, and it is possible that other resonance(s) are created outside the system RX bandwidth (0.5–1.5 GHz). At 0.0 inch separation — electrically joined — it is possible that the RX spectrum is a mix of the individual within-system bandwidth resonances and the out-of-system bandwidth resonances.

and $A(\theta, \tau)$ is the *symmetrical* ambiguity function. Time-frequency distributions can then be classified with respect to the kernel function $\varphi(\theta, \tau)$ (Cohen, 1989, 1995).

With

$$R(\tau) = \int_{-\infty}^{+\infty} s^*(u)s(u + \tau)du$$

defined as the *deterministic auto-correlation function*, a *deterministic generalized local autocorrelation function* can be defined as (Choi & Williams,

FRAME TEST

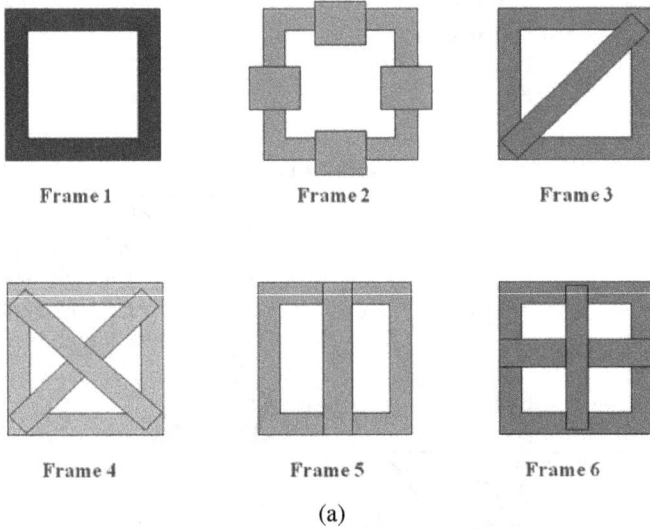

(a)

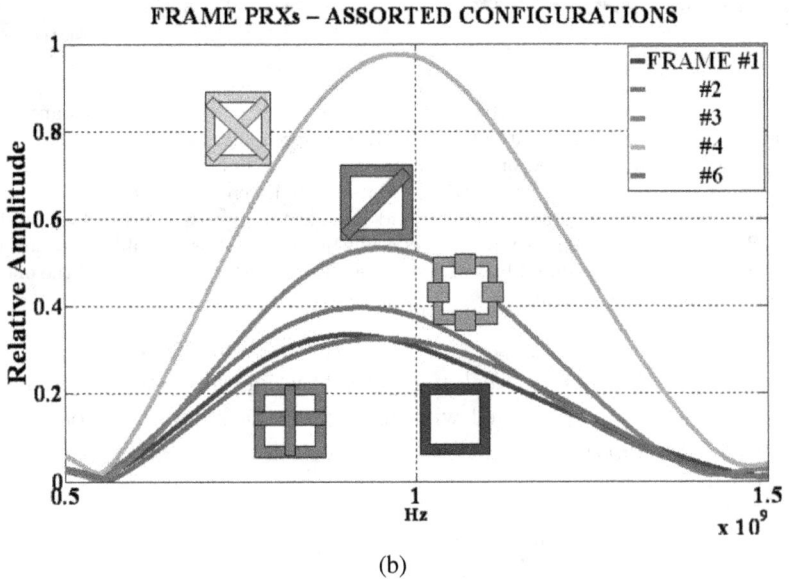

(b)

Fig. 1.9.2   PRX spectra of five of the frame targets. The addition of a diagonal cross to the frame (Frame #4) results in an increase in response amplitude, but the addition of an upright cross (Frame #6) results in a decrease. The decrease could be due to the RX spectrum being a mix of the individual within-system bandwidth resonances and out-of-system bandwidth resonances.

RING TEST

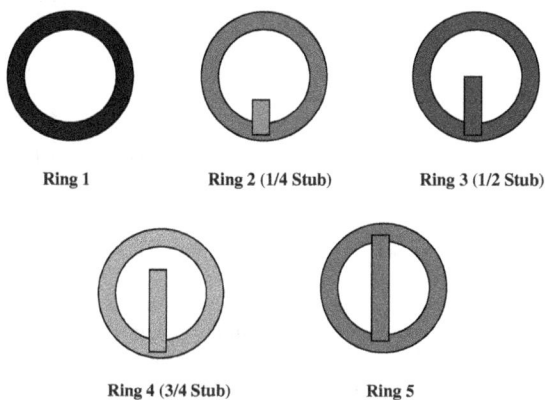

Ring 1   Ring 2 (1/4 Stub)   Ring 3 (1/2 Stub)

Ring 4 (3/4 Stub)   Ring 5

MAP_Data_05-23-04_LocalGlobal_WithData

(a)

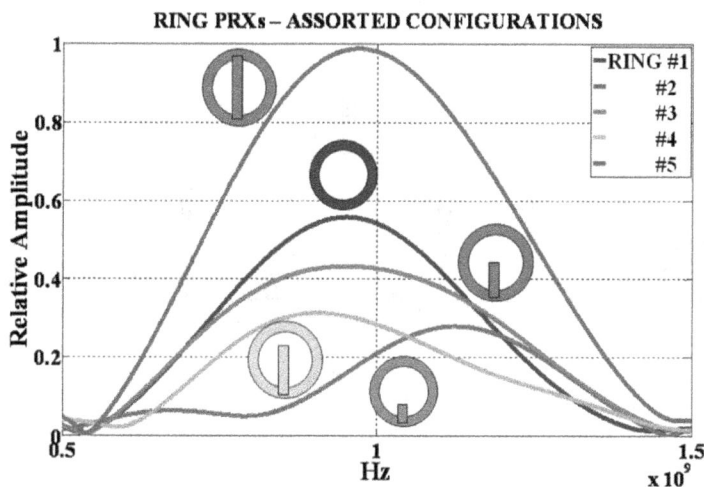

RING PRXs – ASSORTED CONFIGURATIONS

(b)

Fig. 1.9.3   PRX spectra of five ring targets. The smallest addition to the ring (Ring #2) results in the greatest PRX band decrease with an additional change in the peak of the resonance band. A lengthening of the addition (Ring #3) results in an increase in amplitude but still below that of the target without the addition (Ring #1). A further lengthening of the addition (Ring #4) results in another decrease in the amplitude of the band. With the addition finally spanning the ring (Ring #5), there is a major ncrease in amplitude of the spectral band. It should be noted that these band amplitude changes are relative to the system 1 GHz (0.5–1.5 GHz) bandwidth, and a decrease in the amplitude of the band within this bandwidth might signal a generation or an increase in amplitude of another band outside the bandwidth.

UTX?   KRX

Mixing Matrix = Target at All Aspects = Channel

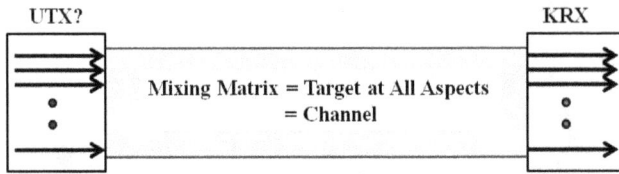

UTX = Unknown Set of target aspect-independent signals.
KRX = Known Set of Received Signals
     = Set of MRXs or PRXs at various target aspect angles.

Fig. 1.10.1 In this model a set, or matrix, of unknown aspect independent signals, UTX, enters a channel, and a known set of signals, KRX, is received (i.e., either PRXs or MRXs at set target aspect angles). By assuming some statistical properties of the channel, the KRX can be decomposed to obtain an estimate of the common aspects of the RX signals, i.e., a set of signals designated: ERX. Therefore the ERX are an estimate of the UTX, or [UTX]. Any later received RX can then be correlated with any of the ERX/[UTX] set for target identification purposes. The role of the target in this model is a mixing matrix role. There are several strategies for extracting the source signals, ERX/[UTX], e.g., PCA, SVD, ICA, GSO, etc, and any method used is only as valid as the assumptions implicit in that method.

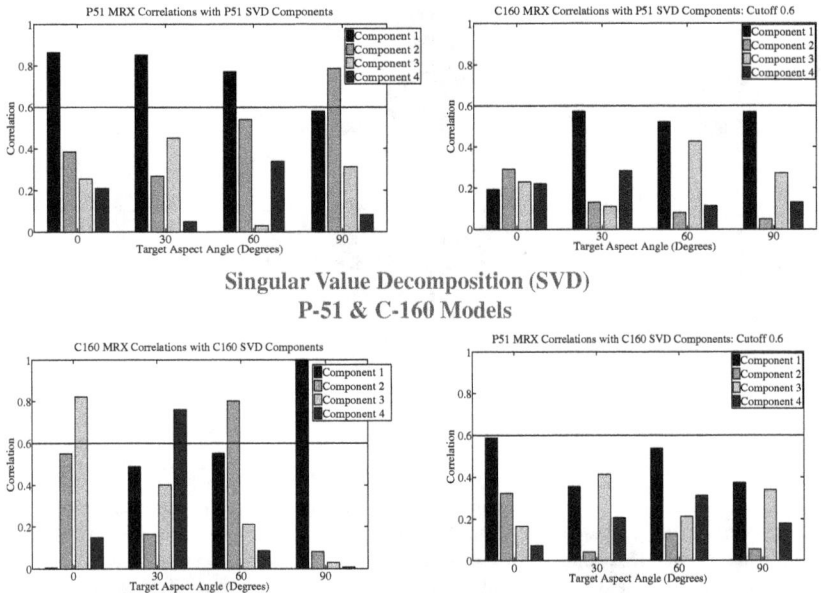

Singular Value Decomposition (SVD)
P-51 & C-160 Models

Fig. 1.10.1.1 SVD Target Identification: Correlations of Target MRXs for 0, 30, 60 & 90° aspect angles, with Target [UTX]s = SVD components.
Upper Left: Correlations of P-51 Model MRXs with P-51 Model [UTX]s. Upper Right: Correlations of C-160 Model MRXs with P-51 Model [UTX]s.
Lower Left: Correlations of C-160 Model MRXs with C-160 Model [UTX]s. Upper Right: Correlations of P-51 Model MRXs with C-160 Model [UTX]s.

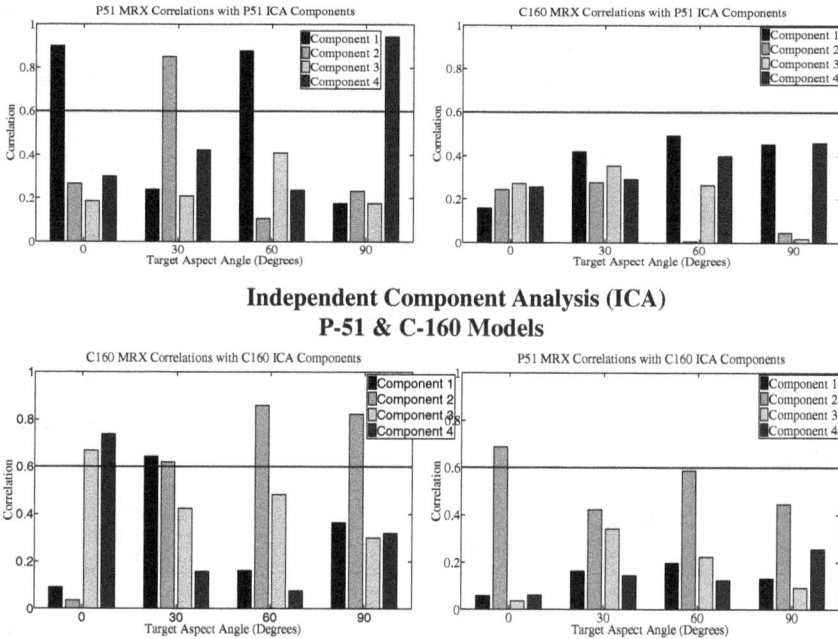

Independent Component Analysis (ICA)
P-51 & C-160 Models

Fig. 1.10.2.1   ICA Target Identification: Correlations of Target MRXs for 0, 30, 60 & 90° aspect angles, with Target [UTX]s = ICA components.
Upper Left: Correlations of P-51 Model MRXs with P-51 Model [UTX]s. Upper Right: Correlations of C-160 Model MRXs with P-51 Model [UTX]s.
Lower Left: Correlations of C-160 Model MRXs with C-160 Model [UTX]s. Upper Right: Correlations of P-51 Model MRXs with C-160 Model [UTX]s.

1989; Cohen, 1995):

$$R_t(\tau) = \iint_{-\infty}^{+\infty} s^*(u - \tau/2)s(u + \tau/2)\varphi(\theta, \tau)\exp[i\theta(u - t)]d\theta du.$$

This auto-correlation is dependent on time.

1.13.1.   *Wigner-Ville distribution*

The WVD (Wigner, 1932; Ville, 1948) is defined as:

$$WVD(t, \omega) = \frac{1}{2\pi} \int_{-\infty}^{+\infty} s^*(t - \tau/2)s(t + \tau/2)\exp[-i\tau\omega]d\tau$$

$$= \frac{1}{2\pi} \int_{-\infty}^{+\infty} S^*(\omega - \theta/2)S(\omega + \theta/2)\exp[-it\theta]d\theta,$$

Independent Component Analysis (ICA)
Ford Truck & Humvee

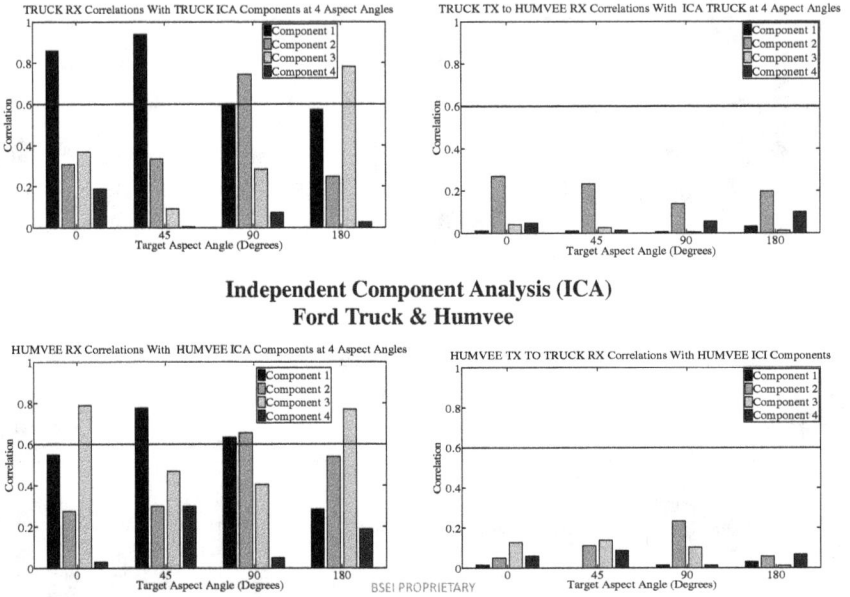

Fig. 1.10.2.2   ICA Target Identification: Correlations of Target MRXs for 0, 30, 60 & 90° aspect angles, with Target [UTX]s = ICA components.
Upper Left: Correlations of Truck MRXs with Truck [UTX]s. Upper Right: Correlations of Humvee MRXs with Truck [UTX]s.
Lower Left: Correlations of Humvee MRXs with Humvee [UTX]s. Upper Right: Correlations of Truck MRXs with Humvee [UTX]s.

where s is a time domain signal and is real. The Wigner-Ville Distribution (WVD) is thus the Fourier transform with respect to τ of an auto-correlation specifically defined as:

$$R(t, \tau) = s^*(t - \tau/2)s(t + \tau/2),$$

or

$$R(t, \tau) = \int_{-\infty}^{+\infty} WVD(t, \omega) \exp[i\tau\omega] d\omega = s^*(t - \tau/2)s(t + \tau/2).$$

As $\varphi(\theta, \tau) = 1$ for the WVD, the WVD characteristic function is:

$$M(\theta, \tau) = A(\theta, \tau).$$

The one-sided WVD time-frequency spectra for a Truck TX and with Truck as target (i.e., Truck TX on Truck), at aspect angles: 0°, 45°, 90° and 180° and for (a): PRX; (b): MRX, are shown in Figs. 1.13.1.1–1.13.1.4, below. For each figure, plots on the left show the WVD frequency marginal

TRUCK TX (BLUE) TO BOTH TARGETS; HUMVEE TX (RED) TO BOTH TARGETS

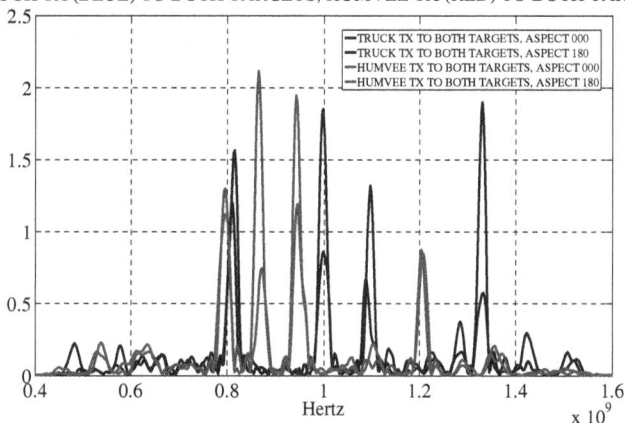

(a)

TRUCK TX (BLUE) TO BOTH TARGETS; HUMVEE TX (RED) TO BOTH TARGETS

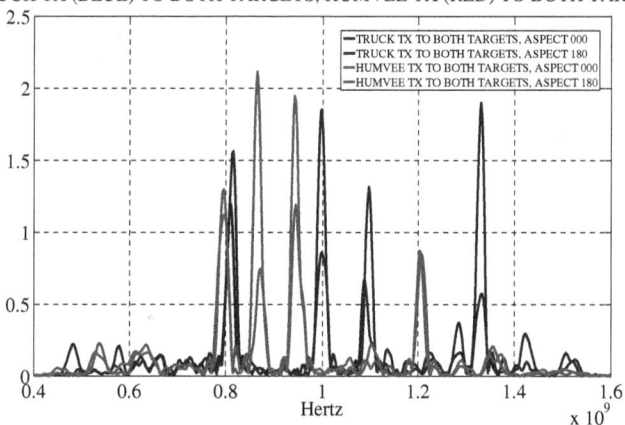

(b)

Fig. 1.10.3.1   MRX spectra of Truck and Humvee at four aspect angles: 000, 045, 090, 180°. a: Truck MRX elicited by Truck MTX *with only the Truck present* and also Humvee MRX elicited by Humvee MTX *with only the Humvee present* and with targets at the aspect angles indicated. b: a: Truck MRX elicited by Truck MTX with both Truck and Humvee present and also Humvee MRX elicited by Humvee *MTX with both Truck and Humvee present* and with targets at the aspect angles indicated. Note that the spectral profiles (spectral location of bands) of a and b are similar while the amplitudes of these bands change with angle.

and the FT spectrum. It is apparent that the WVD contains cross-terms not contained in the FT. It should be noticed that the time symmetry of the WVD for the MRX (b) is due to the convolution of the target with the MTX signal, i.e., the time-reversed impulse response. The convolution of

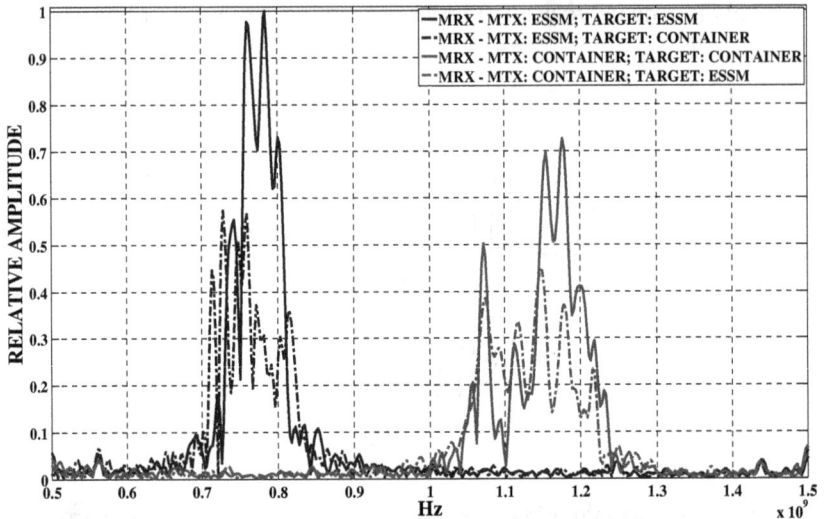

Fig. 1.11.1　MRX spectra obtained for the two targets: the extended Sea Sparrow missile mockup (ESSM) and its container. Using the MTX for the ESSM on the container target, or using the MTX for the container on the ESSM target, results in reduced amplitude MRXs in comparison with using the MTX for the ESSM on the ESSM target, and the MTX for the container on the container target.

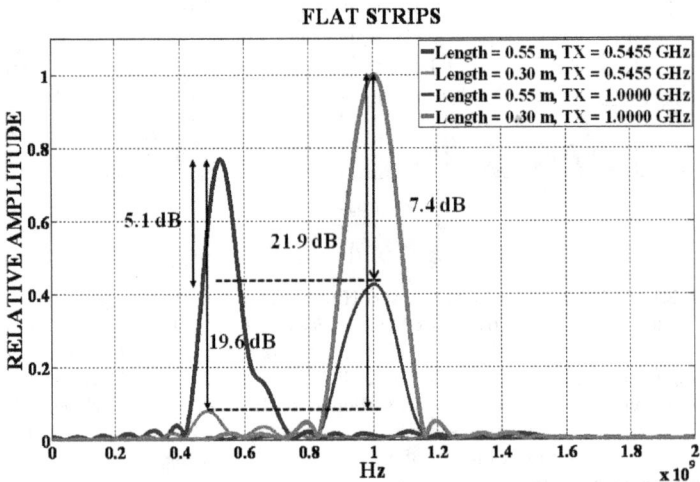

Fig. 1.11.2　MRX spectra for two simple flat aluminum strips, one of length 0.30 m, and the other of length 0.55 m. The resonant frequency of the smaller strip is 1 GHz and of the longer strip, 0.5455 GHz. Changing the frequency of MTX from mismatching to matching the relevant strip's resonant frequency enhances the MRX by 5.1 dB in the case of the 0.55 m strip, and 21.9 dB in the case of the 0.30 m strip.

HUMVEE

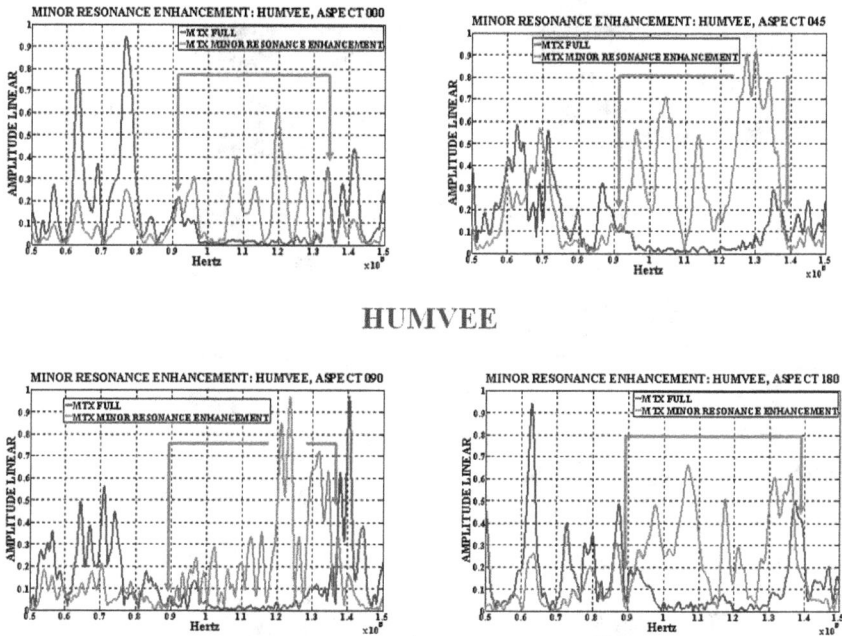

Fig. 1.11.3 Minor resonance enhancement: Spectra for Humvee target at aspect angles 000°, 045°, 090° and 180°. Initially full MRX returns were obtained, and alternative MTXs were designed to address resonances in the 1–1.4 GHz range — indicated by arrows. The amplitude of the MRX minor bands were enhanced, while the amplitude of the major bands were minimized.

the MTX with the target impulse to obtain the symmetrical-in-time MRX was described in the INTRODUCTION.

The one-sided WVD time-frequency spectra for Humvee TX and with Humvee as target, (i.e., humvee TX on humvee) at aspect angles: 0°, 45°, 90° and 180° and for (a): PRX; (b): MRX, are shown in Figs. 1.13.1.5–1.13.1.8, below. Again, for each figure, plots on the left show the WVD frequency marginal and the FT spectrum. It is apparent again that the WVD contains cross terms not contained in the FT. It should be noticed again that the time symmetry of the WVD for the MRX (b) is due to the convolution of the target with the MTX signal, i.e., the time-reversed impulse response. The convolution of the MTX with the target impulse to obtain the symmetrical-in-time MRX was described in the INTRODUCTION.

Figure 1.13.1.9 shows the one-sided WVD time-frequency spectra for a corner reflector TX and with corner reflector as target (i.e., corner reflector

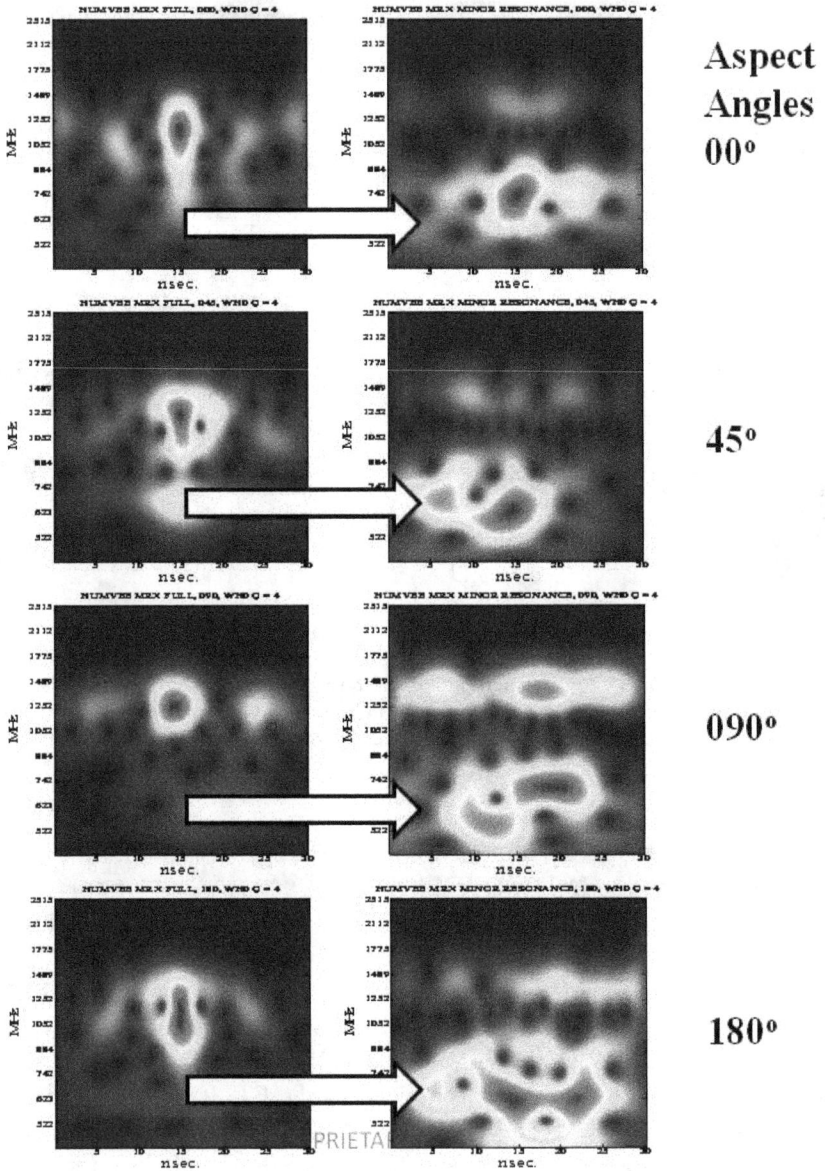

Fig. 1.11.4 Minor Resonance Enhancement, Time-frequency spectra; WH-0 (scaling function) filters. Target, Humvee at aspect angles 000°, 045°, 090° and 180°. Initially full MRX returns were obtained (left column), and alternative MTXs were designed to address resonances in the 1–1.4 GHz range, resulting in the MRX spectra in the right column in which spectral bands barely seen in the left column have been amplified — indicated by arrows. The amplitude of the MRX minor bands were enhanced, while the amplitude of the major bands were minimized.

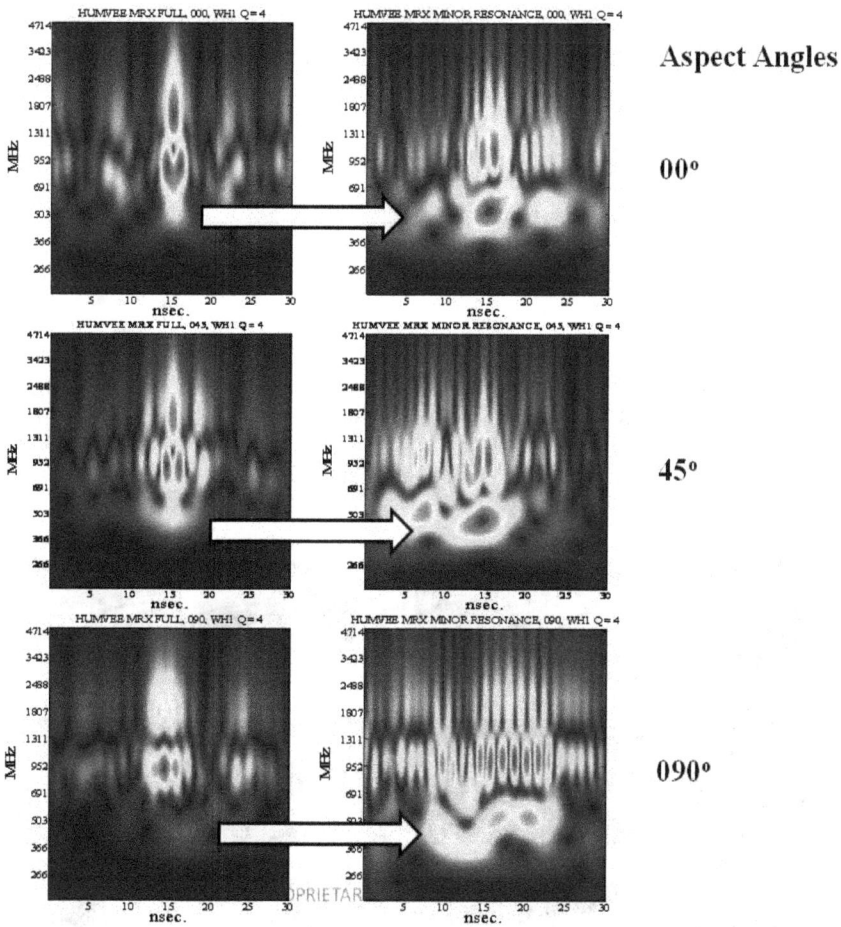

Fig. 1.11.5 Minor Resonance Enhancement, Time-frequency spectra; WH-1 (wavelet) filters. Target, Humvee at aspect angles 000°, 045° and 090°. Initially full MRX returns were obtained (left column), and alternative MTXs were designed to address resonances in the 1–1.4 GHz range, resulting in the MRX spectra in the right column in which spectral bands barely seen in the left column have been amplified — indicated by arrows. The amplitude of the MRX minor bands were enhanced, while the amplitude of the major bands were minimized.

TX on corner reflector), and for (a): PRX; (b): MRX. As expected, there is only a minimal difference between the corner reflector WVD PRX and MRX spectra. As noted in Section 1.5 there is little difference between the RXs elicited by PTX and MTX in the case of the corner reflector. The MAP technique exploits spectral diversity offered by the target. However, a corner reflector, being "mirror-like", returns "all" spectral components in the TX signal. Therefore there is no substantial difference in the PRX and MRX signals elicited by PTX and MTX signals.

1.13.2. *Ambiguity Function*

The symmetric ambiguity function, AF, is defined:

$$AF(\theta, \tau) = \int s^*(t - \tau/2)s(t + \tau/2)\exp[i\theta t]dt.$$

and is generally complex. The AF is thus the Fourier transform with respect to t of an auto-correlation also introduced and used above in defining the WVD (Section 1.13.1), and defined as:

$$R(t, \tau) = s^*(t - \tau/2)s(t + \tau/2).$$

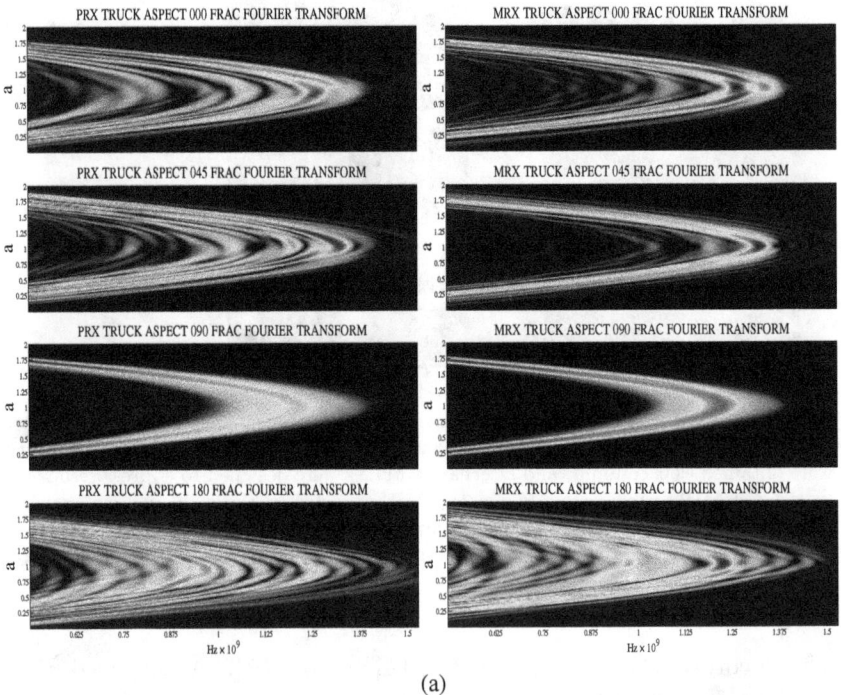

(a)

Fig. 1.12.1   FRFT Spectra $(-2.0 \leq a \leq 0;\ 0 \leq \alpha \leq +\pi)$. Target orientations (degrees): 000, 045, 090, 180°. The conventional Fourier transform is at slice $a = +1$. Left column: PRXs. Right column: MRXs. Notice the higher symmetry and definition in the right column (MRX). a: Truck target; b: Humvee target.

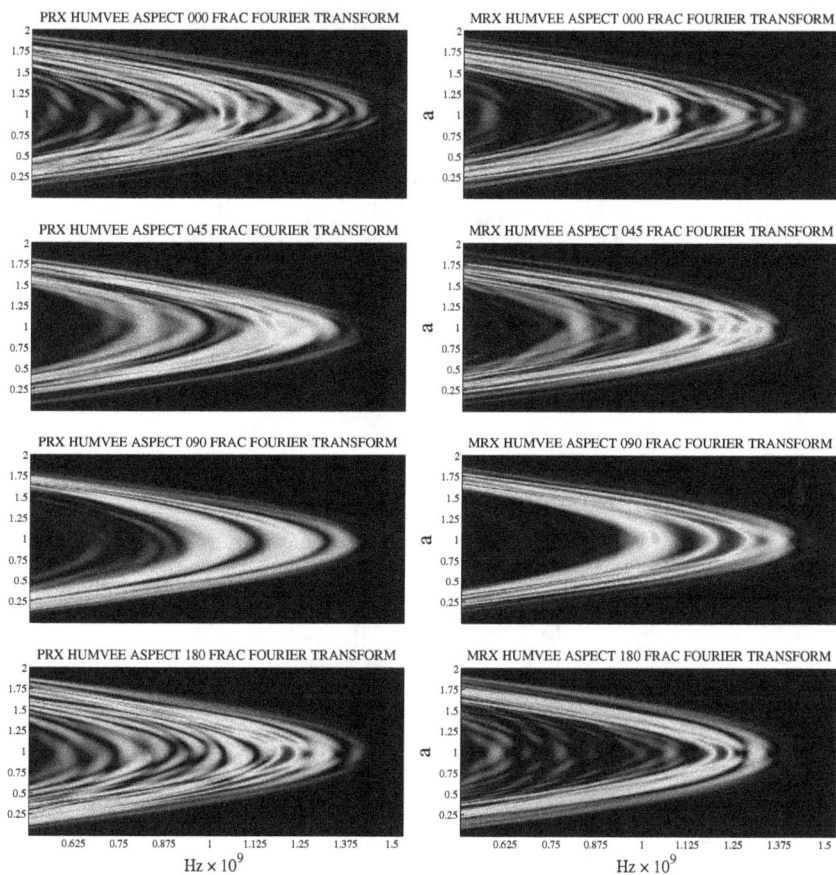

(b)

Fig. 1.12.1 *(Continued)*

or

$$R(t, \tau) = \int_{-\infty}^{+\infty} AF(\theta, \tau) \exp[-i\theta t] d\theta = s^*(t - \tau/2)s(t + \tau/2).$$

The AF was first introduced by Ville (1948) and Moyal (1949), and extensively applied to conventional radar signals by Rihaczek (1969, Rihaczek & Hershkowitz, 1996; 2000). The relation of the AF to matched filters was described by Woodward (1953) and, as mentioned above, AF is the characteristic function of the WVD. Specifically, the WVD is the double

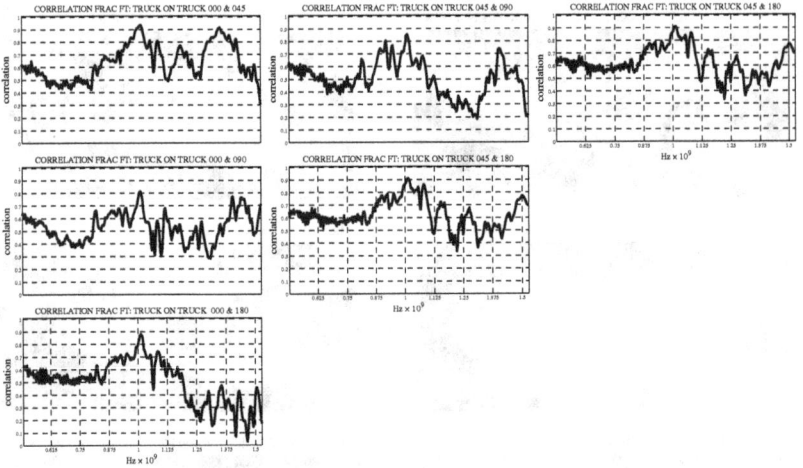

(a)

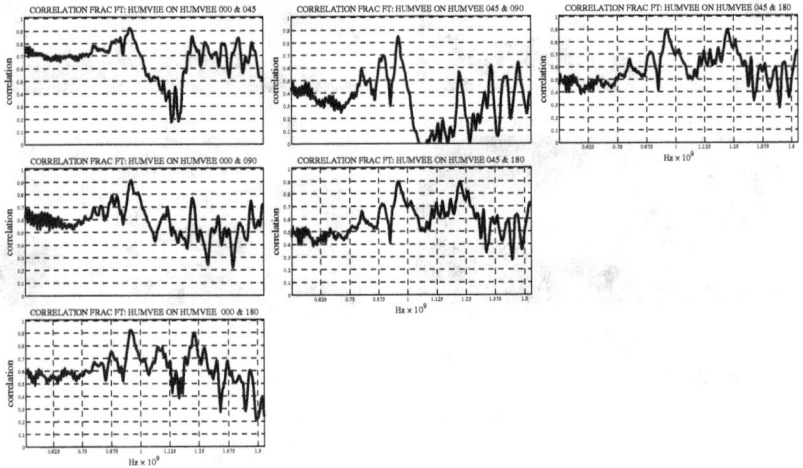

(b)

Fig. 1.12.2   Correlations of the spectra shown in Fig. 1.12.1. a: Truck target; b: Humvee target.

Fourier transform of the symmetric AF:

$$WVD(t, \omega) = \iint_{-\infty}^{+\infty} AF(\theta, \tau) \exp[-i(\omega\tau + \theta t)]d\theta d\tau.$$

However, the AF differs from the WVD in that while the auto-terms are concentrated around the origin, the cross-terms are concentrated away from

CORRELATIONS FRAC FT TRUCK ON TRUCK

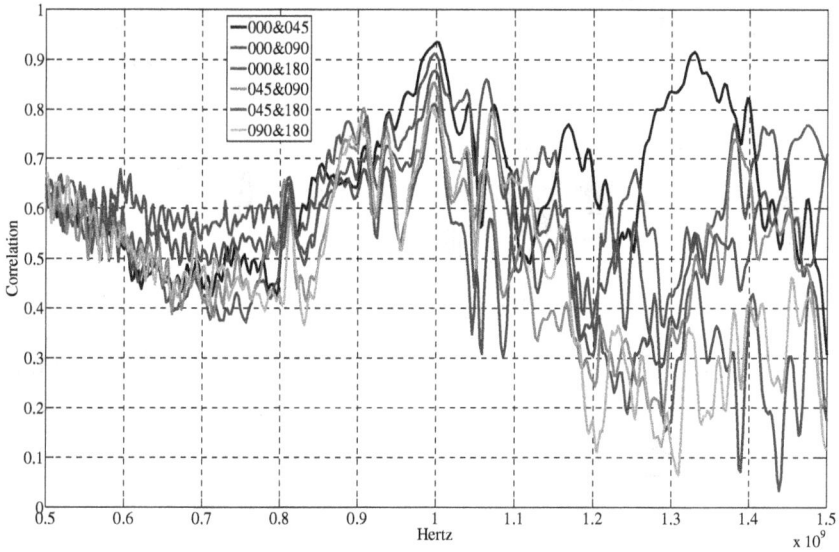

(a)

CORRELATIONS FRAC FT HUMVEE ON HUMVEE

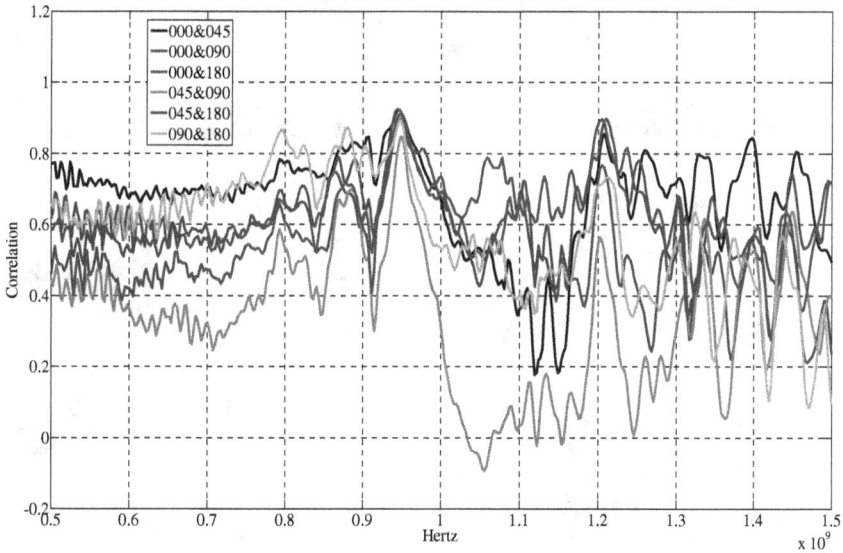

(b)

Fig. 1.12.3   The correlations of Fig. 1.12.2 overlaid. a: Truck target; b: Humvee target.

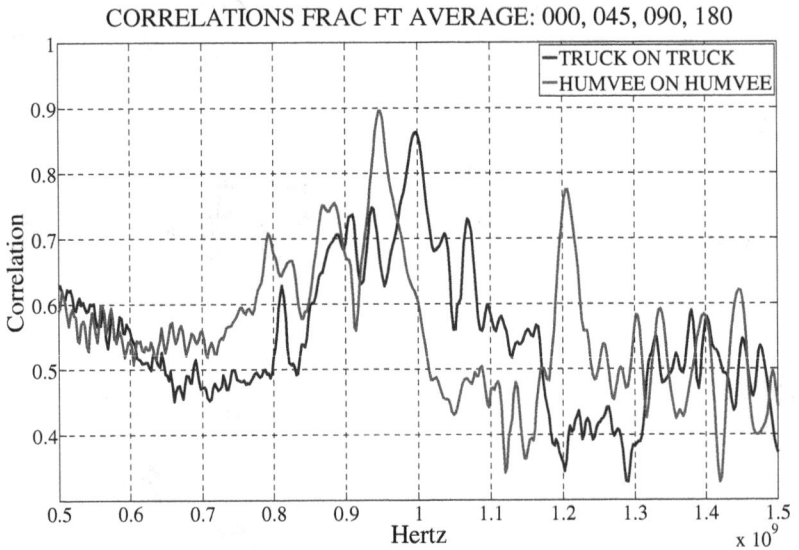

Fig. 1.12.4   Average of the correlations shown in Fig. 1.12.3.

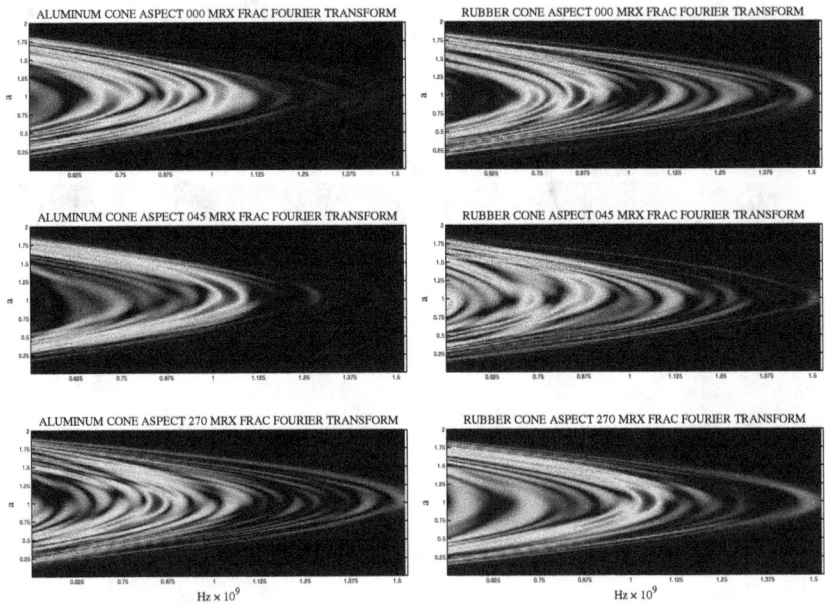

Fig. 1.12.5   MRX FRFT Spectra $(-2.0 \leq a \leq 0;\ 0 \leq \alpha \leq +\pi)$ for traffic cone targets. Target orientations (degrees): 000, 045, 270°. The conventional Fourier transform is at slice $a = +1$. Left column: Cone with aluminum surface. Right column: Cone with rubber surface.

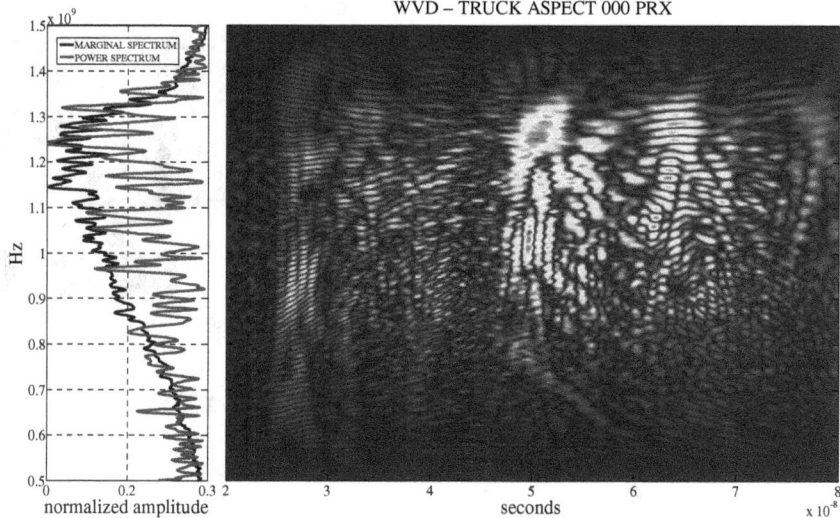

(a)

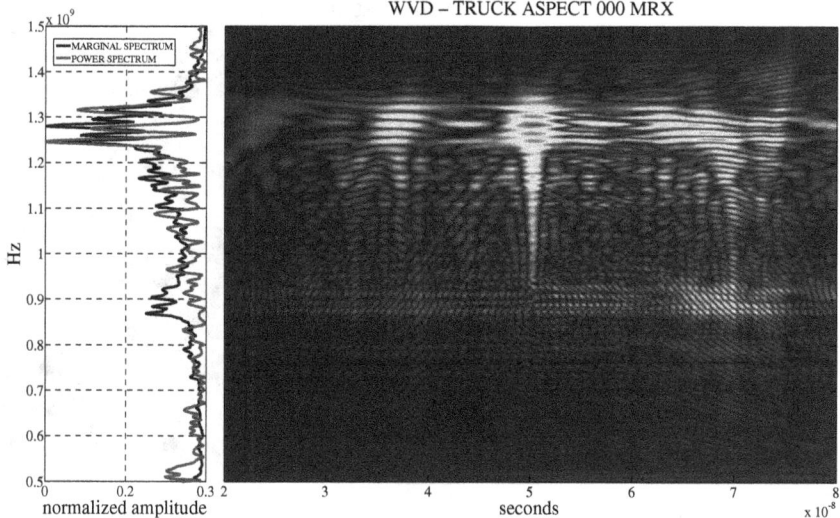

(b)

Fig. 1.13.1.1 One-sided WVD time-frequency spectra: truck TX and with truck as target, Aspect Angle: 000°. a: PRX. b: MRX. Plots on the left show the WVD frequency marginal and the FT spectrum. The WVD contains cross-terms not contained in the FT. The time symmetry of the WVD for the MRX (b) is due to the convolution of the target with the MTX signal, i.e., the time-reversed impulse response.

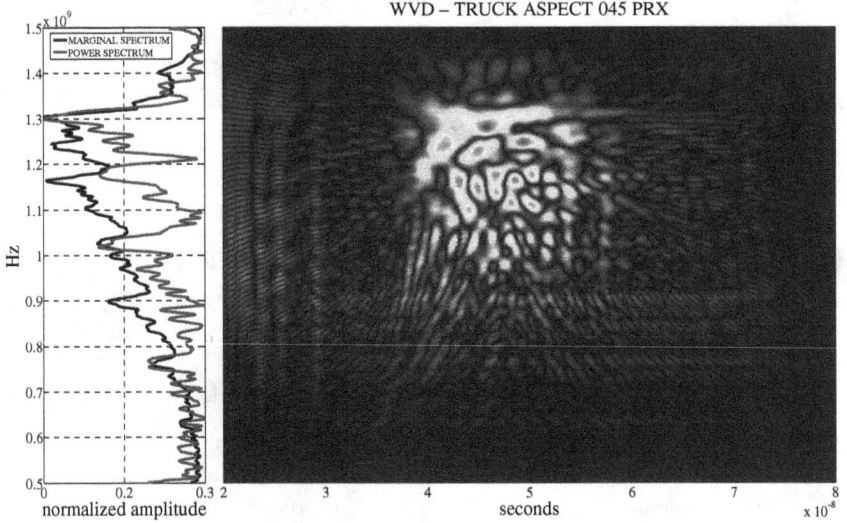

(a)

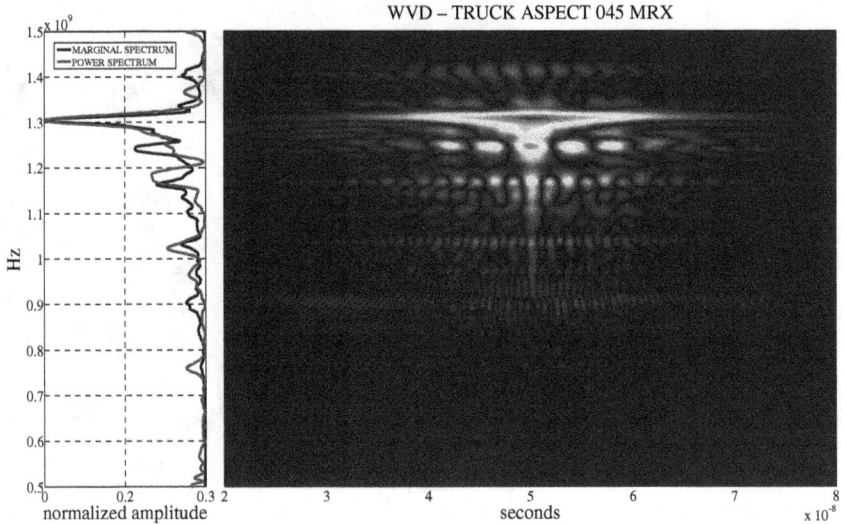

(b)

Fig. 1.13.1.2   One-sided WVD time-frequency spectra: truck TX and with truck as target, Aspect Angle: 045°. a: PRX. b: MRX. Plots on the left show the WVD frequency marginal and the FT spectrum. The WVD contains cross-terms not contained in the FT. The time symmetry of the WVD for the MRX (b) is due to the convolution of the target with the MTX signal, i.e., the time-reversed impulse response.

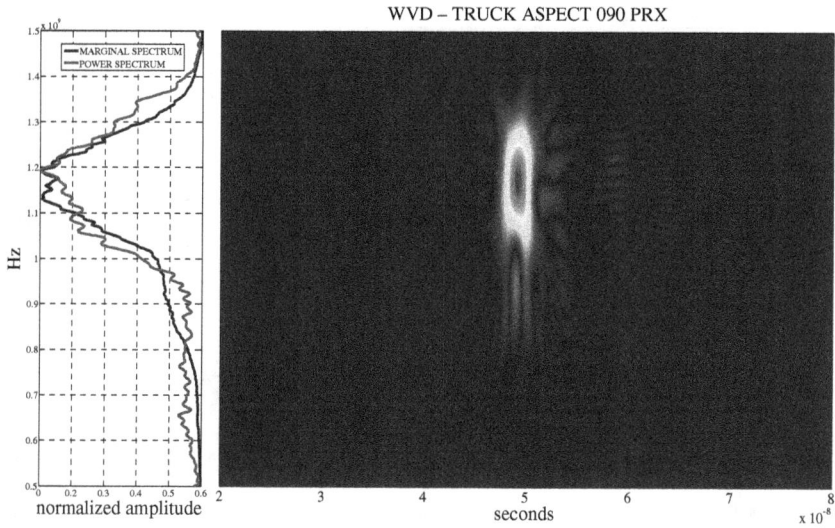

(a)

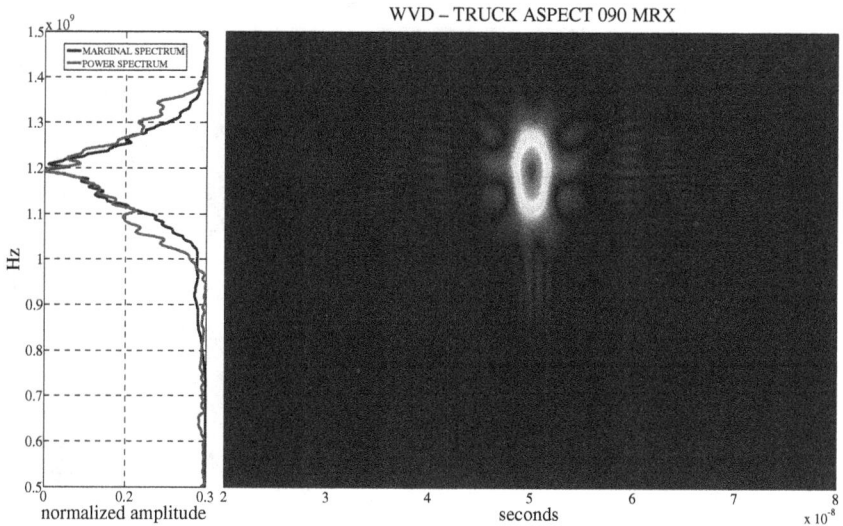

(b)

Fig. 1.13.1.3   One-sided WVD time-frequency spectra: truck TX and with truck as target, Aspect Angle: 090°. a: PRX. b: MRX. Plots on the left show the WVD frequency marginal and the FT spectrum. The WVD contains cross-terms not contained in the FT. The time symmetry of the WVD for the MRX (b) is due to the convolution of the target with the MTX signal, i.e., the time-reversed impulse response.

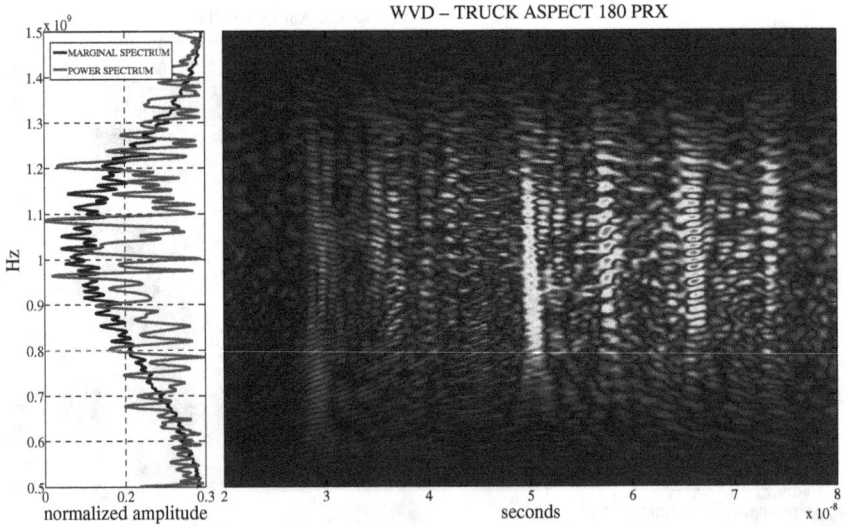

(a)

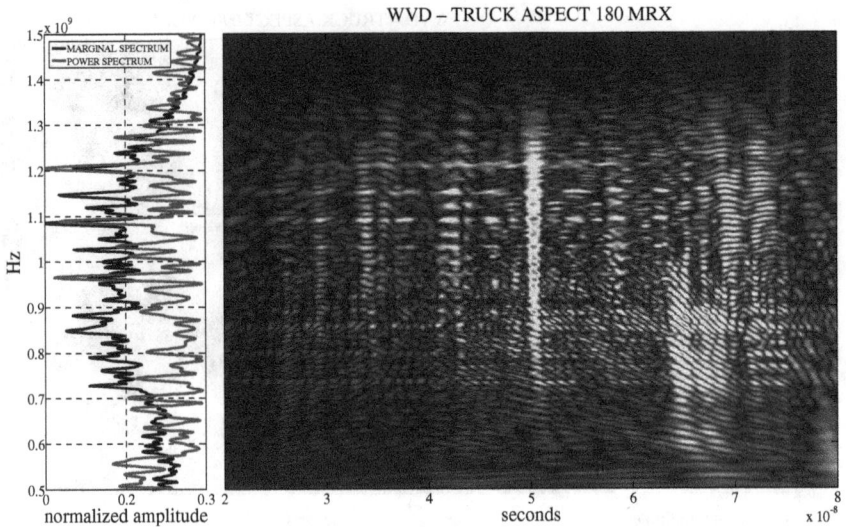

(b)

Fig. 1.13.1.4   One-sided WVD time-frequency spectra: truck TX and with truck as target, Aspect Angle: 180°. a: PRX. b: MRX. Plots on the left show the WVD frequency marginal and the FT spectrum. The WVD contains cross-terms not contained in the FT. The time symmetry of the WVD for the MRX (b) is due to the convolution of the target with the MTX signal, i.e., the time-reversed impulse response.

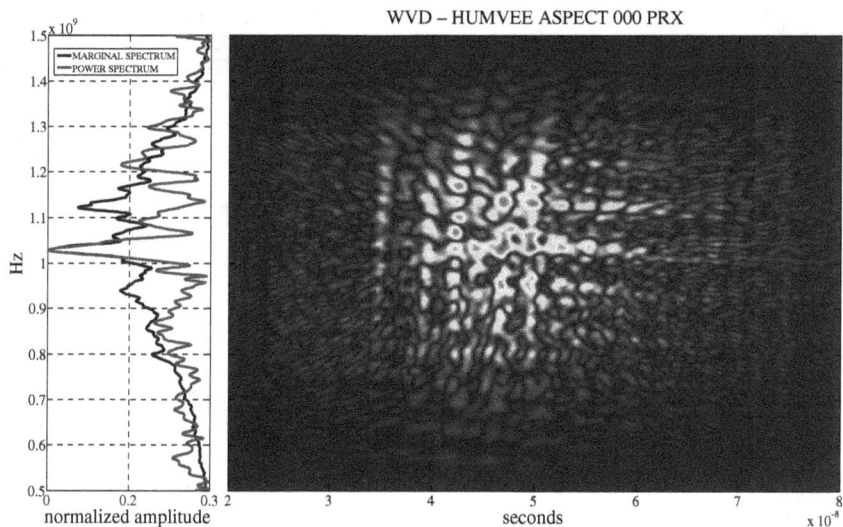

(a)

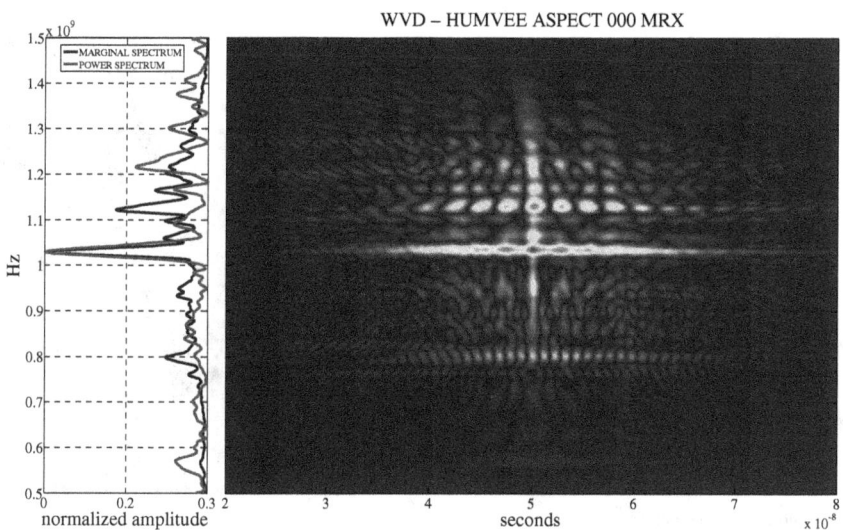

(b)

Fig. 1.13.1.5 One-sided WVD time-frequency spectra: humvee TX and with humvee as target, Aspect Angle: 000°. a: PRX. b: MRX. Plots on the left show the WVD frequency marginal and the FT spectrum. The WVD contains cross-terms not contained in the FT. The time symmetry of the WVD for the MRX (b) is due to the convolution of the target with the MTX signal, i.e., the time-reversed impulse response.

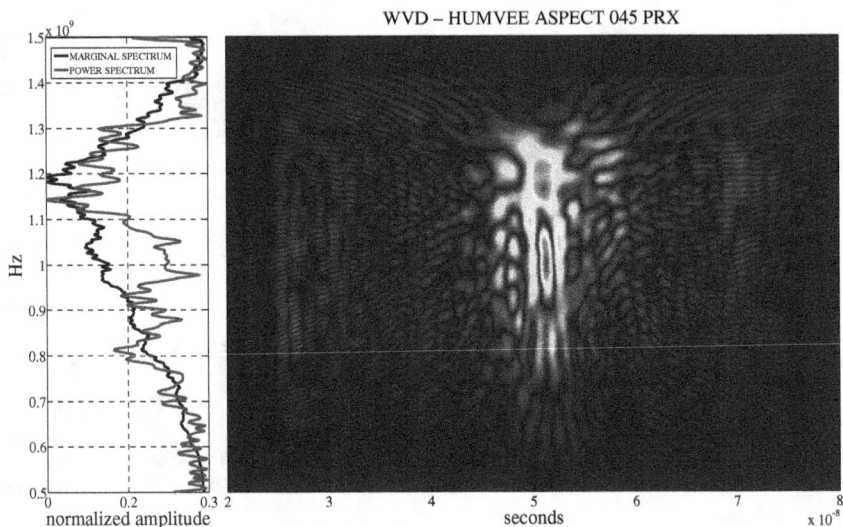

(a)

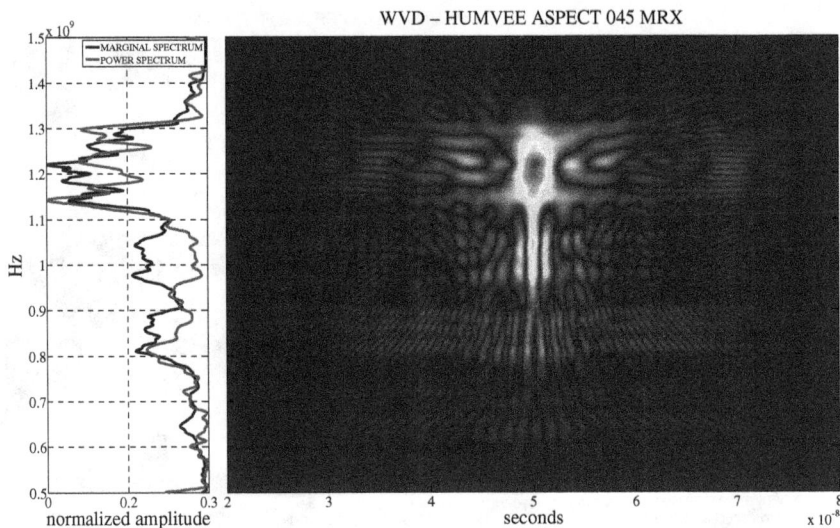

(b)

Fig. 1.13.1.6   One-sided WVD time-frequency spectra: humvee TX and with humvee as target, Aspect Angle: 045°. a: PRX. b: MRX. Plots on the left show the WVD frequency marginal and the FT spectrum. The WVD contains cross-terms not contained in the FT. The time symmetry of the WVD for the MRX (b) is due to the convolution of the target with the MTX signal, i.e., the time-reversed impulse response.

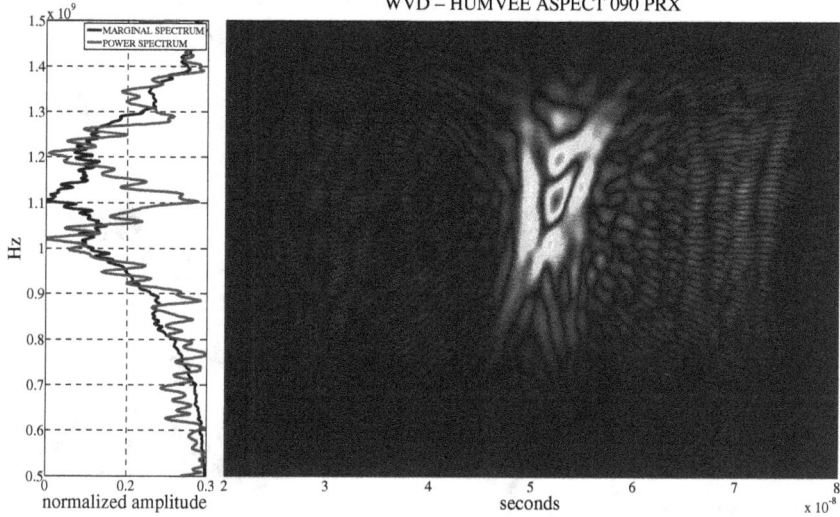

(a)

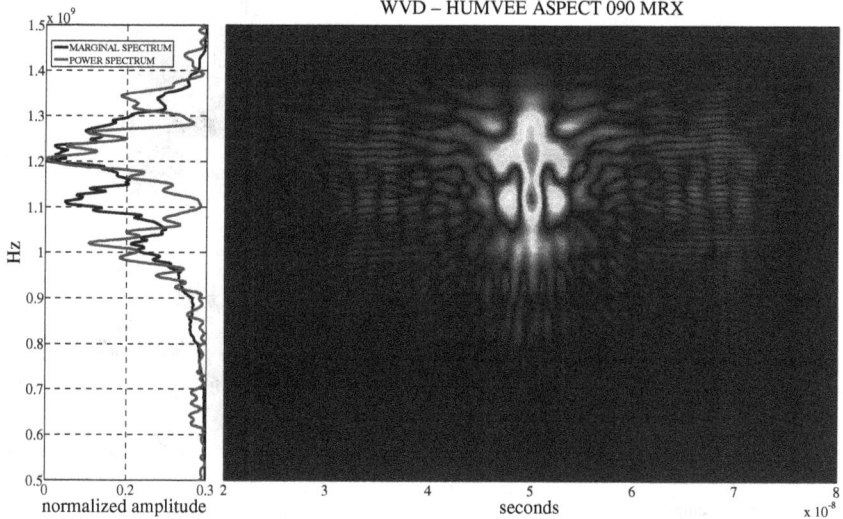

(b)

Fig. 1.13.1.7   One-sided WVD time-frequency spectra: humvee TX and with humvee as target, Aspect Angle: 090°. a: PRX. b: MRX. Plots on the left show the WVD frequency marginal and the FT spectrum. The WVD contains cross-terms not contained in the FT. The time symmetry of the WVD for the MRX (b) is due to the convolution of the target with the MTX signal, i.e., the time-reversed impulse response.

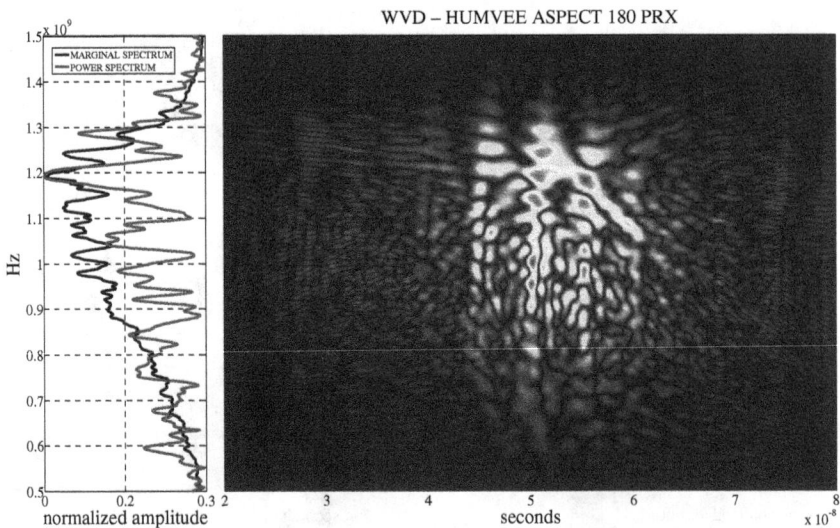

(a)

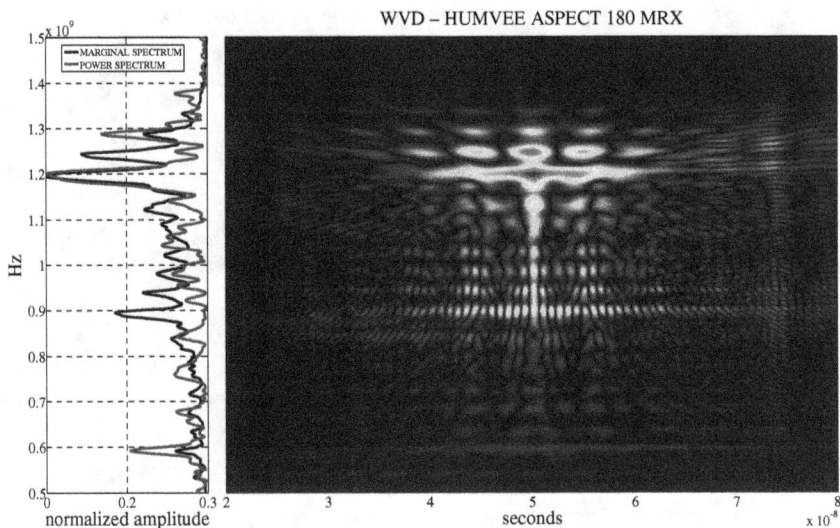

(b)

Fig. 1.13.1.8 One-sided WVD time-frequency spectra: humvee TX and humvee as target, Aspect Angle: 180°. a: PRX. b: MRX. Plots on the left show the WVD frequency marginal and the FT spectrum. The WVD contains cross-terms not contained in the FT. The time symmetry of the WVD for the MRX (b) is due to the convolution of the target with the MTX signal, i.e., the time-reversed impulse response.

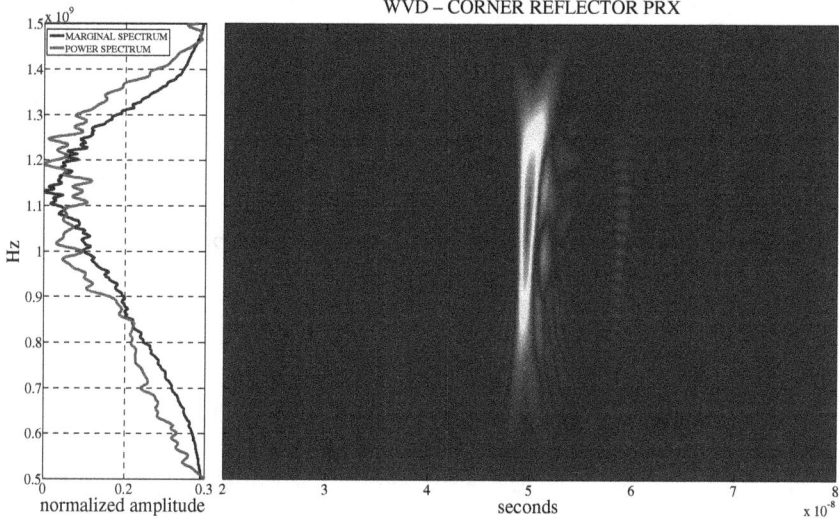

(a)

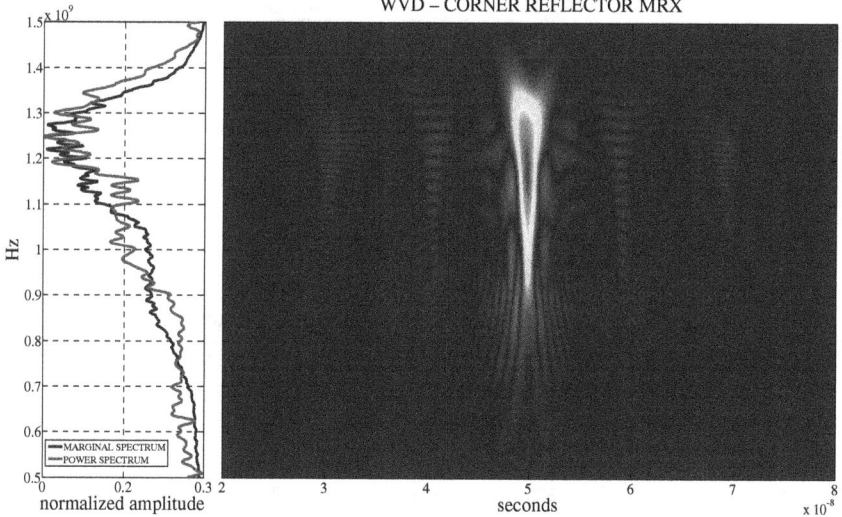

(b)

Fig. 1.13.1.9   One-sided WVD time-frequency spectra: corner ceflector TX and with corner refelector as target. a: PRX. b: MRX. Plots on the left show the WVD frequency marginal and the FT spectrum. There are only minimal cross-terms. The time symmetry of the WVD for the MRX (b) is due to the convolution of the target with the MTX signal, i.e., the time-reversed impulse response. There is only a minimal difference between the corner reflector WVD PRX and MRX spectra, as expected.

the origin. Therefore it is possible to suppress the cross-terms by applying a low-pass 2D filter in the ambiguity domain. The AF is considered to extend the notion of auto-correlation to non-stationary signals (Flandrin, 1998).

The one-sided AF time-frequency spectra for truck TXs and with truck as target, at aspect angles: 0°, 45°, 90° and 180° and for (a): PRX; (b): MRX, are shown in Figs. 1.13.2.1–1.13.2.4, below. For each figure, plots show (i) the AF frequency marginal, (ii) the FT direct power spectrum and (iii) the FT power spectrum calculated from the autocorrelation. The time symmetry of the AF for both the PRX and the MRX is due to the chosen symmetry of the autocorrelation time lag range. The cross-terms are away from the origin.

The one-sided AF time-frequency spectra for humvee TXs and with humvee as target, at aspect angles: 0°, 45°, 90° and 180° and for (a): PRX; (b): MRX, are shown in Figs. 1.13.2.5–1.13.2.8, below. Again, for each figure, plots show (i) the AF frequency marginal, (ii) the FT direct power spectrum and (iii) the FT power spectrum calculated from the autocorrelation. Again, the time symmetry of the AF for both the PRX and the MRX is due to the chosen symmetry of the autocorrelation time lag range. The cross-terms are away from the origin.

Figure 1.13.2.9 shows the one-sided AF time-frequency spectra for a corner reflector TX and with the corner reflector as target, and for (a): PRX; (b): MRX. As expected, there is only a minimal difference between the corner reflector AF PRX and MRX spectra.

1.14.   Nonlocal Transformations: Hilbert-Huang Transform

The Hilbert Transform (HT) provides a signal's instantaneous spectral profile and provides a time versus instantaneous frequency representation of a signal (Fig. 1.14.1). The Hilbert-Huang Transform (HHT) is an *energy-time-frequency* method of signal analysis for nonstationary and nonlinear signals (Huang *et al.*, 1998, 1999, 2003; Huang & Shen, 2005; Huang, & Attoh-Okine, 2005). The HHT is composed of an *empirical mode decomposition* (EMD) together with Hilbert transforms of those EMDs, and is designed for analysis of *signals that are neither linear, nor stationary*. The motivation for the method is that analyses in terms of any *a priori* chosen basis cannot fit the signals arising from all systems (Huang & Shen, 2005). In other words, one basis set doesn't fit all (mathematically or physically). Therefore the HHT adopts an adaptive approach to the analysis of a

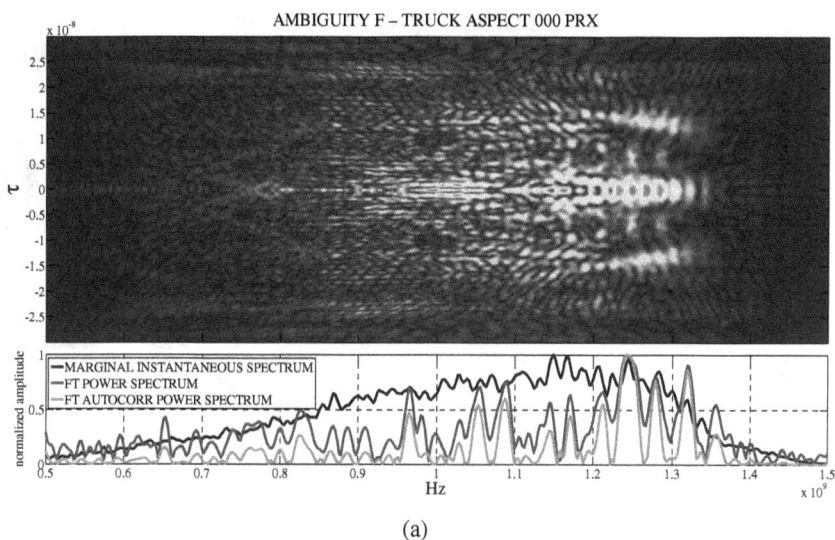

(a)

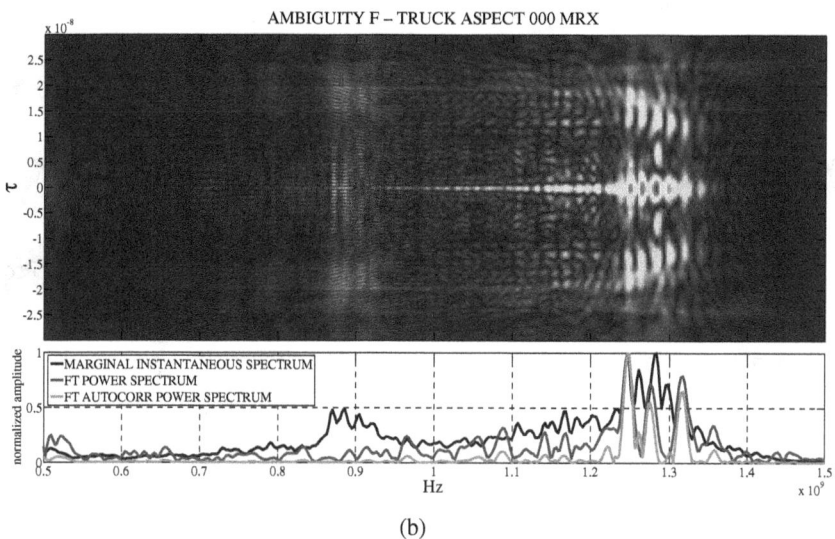

(b)

Fig. 1.13.2.1   One-sided AF time(lag)-frequency spectra: truck TX and truck as target, Aspect Angle: 000°. a: PRX. b: MRX. Plots below show the AF frequency marginal, the FT direct power spectrum and the FT power spectrum calculated from the autocorrelation. The time symmetry of the AF for both the PRX and the MRX is due to the chosen symmetry of the autocorrelation time lag range.

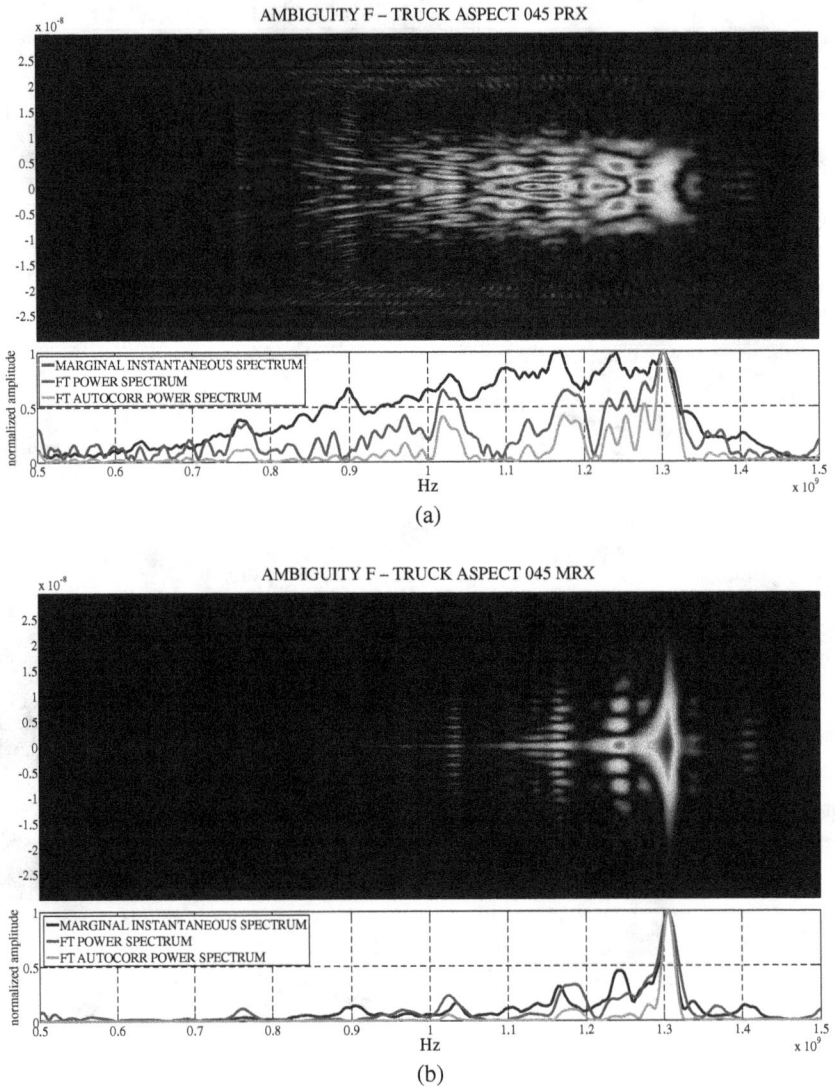

Fig. 1.13.2.2   One-sided AF time(lag)-frequency spectra: truck TX and truck as target, Aspect Angle: 045°. a: PRX. b: MRX. Plots below show the AF frequency marginal, the FT direct power spectrum and the FT power spectrum calculated from the autocorrelation. The time symmetry of the AF for both the PRX and the MRX is due to the chosen symmetry of the autocorrelation time lag range.

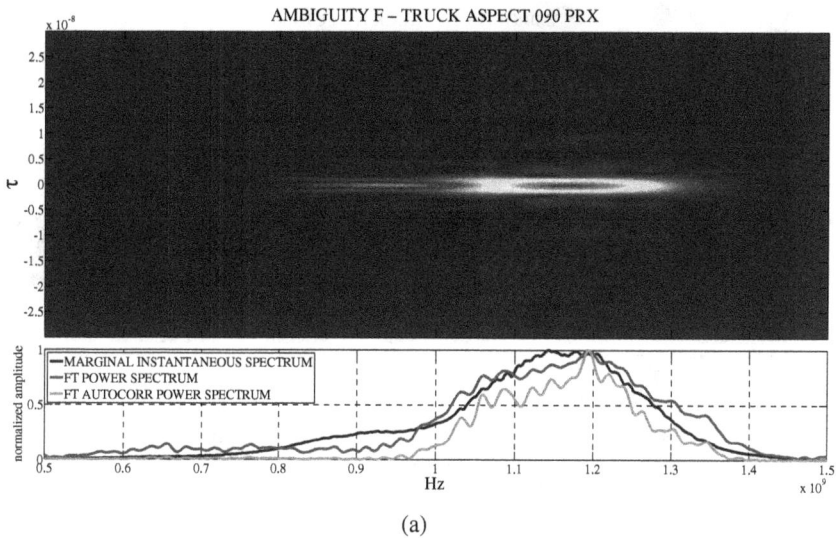

(a)

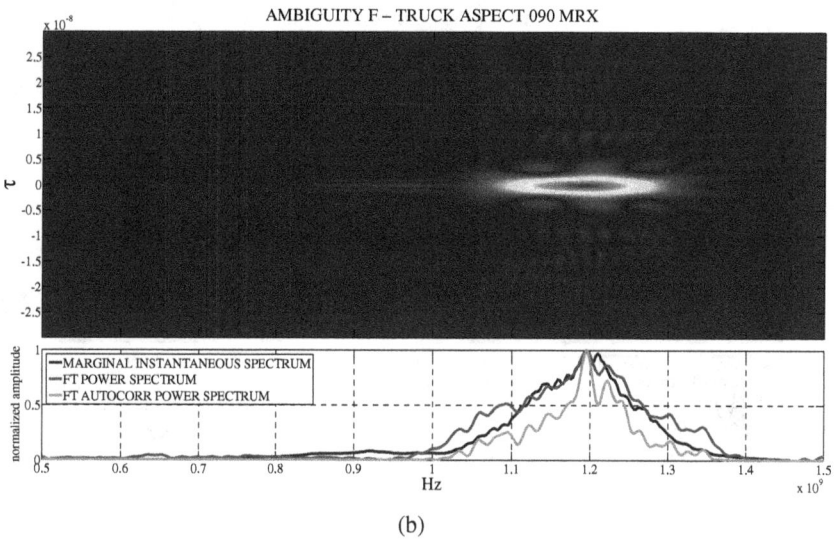

(b)

Fig. 1.13.2.3   One-sided AF time(lag)-frequency spectra: truck TX and truck as target, Aspect Angle: 090°. a: PRX. b: MRX. Plots below show the AF frequency marginal, the FT direct power spectrum and the FT power spectrum calculated from the autocorrelation. The time symmetry of the AF for both the PRX and the MRX is due to the chosen symmetry of the autocorrelation time lag range.

(a)

(b)

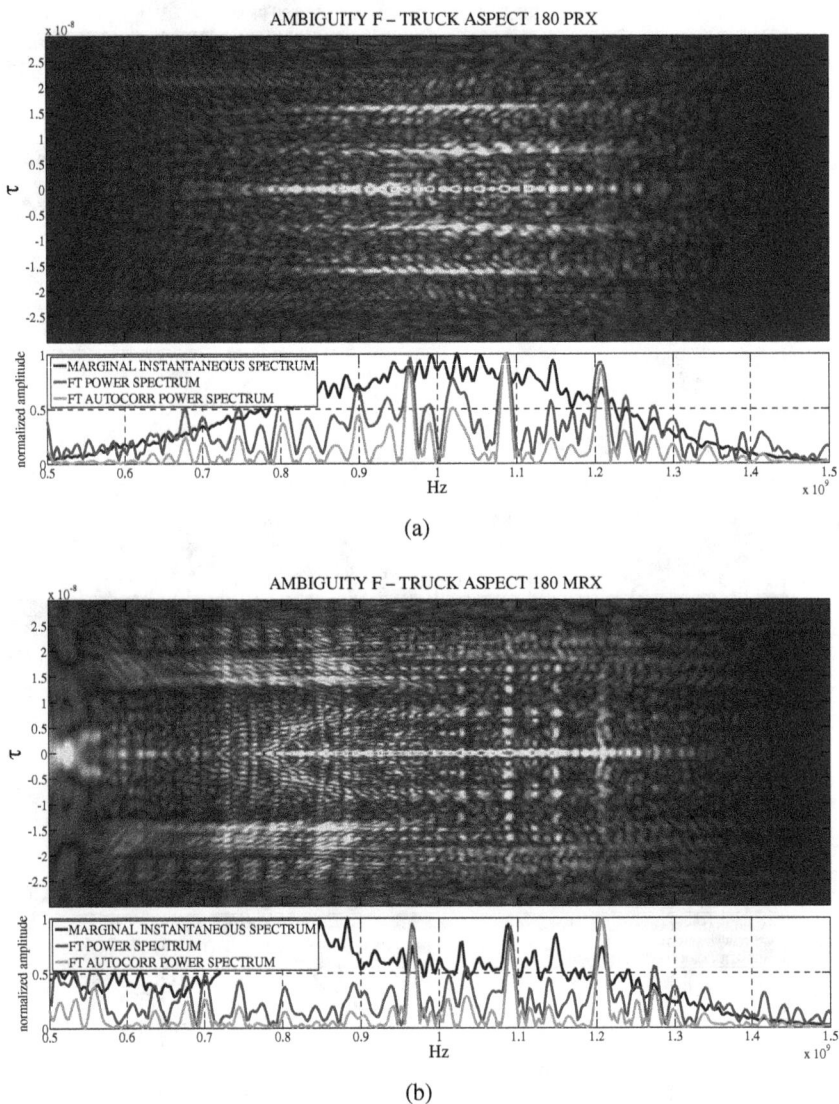

Fig. 1.13.2.4   One-sided AF time(lag)-frequency spectra: truck TX and truck as target, Aspect Angle: 180°. a: PRX. b: MRX. Plots below show the AF frequency marginal, the FT direct power spectrum and the FT power spectrum calculated from the autocorrelation. The time symmetry of the AF for both the PRX and the MRX is due to the chosen symmetry of the autocorrelation time lag range.

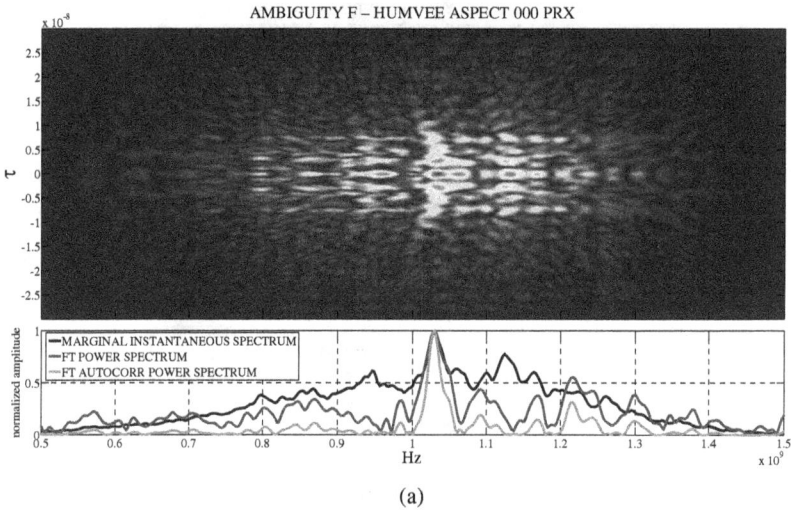

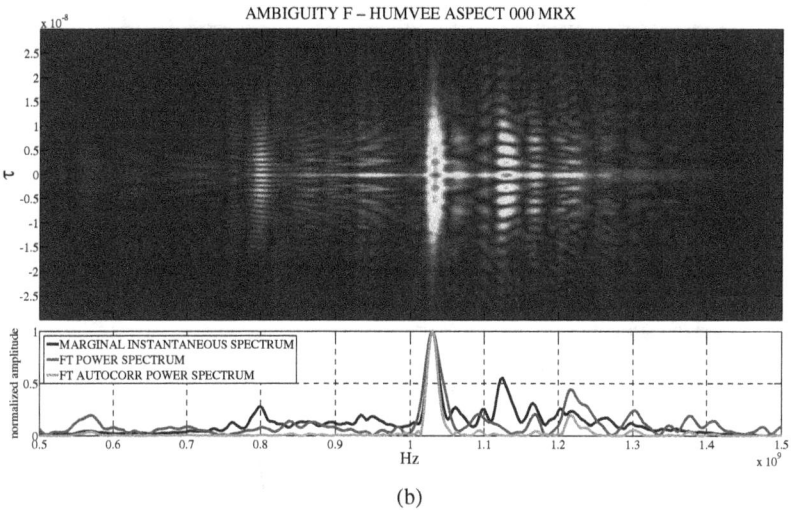

Fig. 1.13.2.5 One-sided AF time(lag)-frequency spectra: humvee TX and humvee as target, Aspect Angle: 000°. a: PRX. b: MRX. Plots below show the AF frequency marginal, the FT direct power spectrum and the FT power spectrum calculated from the autocorrelation. The time symmetry of the AF for both the PRX and the MRX is due to the chosen symmetry of the autocorrelation time lag range.

nonlinear and nonstationary signal into its multiple components, derived empirically. This adaptive approach is called a *sifting process* that yields an empirical basis set of intrinsic mode function (IMF) components. The basis set is derived from the data itself. In some instances, the HHT can

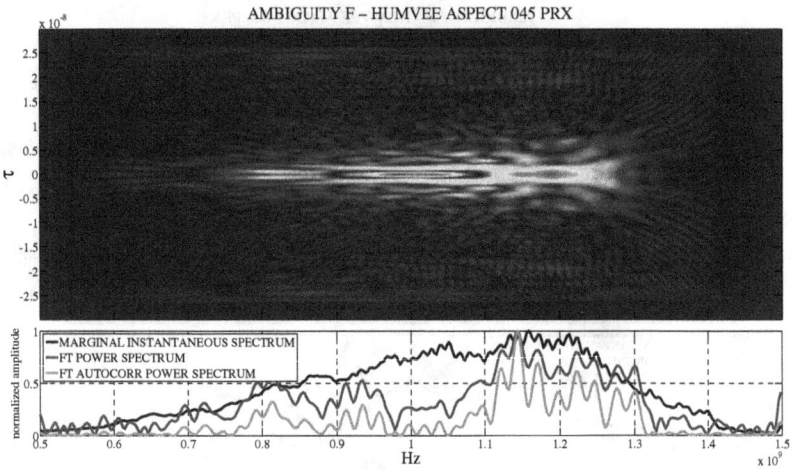

(a)

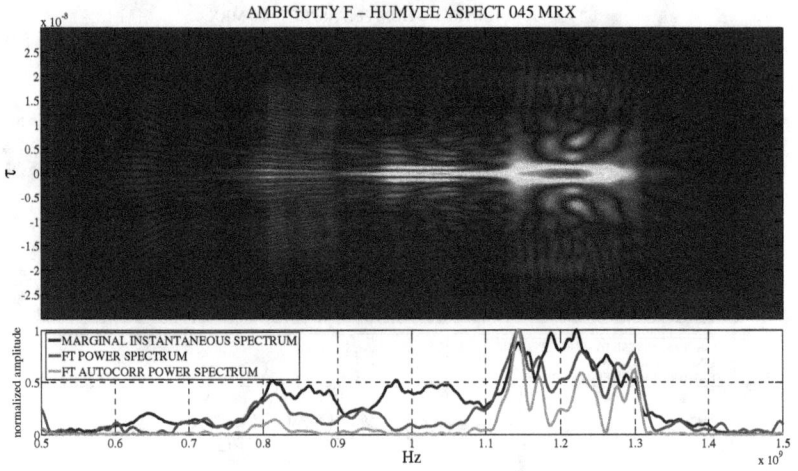

(b)

Fig. 1.13.2.6   One-sided AF time(lag)-frequency spectra: humvee TX and humvee as target, Aspect Angle: 045°. a: PRX. b: MRX. Plots below show the AF frequency marginal, the FT direct power spectrum and the FT power spectrum calculated from the autocorrelation. The time symmetry of the AF for both the PRX and the MRX is due to the chosen symmetry of the autocorrelation time lag range.

provide more accurate identifications of nonstationary systems than wavelet methods (Shan & Li, 2010).

As the HHT is relatively new, there are still problems with the transform that continue to be addressed, such as the choice of signal

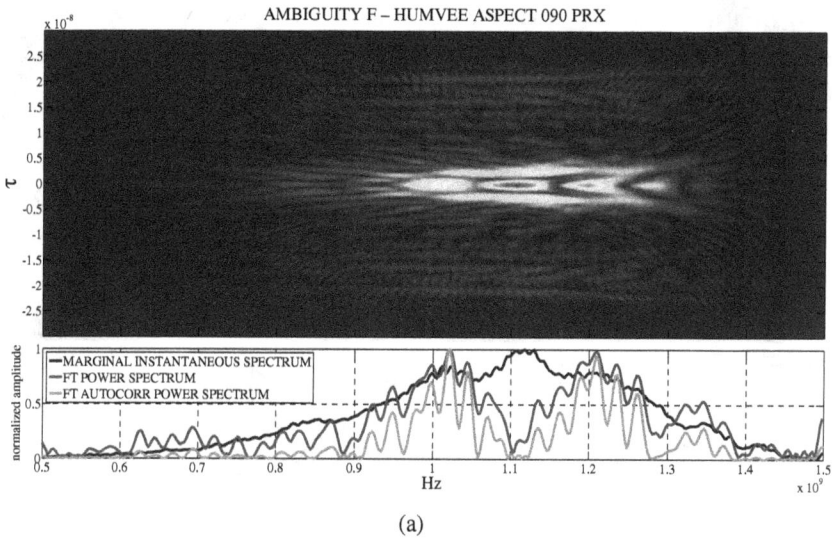

(a)

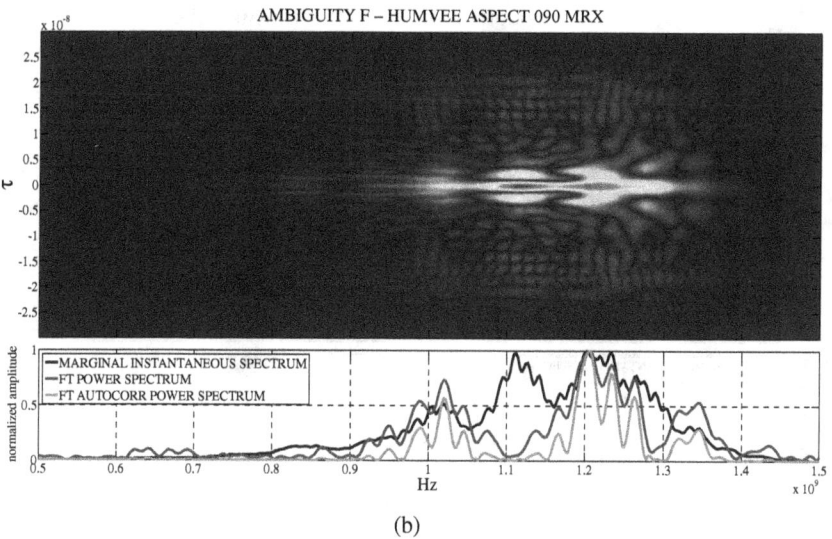

(b)

Fig. 1.13.2.7 One-sided AF time(lag)-frequency spectra: humvee TX and humvee as target, Aspect Angle: 090°. a: PRX. b: MRX. Plots below show the AF frequency marginal, the FT direct power spectrum and the FT power spectrum calculated from the autocorrelation. The time symmetry of the AF for both the PRX and the MRX is due to the chosen symmetry of the autocorrelation time lag range.

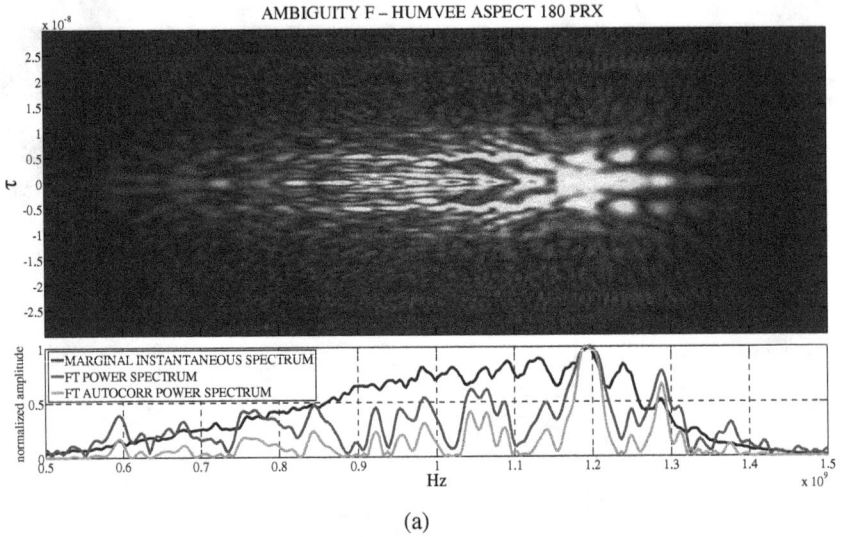

(a)

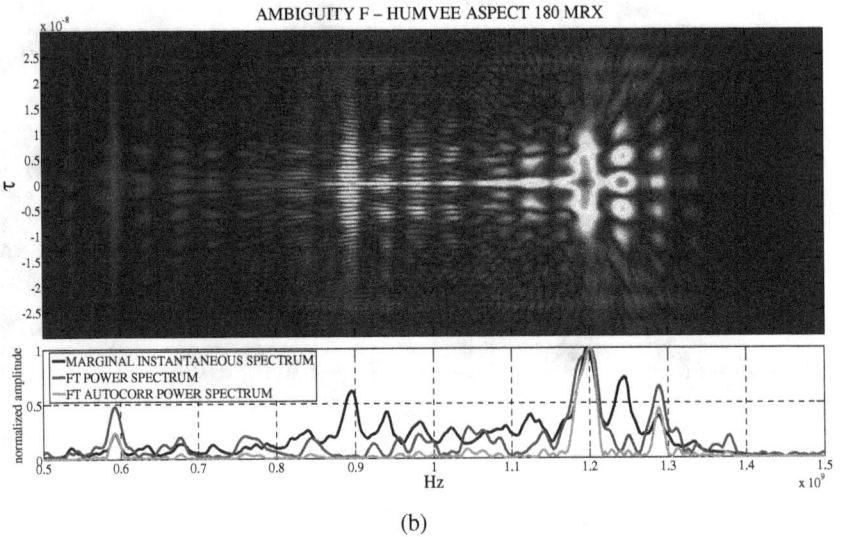

(b)

Fig. 1.13.2.8   One-sided AF time(lag)-frequency spectra: humvee TX and humvee as target, Aspect Angle: 180°. a: PRX. b: MRX. Plots below show the AF frequency marginal, the FT direct power spectrum and the FT power spectrum calculated from the autocorrelation. The time symmetry of the AF for both the PRX and the MRX is due to the chosen symmetry of the autocorrelation time lag range.

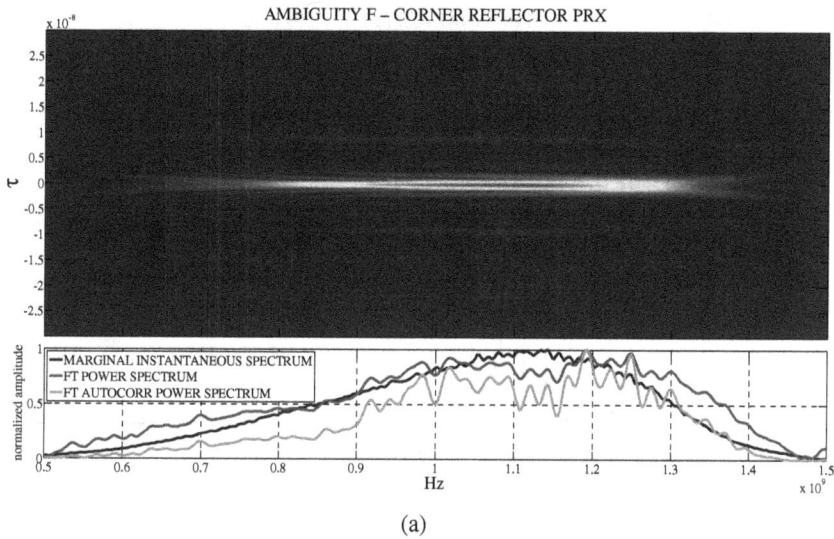

(a)

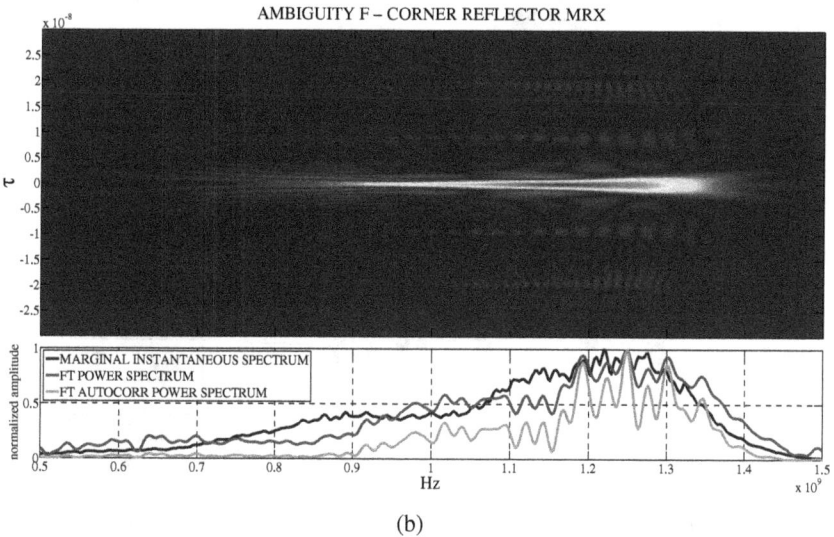

(b)

Fig. 1.13.2.9 One-sided AF time(lag)-frequency spectra: corner reflector TX and corner reflector as target. a: PRX. b: MRX. Plots below show the AF frequency marginal, the FT direct power spectrum and the FT power spectrum calculated from the autocorrelation. The time symmetry of the AF for both the PRX and the MRX is due to the chosen symmetry of the autocorrelation time lag range. There is only a minimal difference between the corner reflector AF PRX and MRX spectra, as expected.

(a)

(b)

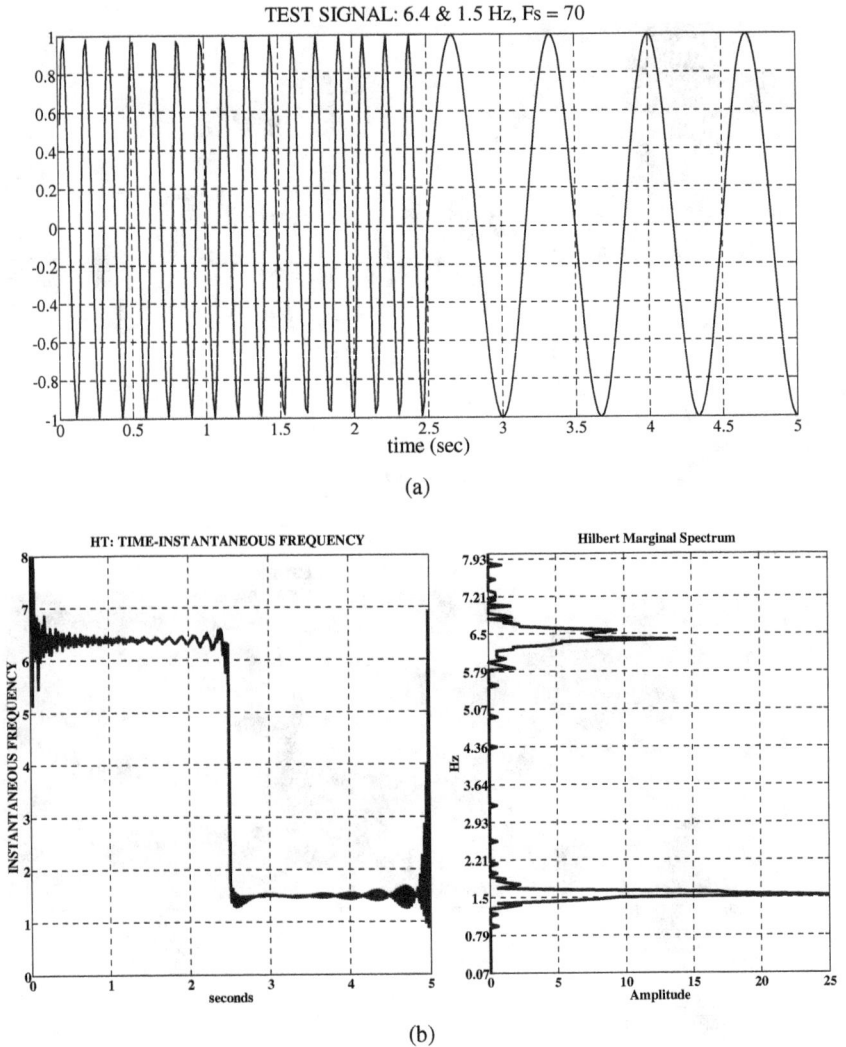

Fig. 1.14.1 Hilbert Transform of a test signal with two frequency components, 6.4 and 1.5 Hz, (Fs = 70 Ss). a: time domain representation of test signal. b: Left — Hilbert transform of test signal. Right — Hilbert marginal spectrum.

envelope generation method (Chen *et al.*, 2006), the stopping criteria for the sifting process, and the production of mode mixing for some signals (Datig & Schlurmann, 2004; Huang & Wu, 2008). The HHT method offers a meaningful insight into physically-defined (as opposed to mathematically-defined) instantaneous frequencies.

The Hilbert Transform, which accesses a signal's instantaneous frequencies is:

$$y(t) = \frac{1}{\pi} P \int_{-\infty}^{+\infty} \frac{x(\tau)}{t - \tau} d\tau,$$

where P is the Cauchy principal value of the singular integral. The analytic signal is defined:

$$z(t) = x(t) + iy(t) = a(t) \exp[i\varphi t],$$

where

$a(t) = \sqrt{x^2 + y^2} =$ the instantaneous amplitude, and
$\varphi(t) = \arctan(y/x) =$ the instantaneous phase function.

The instantaneous frequency is then defined:

$$\omega = \frac{d\varphi}{dt},$$

which provides only physically meaningful results for mono-component signals (Huang *et al.*, 1998), Fig. 1.14.1.

Alternatively, the HHT method involves Hilbert transforming a signal's IMF components, which are extracted by a process called *empirical mode decomposition* (EMD), and seeks results for multi-component signals. The decomposition method is as follows.

The necessary condition for IMF extraction are: (i) the signal must be symmetric with respect to the local zero mean, and (ii) have the same number of zero crossings or extrema. The decomposition is composed of the steps:

(1) Identify the local maxima of $x(t)$ and connect them with a (e.g., cubic) spline, to provide the upper envelope, $e_1(t)$.
(2) Identify the local minima of $x(t)$ and connect them with a (e.g., cubic) spline, to provide the lower envelope $e_1(t)$.
(3) Calculate the local mean $m_1(t) = (e_1(t) - e_1(t))/2$.
(4) Calculate the difference $c_1(t) = x(t) - m_1(t)$.
(5) Repeat steps (1)–(4) with $c_1(t)$ replacing $x(t)$ to obtain $c_{11}(t) = x(t) - m_{11}(t)$.
(6) Repeat this *sifting process* k times:

$$c_{1k}(t) = c_{1(k-1)}(t) - m_{1(k-1)}(t),$$

resulting in the first IMF:

$$C1(t) = c_{1k}(t).$$

An IMF is subject to the following constraints:

(a) The number of extrema and the numbet of zero-crossings must either be equal or differ by at most one.
(b) At any point, the mean value of the envelope defined by the local maxima and the envelope defined by the local minima must be zero.

(7) k is defined by a stopping criterion. One criterion for stopping the sifting process is when $SD = 0.2 - 0.3$, where SD is defined as:

$$SD = \sum_{t=0}^{N} \left[\frac{|c_{1(k-1)}(t) - c_{1k}(t)|^2}{c_{1(k-1)}^2(t)} \right].$$

Another criterion used is that the number of zero-crossings and the number of extrema must be equal.

(8) Subtract $C1(t)$ from the signal according to:

$$r_1(t) = x(t) - C1(t),$$

replace $x(t)$ with $r_1(t)$, and repeat steps (1)–(6).

(9) The result is a series of IMFs, $Ci(t), i = 1, 2, \ldots, n$, with a final residue $r_n(t)$ becoming a monotonic function such that (Huang *et al.*, 1998):

$$x(t) = \sum_{i=1}^{n} Ci(t) + r_n(t).$$

(10) As $r_n(t)$ is monotonic and can be neglected the Hilbert transform of $Ci(t)$ results in the *Hilbert spectrum*:

$$X(t) = Re\left(\sum_{i=1}^{n} a_i(t) \exp\left(i \int \omega_i(t) dt \right) \right).$$

(11) In comparison, the Fourier transform of the original data is:

$$F(\omega, t) = Re\left(\sum_{i=1}^{\infty} a_i \exp(-i\omega_i t) \right).$$

with a_i and ω_i constants. Therefore the IMF is a generalized Fourier expansion.

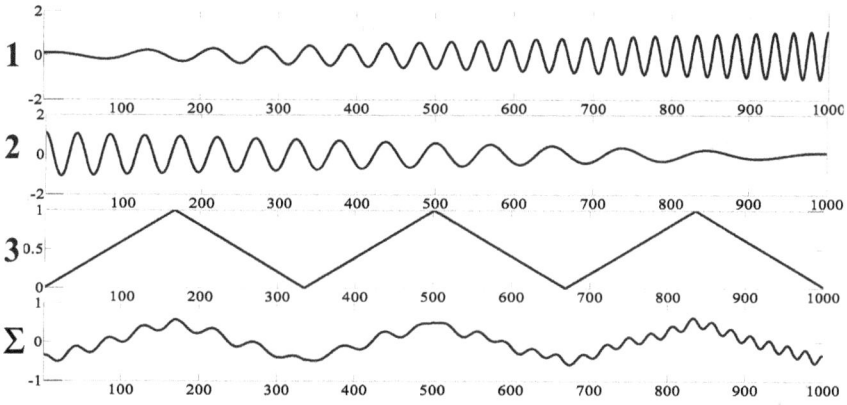

(a)

TEST COMPOSITE SIGNAL

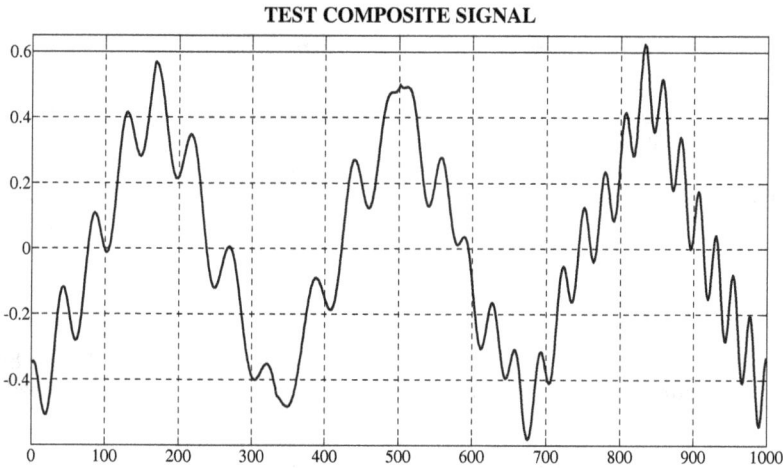

(b)

Fig. 1.14.2   a: The first two signals are oscillatory, frequency modulated signals. The third signal is a slow d.c. component, and the fourth signal is the combination or composite of the three preceding and represents an RX signal, before processing. b: The fourth composite signal amplified. Model signal due to Flandrin & Gonçalvès (2004).

We first demonstrate the method using a test signal as an example (Flandrin & Gonçalvès (2004)). This test signal is a more difficult signal to deconstruct than will be commonly encountered empirically. We construct the test signal as follows (Fig. 1.14.2). The first two component signals are oscillatory, frequency modulated signals. It is these types of signals that one

SPLINE FIT

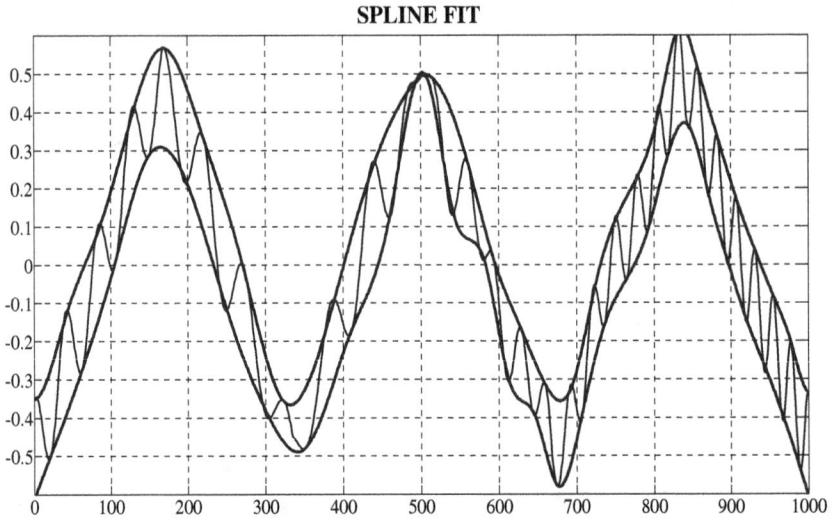

Fig. 1.14.3   The composite signal together with the envelopes calculated by a spline fit. The instantaneous mean is calculated from the difference of the envelopes and subtracted from the original composite signal, which is designated C0. The resultant from the subtraction is the first intrinsic mode function and is designated C1 (or the high frequency component of C0), and the remainder is designated R1 (or the low frequency component of C0).

might want to process further separately using time-frequency methods. The third component signal is a slow d.c. component, and the fourth signal is the combination of the three preceding components and is considered to represent an example RX signal.

Calling this test RX signal: C0, the signal envelopes (Fig. 1.14.3) are calculated and the sifting process begun. After 50 sifting iterations — this choice of 50 is arbitrary and signal-dependent — the first intrinsic mode function, C1 is obtained, and from the relation: C0 = C1 + R1, it should be noticed that R1 captures almost all of the slow d.c. component of the composite signal (Fig. 1.14.4). C1 contains almost all of the high frequency components of the original signal (Figs. 1.14.5–1.14.6), and both C1 and R1 can be separately processed using, e.g., time-frequency methods. This example recovered the signal components after one stage of sifting. In most cases more stages are required.

This example indicates that the HHT, used as a pre-processing method, decomposes the RX signal for separate spectral analysis in a manner similar to a filter bank (Flandrin *et al.*, 2003).

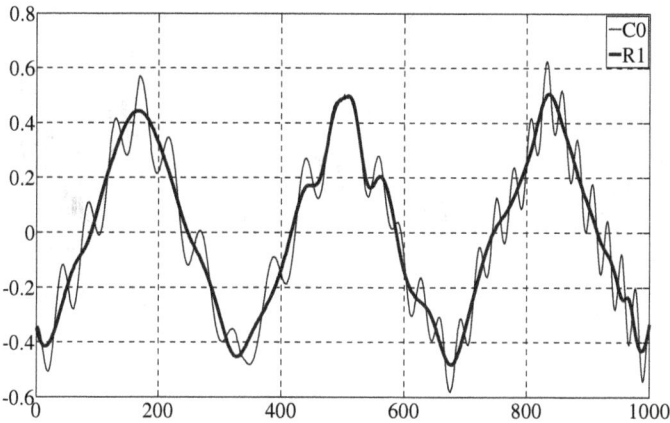

Fig. 1.14.4   The original composite signal, C0, and the first low frequency intrinsic mode functions (IMFs), R1. C1 is obtained by the recursive subtraction of the instantaneous mean, permitting the extraction of the high frequency components in the signal. This process is known as "sifting". The remainder is R1 = R0 − C1. Here, R1 captures almost all of the d.c. component in the C0 signal.

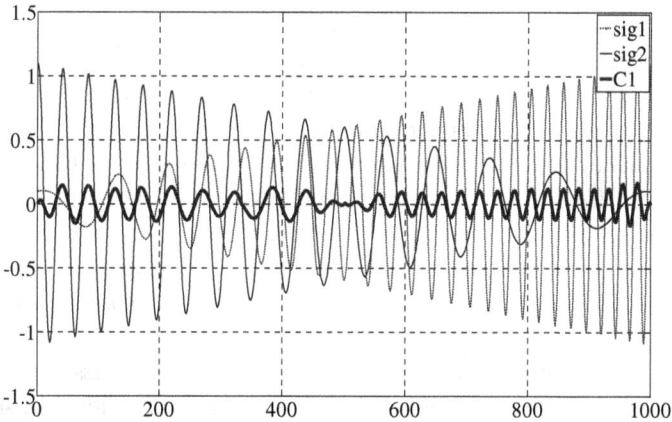

Fig. 1.14.5   The two separate oscillatory components of the original composite signal, signals 1 and 2 in Fig. 1.14.2A, and the high frequency components, C1, of the original signal, C0, obtained by the sifting process. C0 = C1 + R1.

The same procedure can be applied to empirical MAP RX data. In Fig. 1.14.7 is shown the calculated envelopes (upper and lower) for the PRX signal from a barrel target positioned up. The PRX is designated: C0. The instantaneous average is calculated from these envelopes, which is then subtracted from the PRX, or C0, giving C1 (the high frequency IMF).

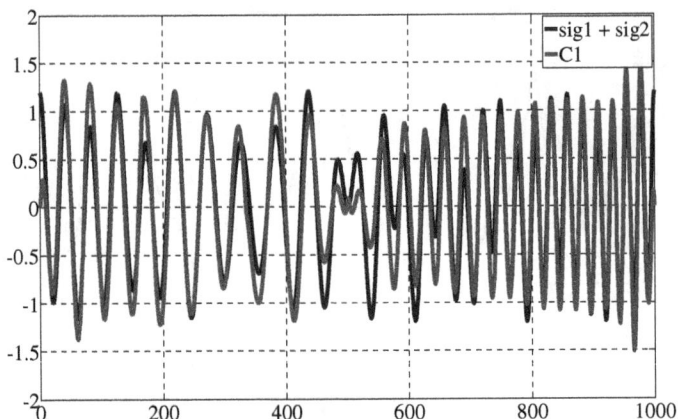

Fig. 1.14.6   The two added oscillatory components of the composite C0 signal: signal 1+ signal 2, and the high frequency IMF signal, C1, obtained by the sifting process. C0 = C1 + R1. It is clear that the sifting process has extracted almost all of the oscillatory, high frequency components, C1, from the composite signal, C0.

The remainder is R1 (the lower frequency remainder IMF). The process is repeated using C1 to obtain C2 and R2, etc.

Figure 1.14.8 shows the time domain C0–C8 IMFs for the target Barrel Up and Barrel Side PRXs. The spectra for the C0–C4 and the R1–R4 IMFs are shown in Fig. 1.14.9. Separation of the high frequency C IMFs and the lower frequency R IMFs progressively shifts downwards — by subtraction of the $C_{(i+1)}$ from the R_i to obtain the $R_{(i+1)}$ IMF — the remaining signal $R_{(i+1)}$ toward the low frequencies. In fact, Fig. 1.14.9 shows that the R's and the C's at each level i, form the output of low-high pass frequency filter pairs. This is reminiscent of wavelet multiresolution analysis (Flandrin *et al.*, 2003).

The correlations of the high-pass C1-4, and the low-pass R1-4 IMFs of Fig. 1.14.9 from the two targets, Barrel Up and Barrel Side, are shown in Table 1.14.1. The original PRXs are from the same barrel target, but in two different aspects: up and side.

1.15.   Nonlocal Transformations: Quadratic Fractional Fourier Transform

The quadratic Fractional Fourier transform (FRFT) is a generalization of the Fourier transform (Namias, 1980; Dickinson & Steiglitz, 1982; McBride & Kerr, 1987; Ozaktas & Mendlovic, 1993; Lohmann, 1993; Ozaktas *et al.*,

PRX: BARREL UP

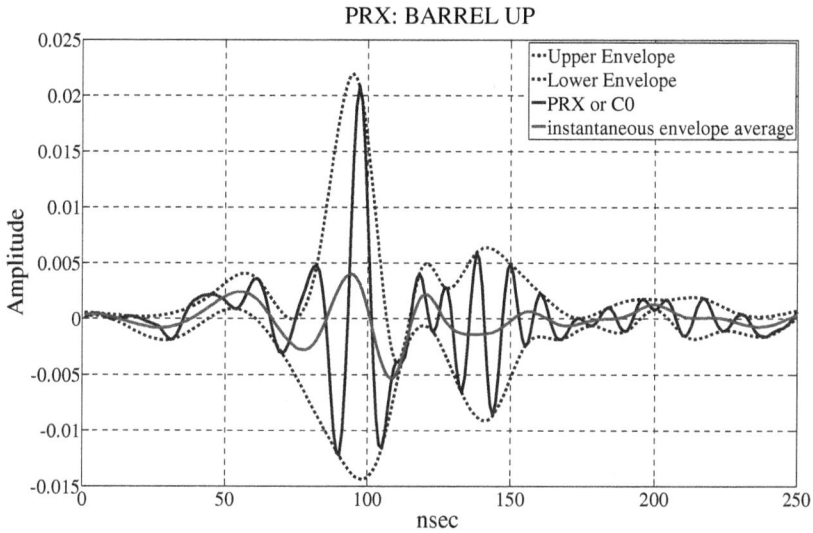

(a)

PRX: BARREL SIDE

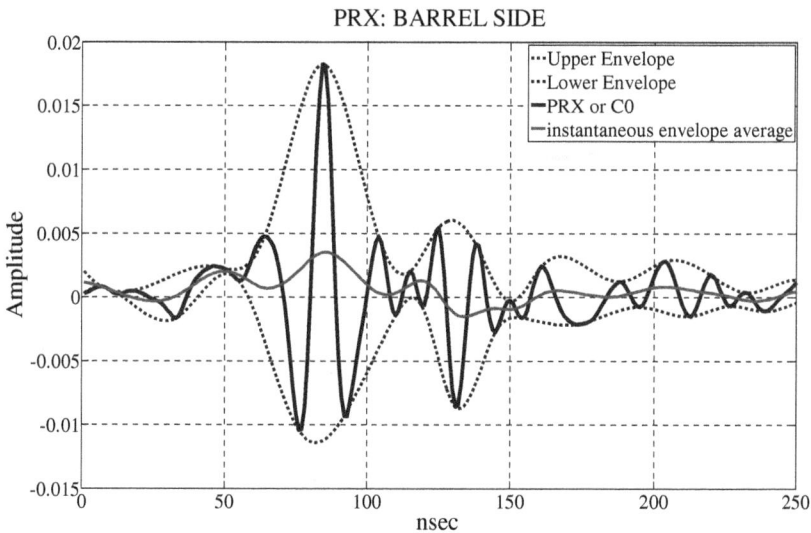

(a)

Fig. 1.14.7   a: $C0 = PRX$ Barrel Up; b: $C0 = PRX$ Barrel Side. A snapshot of the first instantaneous average which is calculated from the difference of the envelopes and updated recursively in the sifting process. The final instantaneous average, not shown here, is then subtracted from the PRX, or $C0$, giving $C1$. The remainder is $R1$. The sifting process is repeated using $C1$, to obtain $C2$ and $R2$, etc.

PRX: BARREL UP

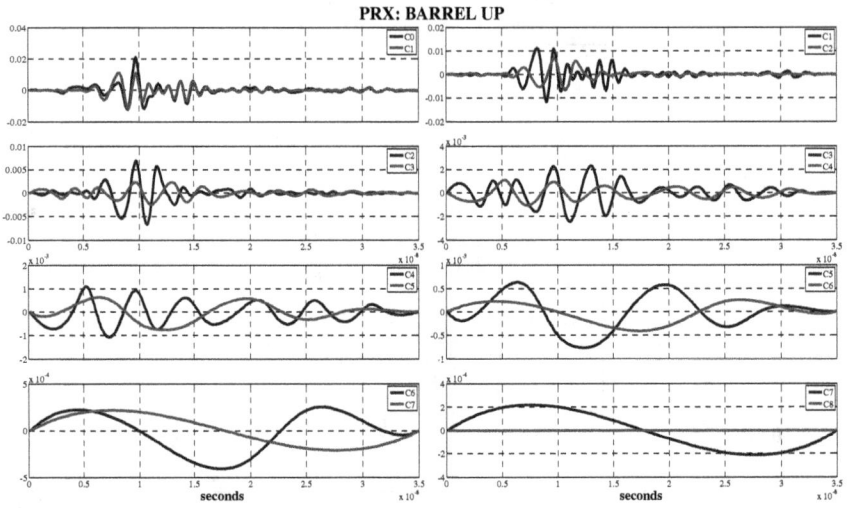

(a)

PRX: BARREL SIDE

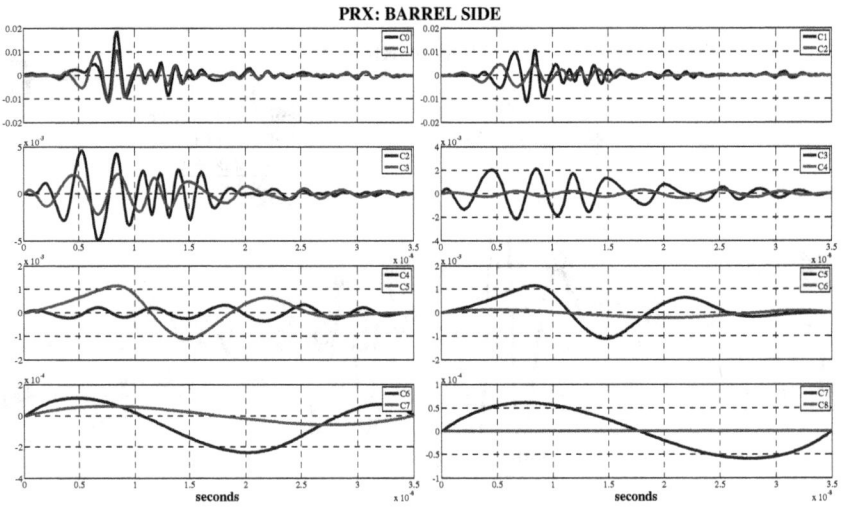

(b)

Fig. 1.14.8 Time Domain: The calculated (sifted) PRX C1, C2, C3, C4, C5, C6, C7 and C8 IMFs for targets: (a): BARREL UP ; and (b): BARREL SIDE.

PRX: BARREL UP

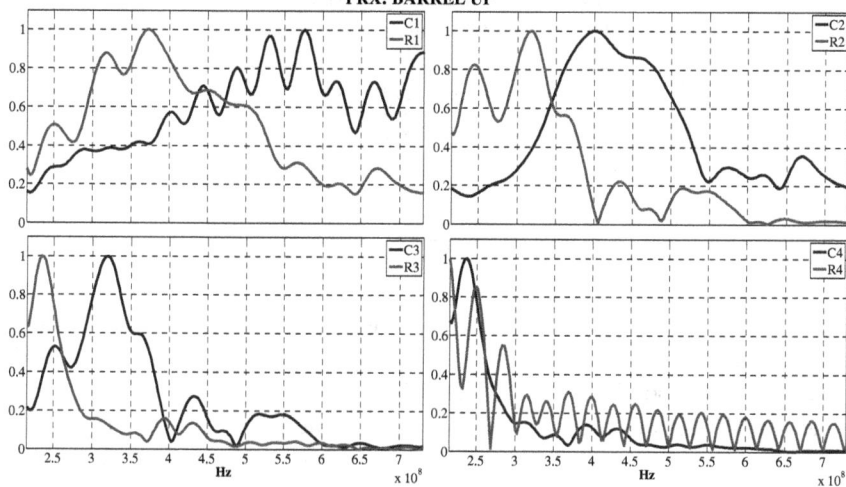

(a)

PRX: BARREL SIDE

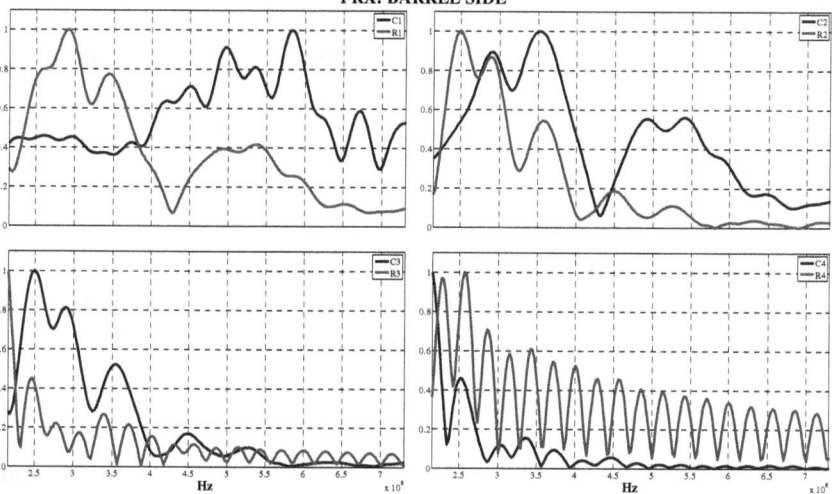

(b)

Fig. 1.14.9 Frequency Domain: The calculated (sifted) PRX C1, R1, C2, R2, C3, R3 and C4, R4 IMFs for (a): BARREL UP; and (b): BARREL SIDE. Notice that the C(i) and R(i) form high-pass, low pass pairs, analyzing the previous R(i-1).

Table 1.14.1　Barrel Up & Barrel Side cross-correlations high-pass & low-pass intrinsic modes.

H-P	Correlation	L-P	Correlation
C1s	0.8627	D1s	0.9683
C2s	0.9122	D2s	0.7735
C3s	0.7894	D3s	0.8443
C4s	0.9030	D4s	0.8290

1994, 2001). Conversely, the linear canonical transforms (LCT), or special affine Fourier transform or ABCD transform (Bernardo, 1996), generalizes the FRFT (Papoulis, 1977; Pei & Ding, 2001; Saxena & Singh, 2005). The FRFT leads to a generalization of the time (or space) and frequency domains. Here, we address the FRFT, but other well-known transforms can be fractionalized (Lohman *et al.*, 1996). A fundamental property of the FRFT is that performing the ath FRFT operation followed by a WVD corresponds to rotating the WVD of the original signal by an angle parameter $\alpha(\alpha = a\pi/2; \ a = \alpha 2/\pi$; both defined below) in the clockwise direction (Lohmann, 1993; Pei & Ding, 2001), or a decomposition of the signal into chirps. Whereas the conventional FT acting on a function can be interpreted as a linear differential operator acting on that function, the FRFT generalizes this definition by having the differential operator depend on the continuous order parameter a. Therefore, the ath order FRFT is the ath power of the FT operator.

The conventional FT is a special case of FRFT, the FRFT adding an additional degree of freedom to signal analysis with introduction of the fraction or order parameter a. The solution to a signal analysis problem can then be optimized depending on the objective at a specific a and the chosen optimization criterion.

A physical picture is conveniently based on diffraction optics. It is well-known that the far-field diffraction pattern is the FT of the diffracting object. The generalization is that FRFTs are the field patterns at closer distances. An alternative picture is an optical path involving many lenses separated by a variety of distances, with the FRFTs being the amplitude distributions of a wave propagating through the total system of lenses. As a wave propagates through the system, the amplitude distribution evolves through FRFTs of increasing order.

Whereas the conventional Fourier transform utilizes cisoidal, and steady state, basis functions, the FRFT (except at $a = 1.0, 2.0, \ldots$) utilizes frequency modulated or chirp basis functions that convolve maximally with signals changing in frequency — as in wave propagation through a dispersive medium — and such wave dispersion can occur on the surface of targets. In the following set of equations, the order parameter variable, a, ranges from -2 to $+2$. When $a = +1$, the transformation is the conventional forward Fourier transformation; when $a = -1$, the transformation is the conventional inverse Fourier transformation. When $a = 0$, the time-domain signal is regained; and when $a = +2$, the time-domain is regained, but parity changed. At other values of a, the ath-order FRFT results with the corresponding chirp basis functions. The relations are described in the following equations:

$$FRFT_a = \int_{-\infty}^{+\infty} K_a(u,t)f(t)dt,$$

where

$$K_a(u,t) = A_a \exp[i\pi(u^2 \cot(\alpha) - 2ut\, csc(\alpha) + t^2 \cot(\alpha))],$$
$$A_a = \sqrt{1 - i\cot(\alpha)},$$
$$\alpha = a\pi/2.$$

Noteworthy is that the kernel, $K_a(u,t)$, functions in the manner of a Green's function in the well-known Schrödinger's equation.

In the case $a = 1$, then $\alpha = \pi/2, u = f$, and

$$FRFT_1(f) = FT(f) = \int_{-\infty}^{+\infty} \exp[-i2\pi ft]f(t)dt,$$

i.e., the conventional forward FT.

In the case $a = -1$, then $\alpha = -\pi/2, u = f$, and

$$FRFT_{-1}(f) = FT(f) = \int_{-\infty}^{+\infty} \exp[+i2\pi ft]f(t)dt,$$

i.e., the conventional inverse FT.

Using operator formalism, the FRFT is generalized to the special affine Fourier transform (SAFT) (Pei & Ding, 2000). For example, let

$$O_F^\alpha(f(t)) = FRFT_\alpha$$

then the SAFT, or canonical transform (Moshinsky & Quesne, 1971; Abe & Sheridan, 1994) is:

$$O_F^{(a,b,c,d)}(f(t)) = \sqrt{\frac{1}{|b|}} \exp[i\pi(u^2(d/b) - ut(1/b))$$

$$+ t^2(a/b)]f(t)dt \quad \text{when } b \neq 0,$$

$$O_F^{(a,b,c,d)}(f(t)) = \sqrt{d}\exp[i\pi(u^2(cd))]f(du)dt \quad \text{when } b = 0,$$

and where $ad - bc = 1$ must be satisfied.

The FT is then described as a rotation by 90° and represented:

$$\begin{bmatrix} a & b \\ c & d \end{bmatrix} = \begin{bmatrix} 0 & 1 \\ -1 & 0 \end{bmatrix},$$

and the FRFT is described as a rotation by an arbitrary angle and represented:

$$\begin{bmatrix} a & b \\ c & d \end{bmatrix} = \begin{bmatrix} \cos\theta & \sin\theta \\ -\sin\theta & \cos\theta \end{bmatrix}.$$

The SAFT has the additive and reversible properties:

$$O_F^{(d,-b,-c,a)}(O_F^{(a,b,c,d)}(f(t))) = f(t).$$

The FRFT is periodic in a (or α) with period 4 (or 2π). Therefore the transform is fully defined for $a \in (-2, 2]$ or $\alpha \in (-\pi, \pi]$. The following relations follow where $\mathcal{F}^i$ is the FRFT for $a = i$, $\mathcal{J}$ is the identity matrix and $\mathcal{P}$ indicates a parity change (Candan *et al.*, 2000; Ozakta *et al.*, 2001):

$$\mathcal{F}^0 = \mathcal{J},$$

$$\mathcal{F}^1 = \mathcal{F},$$

$$\mathcal{F}^2 = \mathcal{P},$$

$$\mathcal{F}^3 = \mathcal{F}\mathcal{P} = \mathcal{P}\mathcal{F},$$

$$\mathcal{F}^4 = \mathcal{F}^0 = \mathcal{J},$$

$$\mathcal{F}^{4j+a} = \mathcal{F}^{4k+a}, \quad \text{where } j \text{ and } k \text{ arbitrary integers.}$$

The FRFT operations are additive. For example, the 0.3th FRFT of the 0.6th FRFT is the 0.9th FRFT. The inverse, $(\mathcal{F}^a)^{-1}$, of the ath order FRFT operator, $\mathcal{F}^a$, is equal to the operator $\mathcal{F}^{-a}$ (because $\mathcal{F}^{-a}\mathcal{F}^a = \mathcal{J}$).

A coordinate multiplication operator, $\mathcal{U}$, can be defined for which impulse signals, δ, are the eigensignals; and also a differentiation operator, $\mathcal{D}$, for which harmonic signals are the eigensignals (Candan *et al.*, 2000). Then $\mathcal{U}$ and $\mathcal{D}$ are Hermitian:

$$[\mathcal{U}, \mathcal{D}] = \frac{i}{2\pi}\mathcal{J}$$

with the following properties:

$$\mathcal{U}f(u) = u(f(u)),$$

$$\mathcal{D}f(u) = \frac{1}{2\pi i}\frac{d}{du}f(u).$$

Under a FRFT, the Hamiltonian Hermitian operator, $\mathcal{H}$, is invariant:

$$\mathcal{H}_{f\mathcal{F}} \equiv 2\pi\frac{1}{2}(\mathcal{D}^2 + \mathcal{D}^2) \equiv 2\mathcal{J}$$

and a chirp multiplication becomes:

$$\mathcal{H}_m \equiv 2\pi\frac{1}{2}\mathcal{U}^2,$$

a chirp convolution is:

$$\mathcal{H}_c \equiv 2\pi\frac{1}{2}\mathcal{D}^2,$$

and the FRFT defined in these terms is:

$$\mathcal{F}^a = \exp[-ia\mathcal{H}] = \exp[-ia[\pi(\mathcal{U}^2 + \mathcal{D}^2) - 1/2]$$

$$= \exp[-i\pi(\csc\alpha - \cot\alpha)\mathcal{U}^2 \exp[-i\pi(\sin\alpha)$$

$$\times \mathcal{D}^2 \exp[-i\pi(\csc\alpha - \cot\alpha)\mathcal{U}^2 \exp[i\alpha/2]$$

Therefore, the FRFT can be described as:

(1) a chirp multiplication, followed by
(2) a chirp convolution, followed by

(3) another chirp multiplication.

(4) a product by a complex amplitude factor.

Alternatively, as the FRFT is a rotation of the WVD (Lohmann, 1993), the FRFT can be described as:

(1) a conversion of an, e.g., 1D signal to a 2D WVD, followed by

(2) a rotation of the WVD, followed by

(3) a 1D FT

With this description, the rotation (2) of the WVD is composed of three shearing processes: left, down and right.

The FRFT can be described in terms of FT eigenfunctions (Pei & Ding, 2002). Let $\psi_n, n = 0, 1, 2, 3 \ldots$ denote eigenfunctions of the ordinary Fourier transform operation, then depending on the order parameter, a, the FRFT will either provide a predominately chirp multiplication (averaging) ($a = 0, 2, 4, 6, 8$, etc.) or predominately chirp convolution (differentiation) ($a = 1, 3, 5, 7, 9$, etc.), and the relation of the averaging, differentiation and Hamiltonian operators is:

$$\pi(\mathcal{D}^2 + \mathcal{U}^2) = \psi_n(u) = (n + 1/2)\psi_n$$
$$H\psi_n(u) = \lambda_n \psi_n(u)$$

Computing the FRFT of a signal corresponds to expressing it in terms of an orthonormal basis formed by chirps — complex exponentials with linearly varying instantaneous frequencies (Almeida, 1994). While there is an FRFT definition for all classes of signals (Cariolario *et al.*, 1998), there are many definitions of the discrete FRFT (DFRFT), and none of these definitions presently satisfy all the properties of the continuous FRFT (Ozaktas *et al.*, 1994; Pei & Ding, 2000). However, Pei & Ding (2000) proposed a closed form under appropriate sampling constraints that provides all properties of the continuous FRFT, including reversibility, except the additive property, but a conversion operation was proposed as a substitute. The sampling operations are:

$$O_{DFRFT}^{-\alpha, \Delta u, \Delta t} = (O_{DFRFT}^{\alpha, \Delta u, \Delta t}(f(t))) = f(t)$$

That is, the DFRFT of order $-\alpha$ with the sampling interval Δu in the input and Δt at the output is the inverse of the DFRFT of order α with the sampling interval Δt in the input and Δu at the output (Erseghe *et al.*, 1999; Pei & Ding, 2000; Li *et al.*, 2007).

The FRFT spectra are *a* order-parameter versus frequency spectra. As the order-parameter, *a*, changes, the orthogonal basis functions change. Therefore, the location of the maximum amplitude in the FRFT spectra will depend on *whether the signal analyzed has frequency modulated components, i.e., a changing instantaneous frequency*. Figure 1.15.1 demonstrates this with three test signals: (i): a 500 MHz, CW; (ii): a 0–1 GHz LFM; and (iii) a square wave. In the case of the first signal, which is CW and has no frequency modulation, the FRFT shows a maximum amplitude at $a = 1.0$, i.e., the conventional FT. In the case of the second signal, which is

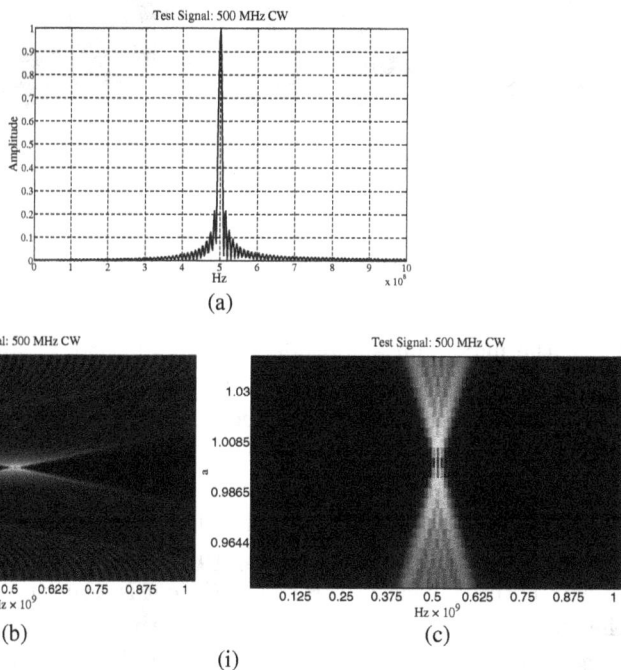

(a)

(b)   (c)

(i)

Fig. 1.15.1   Three test signals: (i): 500 MHz, CW; (ii): 0–1 GHz, LFM; (iii) Square Wave. a: FT of test signal; b: FRFT *a* order-parameter-frequency spectrum of test signal; c: Amplification of the b spectrum. The maximum amplitude in (i) is at $a = 1.0$, i.e., the conventional FT; in (ii) the maximum amplitude is at $a = 1.4386$, which does <u>not</u> give the conventional FT; in (iii) the maximum amplitude is at $a = 1.0$, i.e., again the conventional FT.

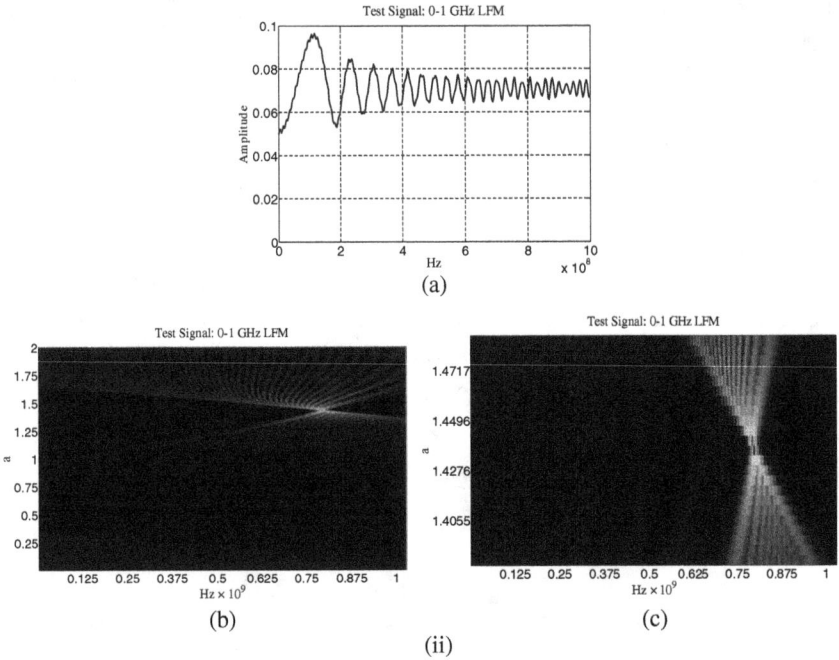

Fig. 1.15.1   (*Continued*)

LFM, the maximum amplitude is at $a = 1.4386$, a value of a which does *not* provide the conventional FT. In the case of the square wave, which, again, has no frequency modulated components, the maximum amplitude is at $a = 1.0$, i.e., again the conventional FT.

Figure 1.15.2(i)–(iii) shows cuts through the FRFT a-frequency spectra of three test signals shown in Fig. 1.15.1 at the position of the maximum amplitude. Here, in Fig. 1.15.2(i)–(iii) the (a) cut is a cut at $a = 1.0$ providing the conventional FT. The (b) cut is a cut at $a = 1.4386$ providing the maximum amplitude but *not* the conventional FT. The (c) cut is a cut at $a = 1.0$ again providing the conventional FT, indicating that the FRFT provides optimum characterization of FM signals.

The question of whether real targets act in a manner similar to frequency dispersive media is addressed in Figs. 1.15.3–1.15.5. The two Figs. 1.15.3–1.15.4 show the FRFT order-parameter-frequency spectra for the PRX (top figure in each), and MRX (bottom figure in each) for a truck

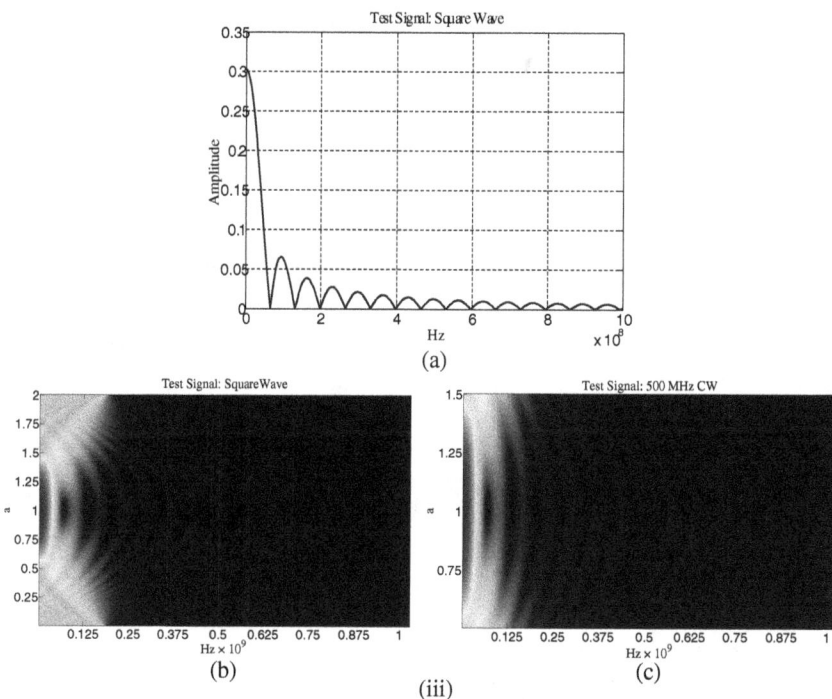

Fig. 1.15.1 (*Continued*)

and humvee target respectively. It should be noted that in each bottom figure (MRX), there is symmetry around the order parameter $a = 1.0$. Previously, MRX time symmetry was also shown in time-frequency spectra. Both time and order-parameter symmetry reflect the convolution operation of the transmitted signal, MTX, with its own complex conjugate, i.e., the target's impulse response, PRX — see INTRODUCTION. It is noteworthy that there is no such symmetry exhibited by the PRX spectrum shown in the top figures in each case.

If the optimum order-parameter providing the maximum amplitude in the FRFT spectra for an RX signal is *not* with $a = 1.0$, i.e., *not* the conventional FT, then the target is dispersive, i.e., the target is *not* non-dispersive. Figure 1.15.5 addresses this result and shows correlations of MTXs with the *a priori* target information, providing means for target identification.

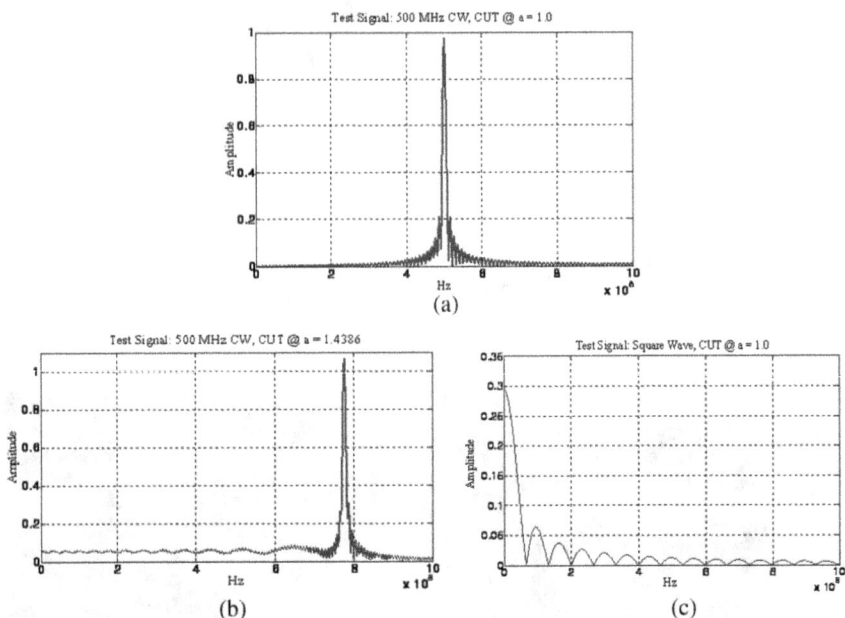

(a)

(b)   (c)

Fig. 1.15.2   Cuts through the FRFT a-frequency spectra of test signals (i)–(iii) of Fig. 1.15.1 at the position of the maximum amplitude. Here, the a cut is a cut at $a = 1.0$ providing the conventional FT. The b cut is a cut at $a = 1.4386$ providing the maximum amplitude but <u>not</u> the conventional FT. The c cut is a cut at $a = 1.0$, again providing the conventional FT.

The order parameter-frequency spectra, Figs. 1.15.3b (MRX, truck) and 1.15.4b (MRX, humvee), were used as the *a priori* information in calculating correlation-order-parameter spectra Fig. 1.15.5. MRXs from both targets were individually correlated against the two spectra, Figs. 1.15.3a and 1.15.4b, at the values of the order parameter, a, shown. In the case of the truck target, Fig. 1.15.5a, the optimum order parameter distinguishing truck from humvee was $a = 0.9674$. In the case of the humvee target, Fig. 1.15.5b, the optimum order parameter distinguishing humvee from hruck was $a = 1.218$. Significantly, in neither case was the optimum order parameter $a = 1.0$, which gives the conventional FT, and indicates a non-dispersive target. The results indicate that these specific targets function in a manner similar to slightly dispersive media. The optimum design of filters to detect these targets would take these dispersions into account (Pei & Ding, 2000).

PRX TRUCK AVERAGE ASPECTS 000/045/090/180 deg FRFT

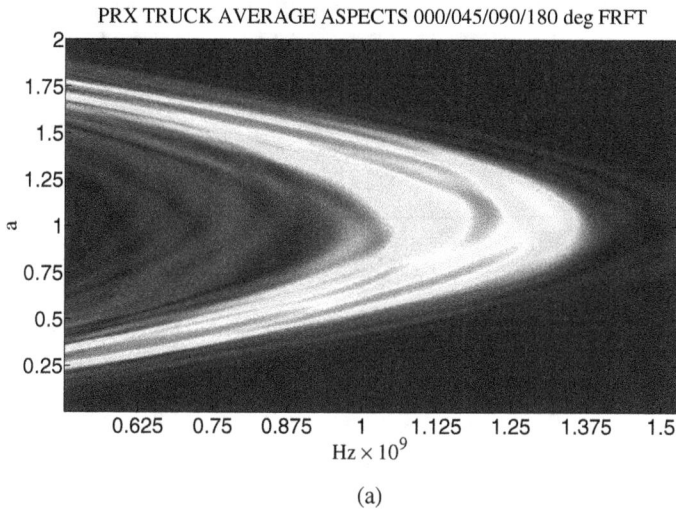

(a)

MRX TRUCK AVERAGE ASPECTS 000/045/090/180 deg FRFT

(b)

Fig. 1.15.3 Order-Parameter-Frequency Spectra for truck target — Average of RX spectra with target at aspect angles: 000°, 045°, 090° and 180°. a: PRX average; b: MRX average. Notice that the spectrum in B is symmetric around the order parameter $a = 1.0$. The MRX time symmetry previously noted and, in this case, order-parameter symmetry reflect the convolution operation of the transmitted signal, MTX, with its own complex conjugate, i.e., the target's impulse response, PRX — see INTRODUCTION. There is no such symmetry exhibited by the PRX spectrum in (a). Compare this spectrum (b), with spectrum (b) of the following Fig. 1.15.4 for a humvee target.

PRX HUMVEE AVERAGE ASPECTS 000/045/090/180 deg FRFT

(a)

MRX HUMVEE AVERAGE ASPECTS 000/045/090/180 deg FRFT

(b)

Fig. 1.15.4 Order-Parameter-Frequency Spectra for humvee target — Average of RX spectra with target at aspect angles: 000°, 045°, 090° and 180°. a: PRX average; b: MRX average. Notice that the spectrum in B is symmetric around the order parameter $a = 1.0$. The MRX time symmetry previously noted and, in this case, order-parameter symmetry reflect the convolution operation of the transmitted signal, MTX, with its own complex conjugate, i.e., the target's impulse response, PRX — see INTRODUCTION. There is no such symmetry exhibited by the PRX spectrum in (a). Compare this spectrum (b), with spectrum (b) of the preceding Fig. 1.15.3 for a truck target.

CORRELATION FRFT : MRX TRUCK A PRIORI WITH HUMVEE MRX

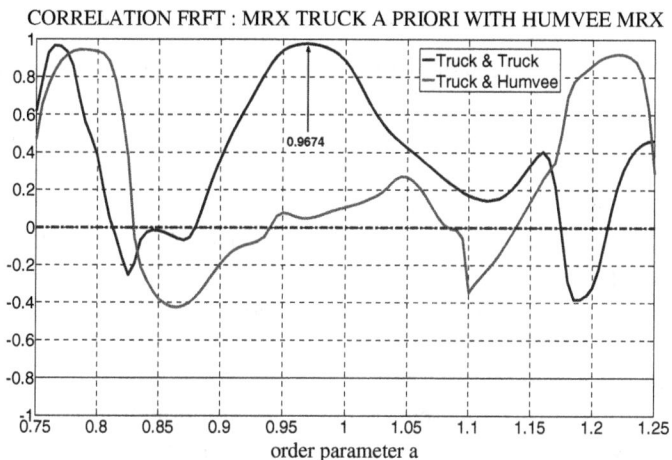

(a)

CORRELATION FRFT : MRX HUMVEE A PRIORI WITH TRUCK MRX

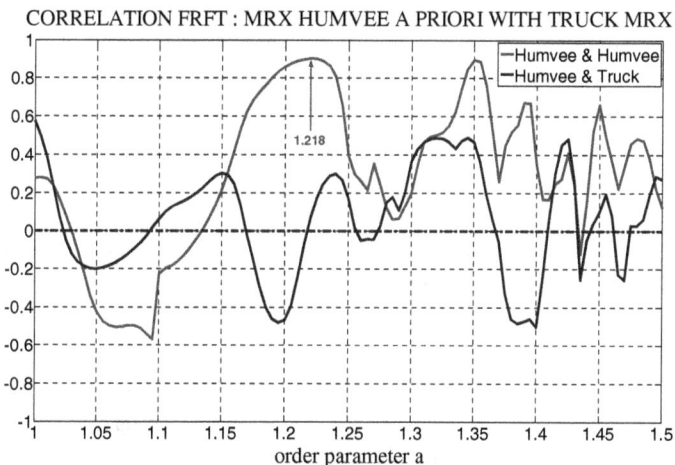

(b)

Fig. 1.15.5 Optimum Order Parameters, a, for Correlations of MTXs with *a priori* target information and providing means for target identification. The order parameter-frequency spectra, Figs. 1.15.3b (MRX, truck) and 1.15.4b (MRX, humvee), were used as the *a priori* information in calculating correlation-order-parameter spectra Fig. 1.15.5. MRXs from both targets were individually correlated against the two spectra, Figs. 1.15.3a and 1.15.4b, at the values of the order parameter, a, shown. In the case of the truck target, Fig. 1.15.5a, the optimum order parameter distinguishing truck from humvee was $a = 0.9674$. In the case of the humvee target, Fig. 1.15.5b, the optimum order parameter distinguishing humvee from truck was $a = 1.218$. Significantly, in neither case was the optimum order parameter $a = 1.0$, which gives the conventional FT, and indicates a non-dispersive target. The results, if that were the case, indicates that these specific targets function in a manner similar to slightly dispersive media.

1.16.   Weber-Hermite Transforms: Local & Global

The FRFT kernels corresponding to different values of the order parameter, a, are related to wavelets (Onural, 1993; Ozaktas *et al.*, 1994). Thus, filtering operations using the FRFT a domain can also be interpreted as filtering in the corresponding wavelet domain. The orthogonal bases of the FRFT are WHWFs[1] (Namias, 1980; Ozaktas *et al.*, 1995; Candan *et al.*, 2000; Cariolaro *et al.*, 1998). As a class, however, FRFTs need not be based on WHWFs, as Li (2008) has recently shown how different varieties of FRFTs can result from using orthogonal bases based on different wavelets.

There are competing methods and approaches to signal analysis and decomposition, all with different assumptions and drawbacks. For example, a Singular Value Decomposition (SVD) eigensignal/eigenimage approach uses basis signals/images which are defined in the signal/image itself. However, this is not an efficient way to treat signals/images, because the basis signals/images change from one signal/image to the next. Therefore the eigenvalues cannot be used for compression and transmission. Another example: the Karhunen-Loève (or Hotelling) transform assumes the signal/image to be processed is ergodic, i.e., that the spatial or temporal statistics of a single signal or image are the same over an ensemble of signals or images. But this is not always the case: signals and images are not the result of simple outcomes of a random process, there being always a deterministic underlying component. And another example: in the case of basis signals/images constructed from the Fourier transform, the assumption is that a signal or image is repeated periodically in all directions, i.e., the assumption is that the signal or image is bandwidth, but not space or time limited. The FRFT and the WH transform, to be discussed, do not require this assumption. Furthermore, while the Fourier transform provides no local information, local information is provided by wavelet transforms, the FRFT and the WH transform.

---

[1]As mentioned in the INTRODUCTION, there are other appellations for the Weber-Hermite wave functions, e.g., Hermite-Gaussian functions. I prefer the designation Weber-Hermite wave function because (a) the name "Hermite-Gaussian" implicates Gaussian in all polynomials, $n = 0, 1, 2, \ldots$ instead of just the first with $n = 0$; (b) Weber's equation is more general than Hermite's equation; (c) the name "Weber-Hermite" follows the *Mathematical Encyclopedia* usage (Hazewinkel, 2002); and (d) other texts, (e.g., Morse & Feshback, 1953, vol. 2, p. 1642; Jones, 1964, p. 86), have used the name: "Weber-Hermite".

The relationship of WHWFs to the FRFT is as follows. Let $\psi_i(u), i = 0, 1, 2, 3, \ldots$ denote WHWFS that are eigensignals (or eigenfunctions) of the ordinary Fourier transform operation with respect to the eigenvalues, λ_i, (Wiener, 1933; Dym & McKean, 1972; McBride & Kerr, 1987), and let these functions constitute an orthonormal basis for the space of well-behaved finite-energy signals (functions). The FRFT is then given by (Ozaktas *et al.*, 2001):

$$\mathbb{F}^\alpha \psi_i = \lambda_i^\alpha \psi_i(u) = \exp(-i\alpha i)\psi_i(u) = \exp(-i\alpha i\pi/2)\psi_i(u),$$

and a given function, $f(u)$, can be expanded as a linear superposition of WHWFs:

$$f(u) = \sum_{i=0}^{\infty} C_i \psi_i(u),$$

with

$$C_i = \int \psi_i(v)f(v)dv.$$

Applying $\mathbb{F}^\alpha$ to both sides:

$$\mathbb{F}^\alpha f(u) = \sum_{i=0}^{\infty} \exp\left(-\frac{iai\pi}{2}\right) C_i \psi_i(u)$$

$$= \int \sum_{i=0}^{\infty} \exp\left(-\frac{iai\pi}{2}\right) \psi_i(u)\psi_i(v)f(v)dv,$$

and

$$K_a(u, v) = \sum_{i=0}^{\infty} \exp\left(-\frac{iai\pi}{2}\right) \psi_i(u)\psi_i(v).$$

Referring back to Section 1.15, we see the relation of WHWFs to the FRFT.

Turning now to derivations of the WHWFs: there are a number of slightly different derivations, but each is noteworthy in presenting a different insight into the physical nature of these functions. We examine 7 here:

(1) A classical derivation is as follows. The parabolic cylinder functions, or Weber Hermite functions, are solutions to Weber's equation (Weber, 1869):

$$\frac{d^2\psi_n(x)}{dx^2} + \left(n + \frac{1}{2} - \frac{1}{4}x^2\right)\psi_n(x) = 0,$$

for which there is a general Weber equation, or parabolic cylinder differential equation (Abramowitz & Stegun, 1972, p. 686):

$$\frac{d^2\psi_n(x)}{dx^2} + (ax^2 + bx + c)\psi_n(x) = 0,$$

with the point $x = \infty$ strongly singular.

This equation permits two solutions derived as follows (Whittaker, 1902; Whittaker & Watson, 1927, p. 347). The substitutions

$$\psi = x^{-1/2}W_{k,m}, \quad z = x^2/2,$$

where $W_{k,m}$ is the Whittaker function (Whittaker & Watson, 1927, p. 347; Abramowitz & Stegun, 1972, p. 505):

$$\frac{d}{zdz}\left[\frac{d(wz^{1/2})}{zdz}\right] + \left(-\frac{1}{4} + \frac{2k}{z^2} + \frac{3}{4z^4}\right)wz^{1/2} = 0,$$

for which:

$$\frac{\partial^2 w}{\partial z^2} + \left(2k - \frac{1}{4}z^2\right)w = 0,$$

converts the Weber equation to the Whittaker equation, which is a special case of the confluent hypergeometric equation.[2] In particular, taking the solution for which $R(z) > 0$ the solution is:

$$\psi_n = (2^{\frac{n}{2}+\frac{1}{4}})(z^{-1/2})(W_{\frac{n}{2}+\frac{1}{4},-\frac{1}{4}}(1/2z^2))$$

$$= \frac{1}{\sqrt{z}}2^{n/2}\exp[-z^2/4](-iz)^{1/4}(iz)^{1/4}(iz)^{1/4}{}_1F_1\left(\frac{1}{2}n + \frac{1}{4};\frac{1}{2};\frac{1}{2}z^2\right)$$

where ${}_1F_1(a;b;z)$ is a confluent hypergeometric function (Gauss, 1812; Abramowitz & Stegun, 1972, p. 503). With

$$w = z^{-1/2}W_{k-1/4}\left(\frac{1}{2}z^2\right),$$

where W is a Whittaker function defined above, Weber's equation can be separated into:

$$\frac{d^2U}{du^2} - (c + k^2u^2)U = 0,$$

---

[2]The hypergeometric differential equation is a second-order linear ordinary differential equation whose solutions are given by the hypergeometric series. The hypergeometric series have the form: $(\sum_{n=0}^{\infty} a_n)(\sum_{n=0}^{\infty} b_n) = \sum_{n=0}^{\infty} c_n$, where $c_n = a_k b_{n-k}$. The confluent hypergeometric equation is a degenerate form of the hypergeometric equation.

or Weber's first derived equation,

$$\frac{d^2V}{dv^2} + (c - k^2v^2)V = 0,$$

or Weber's second derived equation.

For non-negative n, and after renormalization, the solution to Weber's first derived equation reduces to:

$$U_n(x) = 2^{-n/2}\exp[-x^2/4]H_n(x/\sqrt{2}), \quad n = 0, 1, 2, \ldots,$$

which are parabolic cylinder functions or Weber-Hermite functions (WHWFs), and where H_n is a Hermite polynomial.

Similarly, completing the square, Weber's second derived equation can be rewritten as:

$$\frac{\partial^2\psi}{\partial x^2} + \left[a\left(x + \frac{b}{2a}\right)^2 - \frac{b^2}{4a} + c\right]\psi = 0.$$

Defining:
$u = x + b/2a; du = dx$, and substituting, gives:

$$\frac{\partial^2\psi}{\partial u^2} + [au^2 + d]\psi = 0,$$

where $d = -b^2/4a + c$.

Again, this equation admits of two solutions, an even and an odd. Continuing with the even solution, the solution is:

$$\psi(x) = \exp[-x^2/4]\,_1F_1\left(\frac{1}{2}a + \frac{1}{4}; \frac{1}{2}; \frac{1}{2}x^2\right),$$

where $_1F_1(a; b; z)$ is, as before, the confluent hypergeometric function, and the solutions of this equation are, again, the parabolic cylinder or Weber-Hermite functions (WHWFs).

(2) A second parallel derivation commences with the one-dimensional wave equation:

$$-\frac{1}{2m}\frac{\partial^2\psi}{\partial x^2} + V(x)\psi = E\psi,$$

with spring potential:

$$V(x) = \frac{1}{2}kx^2 = \frac{1}{2}m\omega^2x^2,$$

where $= \sqrt{\frac{k}{m}}$, the angular frequency, k is the stiffness constant, m is the mass, and x is the field deflection of the oscillator, or

$$-\frac{1}{2m}\frac{\partial^2 \psi}{\partial x^2} + \frac{1}{2}kx^2\psi = E\psi.$$

This wave equation can be written in dimensionless form by defining the independent variables $\xi = \alpha x$ and an eigenvalue, λ, and requiring:

$$\alpha^4 = mk, \quad \lambda = 2E\left(\frac{m}{k}\right)^{1/2} = \frac{2E}{\omega}.$$

The dimensionless form is then:

$$\frac{\partial^2 \psi}{\partial \xi^2} + (\lambda - \xi^2)\psi = 0.$$

which is a form of Weber's equation.

(3) A derivation based on a familiar model is as follows. The wave equation for a vibrating string is associated with the difference between the total kinetic energy of the string and its potential energy being as small as possible. If a string vibrates with simple harmonic motion, then the time dependence is expressed as:

$$\psi(x,t) = \psi(x)\exp[-i\varepsilon\alpha^2 t],$$

and the function ψ must satisfy Helmholtz's equation:

$$\frac{\partial^2 \psi}{\partial x^2} + k^2\psi = 0,$$

with k a real constant. When $k = 0$, the equation is a one-dimensional Laplace equation; and when k^2 is a function of the coordinates and

$$\varepsilon = (2M/(h/2\pi)^2)E.$$

This equation is Schrödinger's equation for a particle with constant E (Morse & Feshbach, 1953, p. 494).

When ψ is space-dependent, the wave equation is:

$$\frac{\partial^2 \psi}{\partial x^2} + (\varepsilon - \alpha^2 x^2)\psi = 0,$$

where $\alpha = M\omega/(h/2\pi)$ and which is yet another form of Weber's equation. This equation permits solutions as a function of

$$n = \frac{\varepsilon}{2\beta} - \frac{1}{2} = \frac{E}{(h/2\pi)\omega} - \frac{1}{2}.$$

In order for the solutions to be quadratically integrable, it is necessary that n take on integer values: $n = 0, 1, 2, \ldots$ (Morse & Feshbach, 1953, p. 1641). With normalization factors, the solutions are the Weber-Hermite or parabolic cylinder functions (Morse & Feshbach, 1953, p. 1642):

$$\psi_n(t) = \frac{1}{\sqrt{2^n n!}} \left(\frac{\alpha}{\pi}\right)^{1/4} \exp[-\alpha t^2/2] H_n(t\sqrt{\alpha})$$

where $\alpha = M\omega/(h/2\pi)$. For the classical result, we substituted $(h/2\pi) \to 1$, and α becomes a time-frequency trade parameter/variable. This is the general form of the Weber-Hermite functions (WHWFs).

If for a function, $f(x)$, an expansion of the form:

$$f(x) = a_0\psi_0 + a_1\psi_1 + \cdots + a_n\psi_n + \cdots$$

exists, and if it is legitimate to integrate term-by-term between the limits $+\infty$ and $-\infty$ then:

$$a_n = \frac{1}{\sqrt{(2\pi)^{1/2}n!}} \int_{-\infty}^{+\infty} \psi_n(t)f(t)dt,$$

and such a function can be expanded in terms of n WHWFs with coefficients a_n.

(4) There is a derivation based on the Helmholtz equation. This derivation commences with the Helmholtz equation in parabolic cylinder coordinates:

$$\frac{1}{u^2 + v^2}\left(\frac{\partial^2\psi}{\partial u^2} + \frac{\partial^2\psi}{\partial v^2}\right) + \frac{\partial^2\psi}{\partial z^2} + k^2\psi = 0.$$

The equation can be separated with:

$$\psi(u, v, z) = U(u)V(v)Z(z),$$

resulting in:

$$\frac{1}{u^2 + v^2}\left(VZ\frac{\partial^2 U}{\partial u^2} + UZ\frac{\partial^2 V}{\partial v^2}\right) + UV\frac{\partial^2 Z}{\partial z^2} + k^2 UVZ = 0.$$

Dividing by UVZ and separating out the Z part gives:

$$\frac{\partial^2 Z}{\partial z^2} = -(k^2 + m^2)Z,$$

which can be solved, permitting the derivation:

$$\left(\frac{1}{U}\frac{\partial^2 U}{\partial u^2} - k^2 u^2\right) + \left(\frac{1}{V}\frac{\partial^2 V}{\partial v^2} - k^2 v^2\right) = 0.$$

With

$$\left(\frac{1}{U}\frac{\partial^2 U}{\partial u^2} - k^2 u^2\right) = c,$$

$$\left(\frac{1}{V}\frac{\partial^2 V}{\partial v^2} - k^2 v^2\right) = -c,$$

we have again:

$$\frac{d^2 U}{du^2} - (c + k^2 u^2)U = 0.$$

or Weber's first derived equation,

$$\frac{d^2 V}{dv^2} + (c - k^2 u^2)V = 0,$$

or Weber's second derived equation

The solutions of Weber's first derived equation reduce to:

$$U_n(x) = 2^{-n/2}\exp[-x^2/4]H_n(x/\sqrt{2}), \quad n = 0, 1, 2, \ldots$$

or WHWFs as before.

(5) A derivation can be based on the electric Hertz vector, $\mathbf{\Pi}_e$. Hertz showed that the electromagnetic field can be expressed in terms of a single vector function (Hertz, 1889). This potential is known as *the Hertz (electric) vector, the polarization potential, or a "superpotential"*. A second Hertz vector, the *"Hertz magnetic vector potential"*, $\mathbf{\Pi}_m$, and related to the magnetic polarization, was introduced by Righi (1901). Together, $\mathbf{\Pi}_e$ and $\mathbf{\Pi}_m$ form a six vector (i.e., an antisymmetric tensor of the second rank). $\mathbf{\Pi}_e$ and $\mathbf{\Pi}_m$ have similar transformation properties as $\mathbf{E}$ and $\mathbf{B}$ in the case of $\mathbf{\Pi}_e$, and $\mathbf{D}$ and $\mathbf{H}$ in the case of $\mathbf{\Pi}_m$ (Born & Wolf, 1999, p. 84). Their relationship is as follows.

Assuming the Lorentz gauge, a vector, called *the polarization vector*, $\mathbf{p}$, is defined with respect to the actual charges and currents as (Panofsky & Phillips, 1962, p. 254):

$$\frac{\partial \mathbf{p}}{\partial t} = \mathbf{J}; \quad \nabla \mathbf{p} = -\rho,$$

where $\mathbf{J}$ is the free current density, and ρ is the free charge density. A generalization introduces a magnetic density vector, $\mathbf{m}$, so that these equations become (*cf.* Chapou Fernández *et al.*, 2009):

$$\frac{\partial \mathbf{p}}{\partial t} = \mathbf{J} - \frac{1}{\mu\epsilon}(\nabla \times \mathbf{m}); \quad \nabla \mathbf{p} = -\rho.$$

$\mathbf{\Pi}_e$ and $\mathbf{\Pi}_m$ are then defined as two *retarded potentials* (Born & Wolf, 1999, p. 84):

$$\mathbf{\Pi}_e = \frac{1}{\epsilon} \int_V \frac{[\boldsymbol{p}]}{R} dv',$$

$$\mathbf{\Pi}_m = \mu \int_V \frac{[\boldsymbol{m}(r')]}{R} dv',$$

where R is the distance from a point, r, to a volume element, dV', at a point, r', and the brackets [ ] indicate that $\boldsymbol{p}$ is to be evaluated at the retarded time, $t - |r - r'|/v$, where v is the velocity of propagation.

In terms of $\mathbf{\Pi}_e$ and $\mathbf{\Pi}_m$, the electromagnetic field can be defined by the following $\boldsymbol{A}$ vector potential field (Jones, 1964; Born & Wolf, 1999):

$$\boldsymbol{A} = \mu\epsilon\frac{\partial \mathbf{\Pi}_e}{\partial t} + \nabla \times \mathbf{\Pi}_m,$$

$$\varphi = -\nabla \cdot \mathbf{\Pi}_e,$$

$$\nabla^2\mathbf{\Pi}_e - (\mu\epsilon)^2\frac{\partial^2\mathbf{\Pi}_e}{\partial t^2} = -\boldsymbol{p}/\epsilon,$$

$$\nabla^2\mathbf{\Pi}_m - (\mu\epsilon)^2\frac{\partial^2\mathbf{\Pi}_m}{\partial t^2} = -\mu\boldsymbol{m},$$

$$\nabla \cdot (\nabla^2\mathbf{\Pi}_e) = \nabla^2(\nabla \cdot \mathbf{\Pi}_e).$$

Furthermore, in a source-free region, in vacuo, an electromagnetic field can be described in terms of either the electric or the magnetic Hertzian potential.

For a field in the absence of magnetic polarization:

$$\boldsymbol{A} = \mu\epsilon\frac{\partial \mathbf{\Pi}_e}{\partial t},$$

$$\varphi = -\nabla \cdot \mathbf{\Pi}_e,$$

$$\boldsymbol{E} = \nabla(\nabla \cdot \mathbf{\Pi}_e) - \mu\epsilon\frac{\partial^2\mathbf{\Pi}_e}{\partial t^2},$$

$$\boldsymbol{B} = \mu\epsilon\left(\nabla \times \frac{\partial^2\mathbf{\Pi}_e}{\partial t}\right),$$

$$\nabla^2\mathbf{\Pi}_e - (\mu\epsilon)^2\frac{\partial^2\mathbf{\Pi}_e}{\partial t^2} = -\boldsymbol{p}/\epsilon.$$

For a field in the absence of electric polarization:

$$A = \nabla \times \mathbf{\Pi}_m$$

$$\varphi = 0,$$

$$E = -\mu\epsilon \frac{\partial(\nabla \times \mathbf{\Pi}_m)}{\partial t},$$

$$B = \nabla \times (\nabla \times \mathbf{\Pi}_m),$$

$$\nabla^2 \mathbf{\Pi}_m - (\mu\epsilon)^2 \frac{\partial^2 \mathbf{\Pi}_m}{\partial t^2} = -\mu m.$$

For a field with both electric and magnetic polarization:

$$A = \mu\epsilon \frac{\partial \mathbf{\Pi}_e}{\partial t} + \nabla \times \mathbf{\Pi}_m,$$

$$\varphi = -\nabla \cdot \mathbf{\Pi}_e,$$

$$E = \nabla(\nabla \cdot \mathbf{\Pi}_e) - \mu\epsilon \frac{\partial^2 \mathbf{\Pi}_e}{\partial t^2} - \mu\epsilon \frac{\partial(\nabla \times \mathbf{\Pi}_m)}{\partial t},$$

$$B = \mu\epsilon \left(\nabla \times \frac{\partial^2 \mathbf{\Pi}_e}{\partial t} \right) + \nabla \times (\nabla \times \mathbf{\Pi}_m).$$

For a conductor without sources:

$$E = \nabla(\nabla \cdot \mathbf{\Pi}_e) - \mu\epsilon \frac{\partial^2 \mathbf{\Pi}_e}{\partial t^2} - \mu\sigma \frac{\partial \mathbf{\Pi}_e}{\partial t} = \nabla \times (\nabla \times \mathbf{\Pi}_e),$$

$$B = \mu\nabla \times \left(\epsilon \frac{\partial \mathbf{\Pi}_e}{\partial t} + \sigma \mathbf{\Pi}_e \right),$$

$$\nabla^2 \mathbf{\Pi}_e - \mu\epsilon \frac{\partial^2 \mathbf{\Pi}_e}{\partial t^2} - \mu\sigma \frac{\partial \mathbf{\Pi}_e}{\partial t} = 0.$$

Invariance is then obtained for the transformations:

$$\mathbf{\Pi'}_e = \mathbf{\Pi}_e + \nabla \times F - \nabla G,$$

$$\mathbf{\Pi'}_m = \mathbf{\Pi}_e - \mu\epsilon \frac{\partial F}{\partial t},$$

where the vector $F = F_x \mathbf{i} + F_y \mathbf{j} + F_z \mathbf{k}$, and the scalar function, G, defined over three dimensions of Cartesian coordinates, and with $\mathbf{i}$, $\mathbf{j}$, and $\mathbf{k}$ the unit vectors for the x-, y-, and z-axes, respectively, are solutions to the wave equations:

$$\nabla^2 F - (\mu\epsilon)^2 \frac{\partial^2 F}{\partial t^2} = 0,$$

$$\nabla^2 G - (\mu\epsilon)^2 \frac{\partial^2 G}{\partial t^2} = 0.$$

for which the usual definitions apply:

∇ as the gradient operator (grad):

$$\nabla F = \frac{\partial F_x}{\partial x}\mathbf{i} + \frac{\partial F_y}{\partial y}\mathbf{j} + \frac{\partial F_z}{\partial z}\mathbf{k}.$$

$\nabla\cdot$ as the divergence operator (div):

$$\nabla \cdot F = \frac{\partial F_x}{\partial x} + \frac{\partial F_y}{\partial y} + \frac{\partial F_z}{\partial z}.$$

$\nabla\times$ as the curl operator (curl):

$$\nabla \times F = \left(\frac{\partial F_z}{\partial y} - \frac{\partial F_y}{\partial z}\right)\mathbf{i} + \left(\frac{\partial F_x}{\partial z} - \frac{\partial F_z}{\partial x}\right)\mathbf{j} + \left(\frac{\partial F_y}{\partial x} - \frac{\partial F_x}{\partial y}\right)\mathbf{k}.$$

∇^2, sometimes ∇, as the Laplace operator, or the divergence of the gradient (del):

$$\nabla F = \nabla \cdot (\nabla F) = \nabla^2 F = \frac{\partial F_x^2}{\partial x^2} + \frac{\partial F_y^2}{\partial y^2} + \frac{\partial F_z^2}{\partial z^2}.$$

and where SI units are implied:

μ is the magnetic permeability $(kg \cdot m)/(s^2 \cdot A^2)$,
ϵ is the dielectric constant $(A^2 \cdot s^4)/(kg \cdot m^3)$,
σ is the conductivity $((A^2 \cdot s^3)/(kg \cdot m^3))$,
ρ is the charge density $(A \cdot s)/m^3$,
$\boldsymbol{m}$ is the magnetization or magnetic density, A/m,
$\boldsymbol{p}$ is the electric polarization density $(A \cdot s)/m^2$,
$\boldsymbol{J}$ is the current density A/m^2,
$\boldsymbol{E}$ is the electric intensity or electric field strength $(kg \cdot m)/(A \cdot s^3)$, with A = ampere,
$\boldsymbol{B}$ is the magnetic flux density or magnetic induction, $kg/(A \cdot s^2)$
$\boldsymbol{A}$ is the vector potential $(kg \cdot m)/(A \cdot s^2)$,
φ is the scalar potential $(kg \cdot m^2)/(A \cdot s^3)$.

In parabolic cylinder coordinates, u, v, z, the electric Hertz vector, $\boldsymbol{\Pi}_e$ is (Jones, 1964, p. 85):

$$\boldsymbol{\Pi}_e(u, v, z) = U(u)V(v)Z(z),$$

and

$$\frac{\partial^2 Z}{\partial z^2} = m^2 Z,$$

$$\frac{\partial^2 U}{\partial u^2} + [(m^2 + k^2)u^2 - h]U = 0,$$

$$\frac{\partial^2 V}{\partial v^2} + [(m^2 + k^2)v^2 + h]V = 0,$$

where m and h are separation constants. Both the equations in U and V can be treated similarly. Substitutions of:

$$u = [4(m^2 + k^2)]^{-1/4} \boldsymbol{X}, \quad \text{and} \quad i(v + 1/2) = (4(m^2 + k^2)^{-1/2}) \boldsymbol{X},$$

gives:

$$\frac{\partial^2 U}{\partial u^2} + \left(\frac{1}{4}\boldsymbol{X}^2 - i\left(v + \frac{1}{2} \right) \right) U = 0.$$

A further substitution of $\boldsymbol{X} = x \exp[(1/4)\pi i]$ gives:

$$\frac{\partial^2 U}{\partial u^2} + \left(v + \frac{1}{2} - \frac{1}{4}x^2 \right) U = 0,$$

which we recognize again as Weber's equation. As before, the solutions are WHWFs:

$$U_n(x) = 2^{-n/2} \exp[-x^2/4] H_n(x/\sqrt{2}) \quad n = 0, 1, 2, \ldots$$

The Hertzian vectors, $\boldsymbol{\Pi}_e$ and $\boldsymbol{\Pi}_m$, together form a six vector (Nisbet, 1955), and TE and TM waves in a waveguide can be defined in terms of $\boldsymbol{\Pi}_e$ and $\boldsymbol{\Pi}_m$, and shown to be generated by exciter systems that are equivalent respectively to electric and magnetic oscillating dipoles parallel to the direction of propagation (Essex, 1977). The two Hertzian vector approach is related to Whittaker's (1904, 1951) expression of an electromagnetic field in terms of two scalar functions.

The vectors, $\boldsymbol{\Pi}_e$ and $\boldsymbol{\Pi}_m$, can be united in a covariant formulation resulting in a skew tensor of rank two (Nisbet, 1955; McCrea, 1957; Chapou Fernández et al., 2009). This Hertz's tensor is defined:

$$\boldsymbol{\Pi}^{0i} = (\boldsymbol{\Pi}_e)_i, \quad \boldsymbol{\Pi}^{ij} = \varepsilon^{ijk}(\boldsymbol{\Pi}_m)_k.$$

where ε^{ijk} is the Levi-Civita permutation symbol. This potential tensor, $\boldsymbol{\Pi}^{\mu v}, \mu, v = 0, 1, 2, 3$, has space-time components corresponding to the electric Hertz vector, $\boldsymbol{\Pi}_e$, components, and purely spatial components corresponding to the magnetic Hertz vector, $\mathbf{\Pi}_m$, components. The Hertz tensor transforms according to the gauge chosen.

It should also be noticed that $\mathbf{\Pi}^{\mu\nu}$ has the characteristic of the 8-dimensional algebra of octonions (Conway & Smith, 2003; Baez, 2002, 2005; Baez & Huerta, 2011).

An alternative approach to the treatment of Hertz vectors is the so-called "source scalarization" method, whereby a any given distribution of arbitrarily oriented sources is reduced to an equivalent distribution of single-component parallel electric and magnetic sources (Weigelhofer, 2000; Georgieva & Weiglhofer, 2002; Weigelhofer & Georgieva, 2003). The objective is to produce a complete description of the EM field in terms of two so-called "scalar wave" potentials — actually vector potentials. The method has been extended to the case of propagation in stratified gyrotropic media (De Visschere, 2009).

(6) Turning from classical mechanics to quantum mechanics, a quantum mechanical derivation commences with the Hamiltonian of a particle:

$$H = \frac{p^2}{2m} + \frac{1}{2}m\omega^2 x^2,$$

where x is the position operator, p is the momentum operator: $p = -i\hbar\frac{\partial}{\partial x}$; and where the first term is the kinetic energy of the particle, and the second, the potential energy.

The one-dimensional Schrödinger wave equation, inspired by an optical analogy, is:

$$-\frac{\hbar^2}{2m}\frac{d^2\psi}{dx^2} + V(x)\psi = E\psi,$$

or

$$-\frac{\hbar^2}{2m}\frac{d^2\psi}{dx^2} + \frac{kx^2\psi}{2} = E\psi,$$

This equation can be written in dimensionless form using the following substitutions:

$$\xi = \alpha x; \quad \alpha^4 = mk/\hbar^2; \quad \lambda = (2E/\hbar)(m/k)^{1/2} = (2E)/\hbar\omega.$$

The wave equation in this dimensionless form becomes Weber's equation:

$$\frac{d^2\psi}{d\xi^2} + (\lambda - \xi^2)\psi = 0.$$

Alternatively, by a coordinate transformation:

$$x = \left(\frac{\hbar}{m\omega}\right)^{1/2} \xi$$

the wave equation becomes:

$$\left(\frac{d^2}{d\xi^2} + \frac{2E}{\hbar\omega} - \xi^2\right)\psi = 0,$$

which is also Weber's equation. Except when $\xi = \infty$, the solutions are:

$$\psi = \exp(\pm(1/2)\xi^2)^{1/2}y,$$

for which

$$\left(\frac{d^2}{d\xi^2} - 2\xi\left(\frac{d}{d\xi}\right) + \frac{2E}{\hbar\omega} - 1\right)y = 0,$$

with solutions for polynomials of degree n:

$$-2n + \frac{2E}{\hbar\omega} - 1 = 0,$$

resulting in eigenvalues:

$$E_n = (n + 1/2)\hbar\omega.$$

The solutions for y are:

$$y_n = H_n(\xi) = (-1)^n \exp[\xi^2]\frac{d^n}{d\xi^n}\exp[-\xi^2],$$

where $H_n(\xi), n = 0, 1, 2, \ldots$ are Hermite polynomials, satisfying:

$$\left(\frac{d^2}{d\xi^2} - 2\xi\frac{d}{d} + 2n\right)H_n(\xi) = 0,$$

for which $2n = \lambda - 1$, or $\lambda = 2n + 1$. Therefore, substituting, the solutions for ψ are:

$$\psi = \exp\left(\pm\frac{1}{2}\xi^2\right)^{1/2} H_n(\xi), \quad n = 0, 1, 2, \ldots$$

which are WHWFs.

(7) An alternative quantum mechanical view is to commence with the time-independent one-dimensional Schrödinger equation in the bra-ket form:

$$H|\psi\rangle = E|\psi\rangle,$$

which is seen to be a Weber's equation:

$$\frac{-\hbar^2}{2m}\frac{d^2\psi(x)}{dx^2} + \frac{1}{2}m\omega^2 x^2\psi(x) = E\psi(x).$$

The general solution is:

$$\langle x|\psi_n\rangle = \frac{1}{\sqrt{2^n n!}}\left(\frac{m\omega}{\pi\hbar}\right)^{1/4}\exp\left(-\frac{m\omega x^2}{2\hbar}\right)H_n\left(\sqrt{\frac{m\omega}{\hbar}}x\right), \quad n = 0, 1, 2, \ldots$$

where $H_n(x) = (-1)^n\exp(x^2)\frac{d^n}{dx^n}\exp(-x^2)$ are Hermite polynomials; or

$$\psi_n(z) = \left(\frac{\alpha}{\pi}\right)^{1/4}\frac{1}{\sqrt{2^n n!}}H_n(z)\exp(-z^2/2), \quad n = 0, 1, 2, \ldots$$

for $z = \sqrt{\alpha}x$ and $\alpha = m\omega/\hbar$, which is a normalized form of the Weber-Hermite wave functions.

The corresponding energy levels are again:

$$E_n = \hbar\omega(n + 1/2),$$

and the expectation value for the potential energy is:

$$\langle V_n\rangle = \int_{-\infty}^{+\infty}\left(\frac{1}{2}\right)\psi_n^* kx^2\psi_n(x)dx$$

$$= (1/2)k\frac{2n+1}{2\alpha^2} = (1/2)(n+1/2)\hbar\omega = (1/2)E_n.$$

Therefore:

$$\Delta x\Delta p = \frac{1}{2}(2n+1)\hbar, \quad n = 0, 1, 2, \ldots$$

Substituting $x \to t, p \to f$, and $\hbar \to 1$ for the classical case, gives the time-bandwidth products (TBPs) for the WHWFs:

$$\Delta t\Delta f = \frac{1}{2}(2n+1), \quad n = 0, 1, 2, \ldots$$

It should be noted that these time-bandwidth products refer to one standard deviation of the signal duration, and one standard deviation of the signal bandwidth, i.e., not the 90% or 99% support of signal duration and bandwidth in general use in engineering.

The WHWFs — of increasing TBP — can be given a *globally distributed* or a *locally distributed* matrix form. Both matrices are unitary (Figs. 1.16.1 & 1.16.2) and forward and inverse global and local WH transforms of 1D and 2D signals are possible (Fig. 1.16.3). Figures 1.16.4 & 1.16.5

(a)

(b)

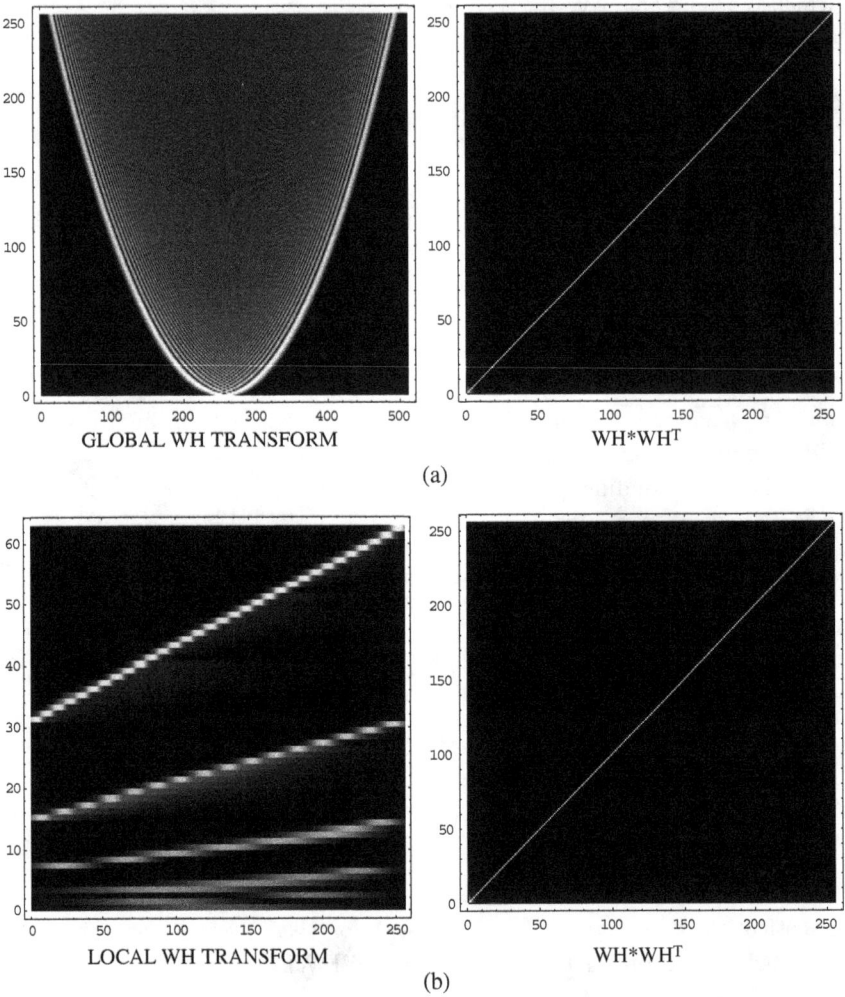

Fig. 1.16.1   WH Transforms. a (Left): *Global* WH transform matrix. b (Left): *Local* WH transform matrix, in both of which the time-bandwidth product (TBP) increases from bottom to top according to $\Delta f \Delta t = \frac{1}{2}(2n+1), n = 0, 1, 2, \ldots$. In the case of both Left matrices, which are unitary, the product of each matrix with its conjugate transpose, results in an identity matrix — both Right matrices. Therefore both Global and Local WH transforms, forward and inverse, of 1D and 2D signals are indicated.

indicate the complementarity of Fourier and WH transforms. As CW signals are the basis functions of the Fourier transform, the Fourier-transform-based power spectrum discriminates well the CW signals and the WH transform does poorly. In contrast, as WH signals are the basis functions

Weber-Hermite Matrix Global 20 Signals

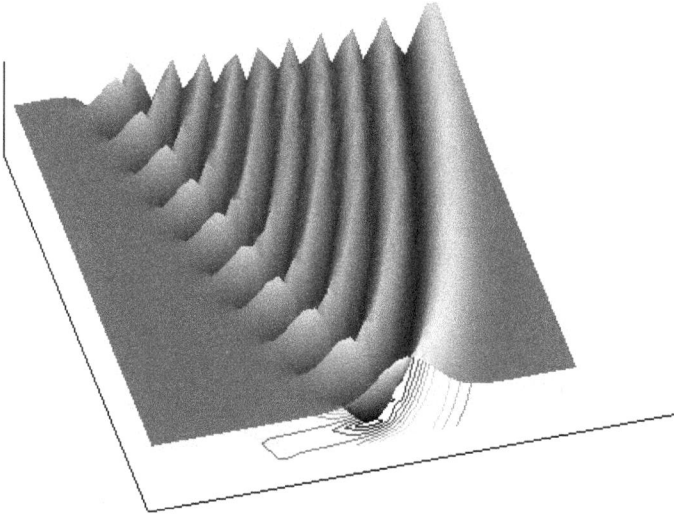

Fig. 1.16.2   A perspective of the Global WH matrix for 20 signals.

of the WH transform, the WH transform discriminates well WH packet signals and the Fourier-transform-based power spectrum does poorly.

The ability to reconstruct corrupted signals by separation of the corruption and the signal and the removal of the corruption in the transform domain is an indication of the appropriateness of a transform for analysis of specific families of signals. Figures 1.16.6–1.16.8 indicate that the WH transform has this capability in the case of wave packet signals, but the Fourier transform does not. Therefore the WH transform has an important role to play in the characterization of radar RX signals.

Figure 1.16.9 shows the Global WH Transform spectra for two 86.5 cm length cone targets of identical shape but different surface composition: one of rubber and the other of aluminum. The objective, here, was to detect surface composition differences of identical targets when the carriers of the surface currents differed. Figure 1.16.10, a comparison of the magnitudes of the two WH transform spectra, indicates a detected difference.

1.17. Radon Transform

Radon (1917) showed how a function can be defined in terms of its integral projections. The mapping to the projections is known as the

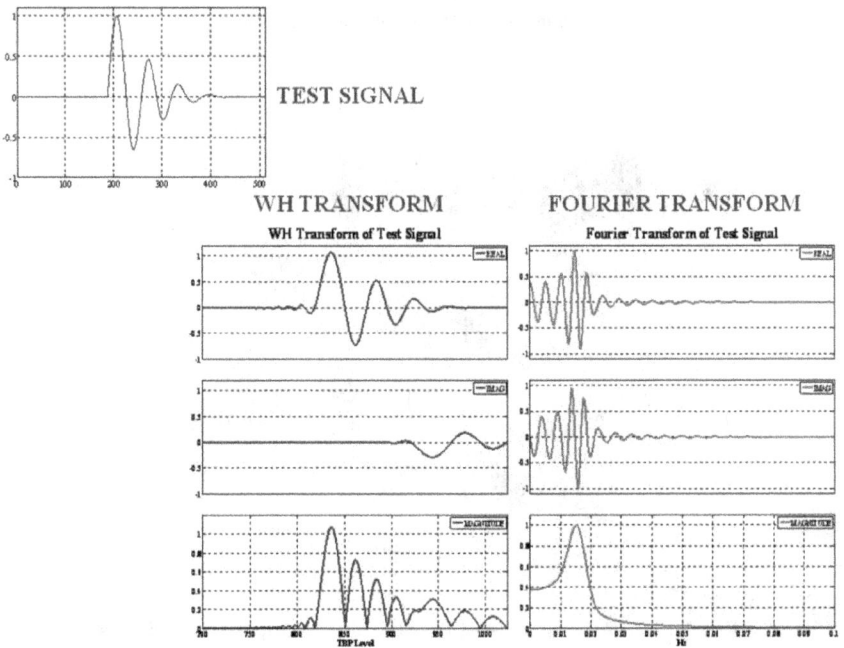

Fig. 1.16.3   A transient test signal (upper), and the real, imaginary and magnitude components of the WH Global transform and the Fourier transform.

Radon transform. Another transform, the Hough transform was designed to detect straight lines in images (Hough, 1962) employing a template. Both are mappings from image space (or source space) to parameter space (or destination space). The two approaches to the mappings can be distinguished by noting that whereas the Radon transform considers how a data point in destination space is obtained from the data in source space, the Hough space considers how a data point in source space is obtained from the data in destination space (Ginkel *et al.*, 2004). The original Hough transform can also be considered as a discrete version of the Radon transform. However, the mathematical formalism for both transforms is identical (Deans, 1981; Illingworth & Kittler, 1988); and Gel'fand *et al.* (1966) formulated the Radon transform in terms of the Dirac delta function, permitting the Radon transform to be treated as a form of template matching. Therefore, the Radon transform and the Hough transform are equivalent and forms of template matching, and henceforth we shall confine ourselves to addressing the Radon tramsform.

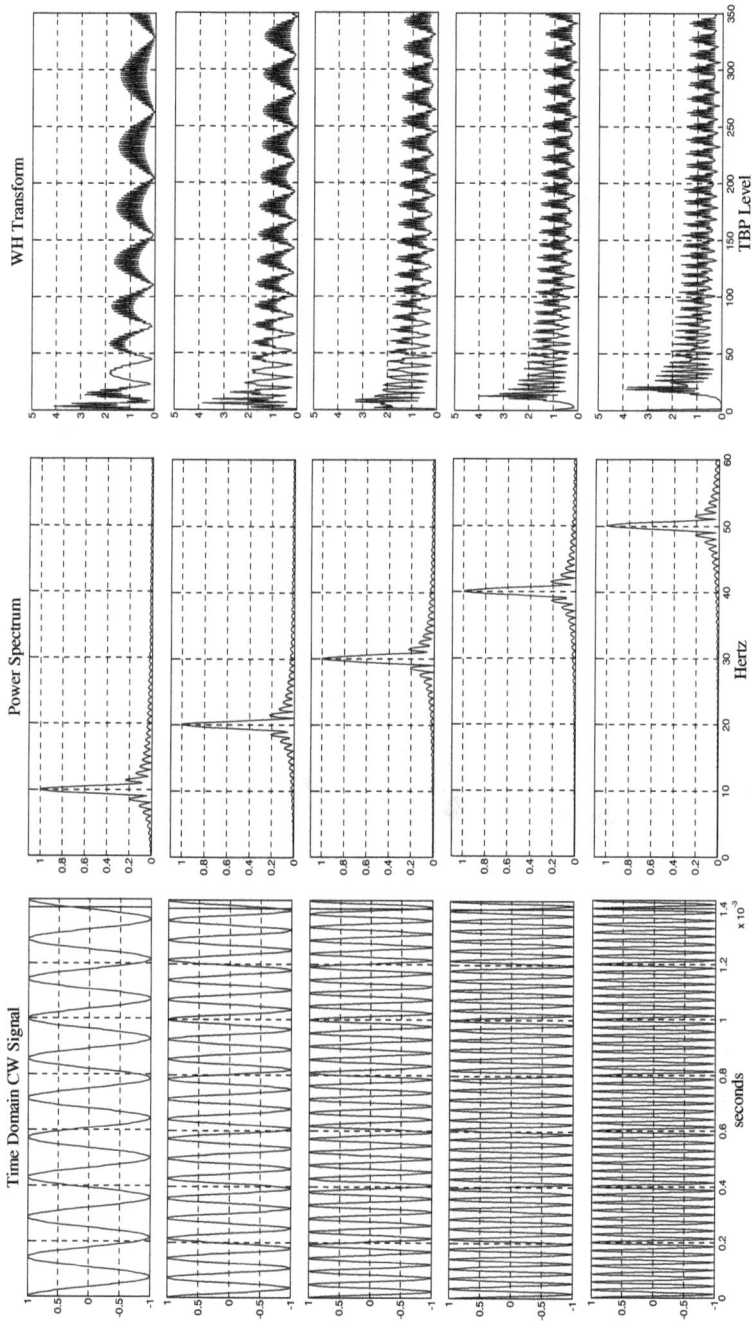

Fig. 1.16.4   Time domain constant wavelength (CW) signals (left), their power spectra (middle) and their Global WH transform (right). As CW signals are the basis functions of the Fourier transform, the Fourier-transform-based power spectra discriminates well CW signals and the WH transform does poorly.

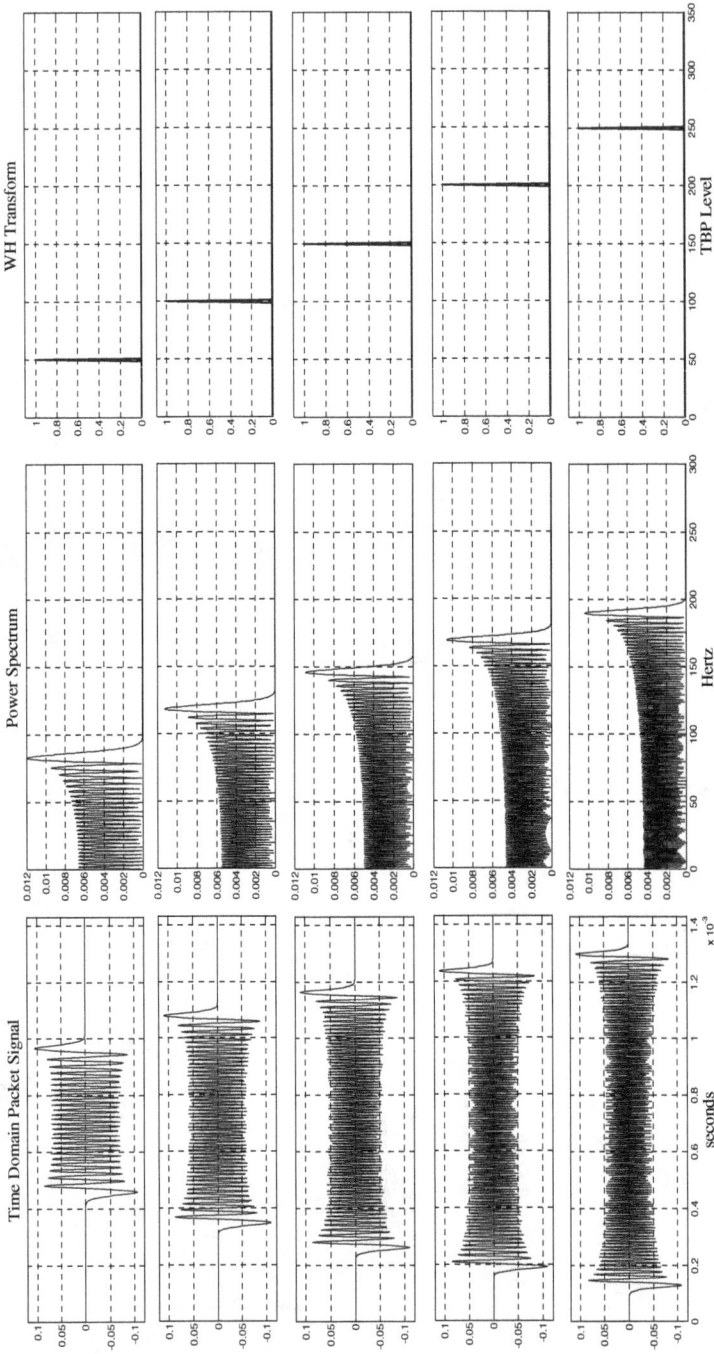

Fig. 1.16.5   Time domain WH packet signals (left), their power spectra (middle) and their Global WH transform (right). As WH signals are the basis functions of the WH transform, the WH transform discriminates well WH packet signals and the Fourier-transform-based power spectrum does poorly.

TEST SIGNAL–THE CORRUPTED SIGNAL

(a)

ONE COMPONENT–THE DESIRED SIGNAL

(b)

OTHER COMPONENT–THE CORRUPTION

(c)

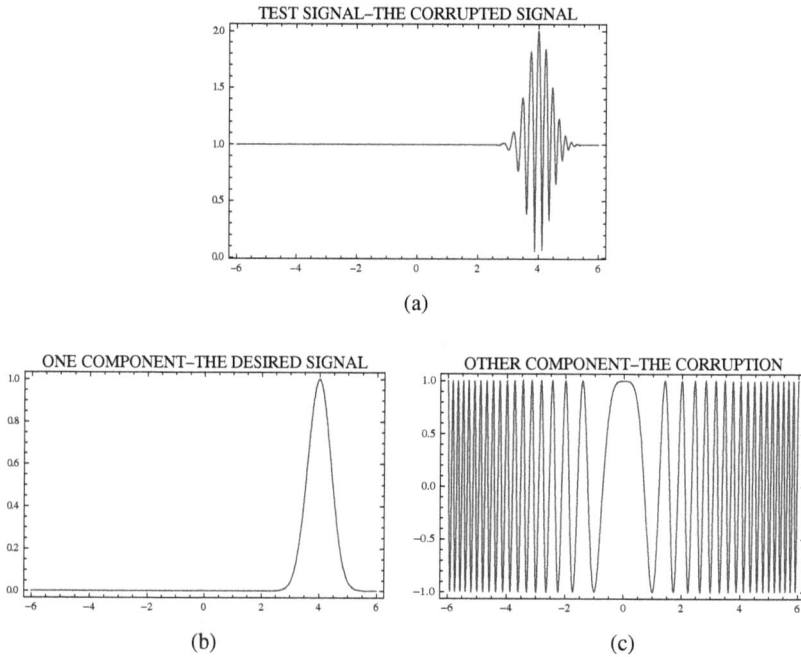

Fig. 1.16.6 (a): A corrupted signal

$$s(t) = \exp[-\pi(x-4)^{(2)}](\exp[-i\pi t^2]rect[t]),$$
$$rect[t] = 1 \quad (-8 \leq t \leq +8),$$
$$rect[t] = 0 \quad (x < -8; x > +8),$$

(b): The desired or designated signal is $[-\pi(x-4)^2]$.
(c): The corruption is $\exp[-i\pi t^2]rect[t]$.

Following Ginkel *et al.* (2004), the Radon transform can be described in terms of generalized functions:

$$(\mathcal{L}_\mathcal{C}\mathcal{J})(p) = \int_\mathcal{R} \mathcal{C}(p,x)\mathcal{J}(x)dx,$$

where:

$\mathcal{L}_\mathcal{C}$ is a linear integral operator with kernel;
$\mathcal{C}$ is a generalized function;
$\mathcal{J}$ is a 2D image;
p is a vector containing parameters;
x are the spatial coordinates;

and the integral is a volume integral.

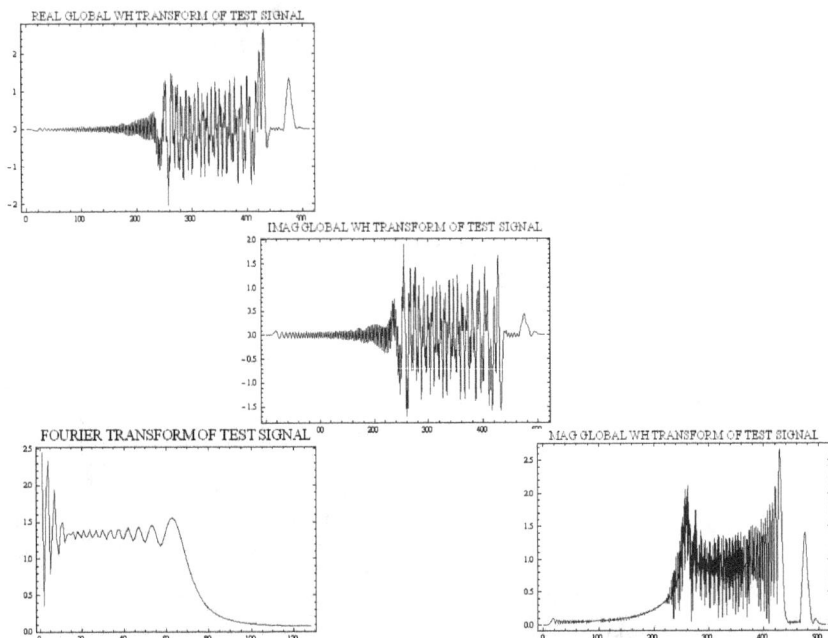

Fig. 1.16.7   The real, imaginary and magnitude of the Global WH transform of the corrupted signal. The signal corruption component and the desired signal component of the transform (indicated by arrow) are separated. However, the Fourier transform of the corrupted signal — lower left — does not permit separation of the two components.

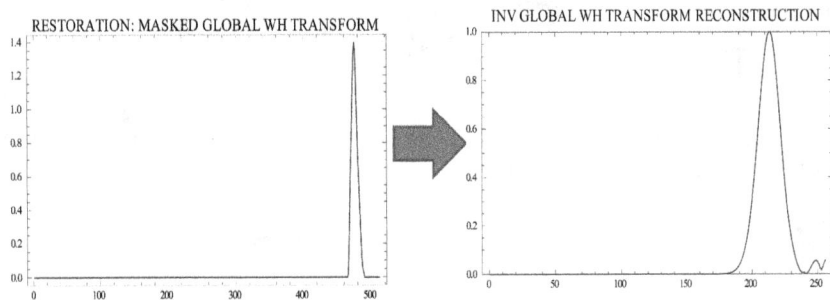

Fig. 1.16.8   The WH Transform permits the corrupted signal component to be masked leaving the desired signal component (left). An inverse WH transform captures the desired signal component without the corruption (right).

The Radon transform of the Cross WVD (or RCWVD) is related to the FRFT (Lohmann & Soffer, 1993). The RCWVD of an angle is equal to the p'th order FRFT of the first function times the p'th order conjugate FRFT of the second function (Raveh & Mendlovic, 1999).

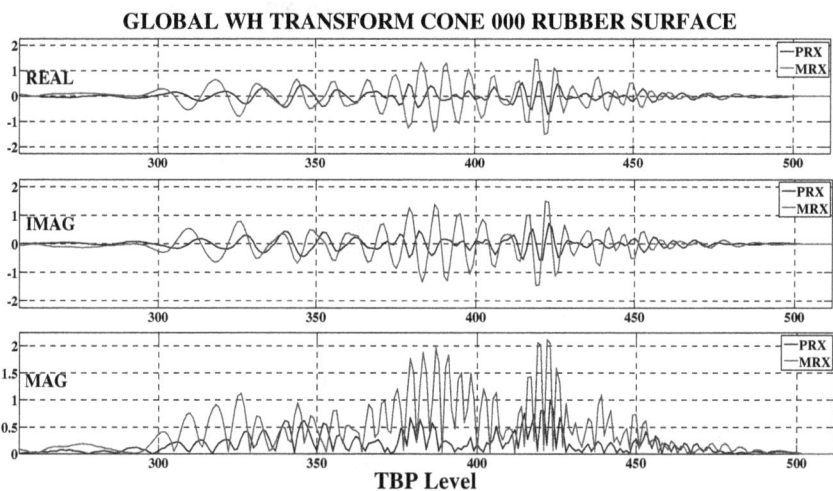

(a)

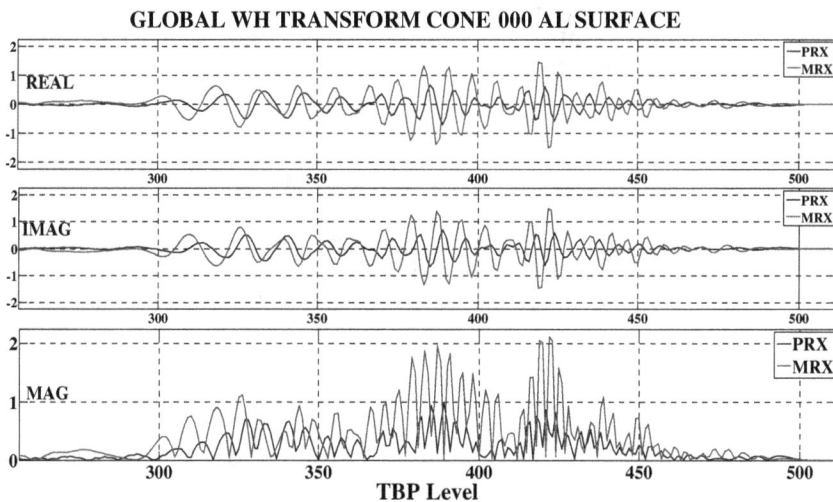

(b)

Fig. 1.16.9 Global WH transforms of PRXs and MRXs for two 86.5 cm length cones of identical shape but different surface composition: one with a rubber surface (a), and on with an aluminum surface (b).

GLOBAL WH TRANSFORM CONE 000 MRX

(a)

GLOBAL WH TRANSFORM CONE 000 MRX
MRX (ALUMINUM SURFACE) – MRX (RUBBER SURFACE)

(b)

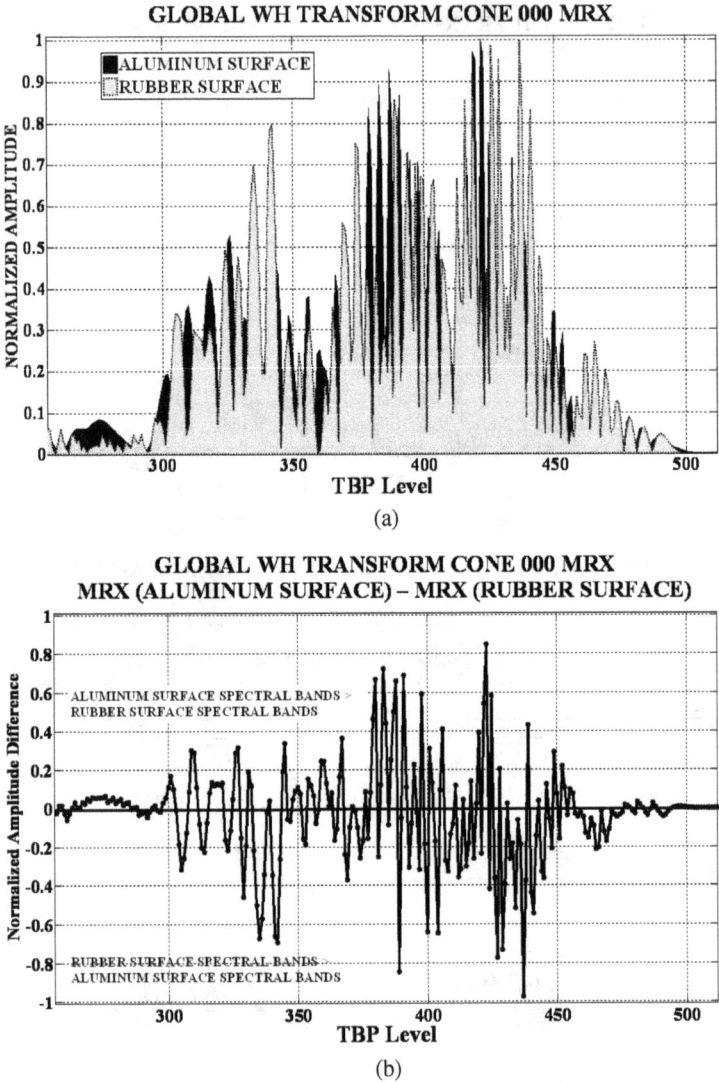

Fig. 1.16.10   Comparisons of the Magnitude of Global WH Transform spectra for a rubber surface and an aluminum surface cone. (a): The Magnitude MRX spectra. (b): The difference of the two spectra shown in A.

The Radon transform and its inverse are well-known (Deans, 1983; Barrett, 1984; Helgason, 1999; Ramm & Katsevich, 1996; Kak & Slaney, 2001). A general description is as follows: the omnidirectional (0-180 degrees) Radon transform of an (x, y) image is the collection of line

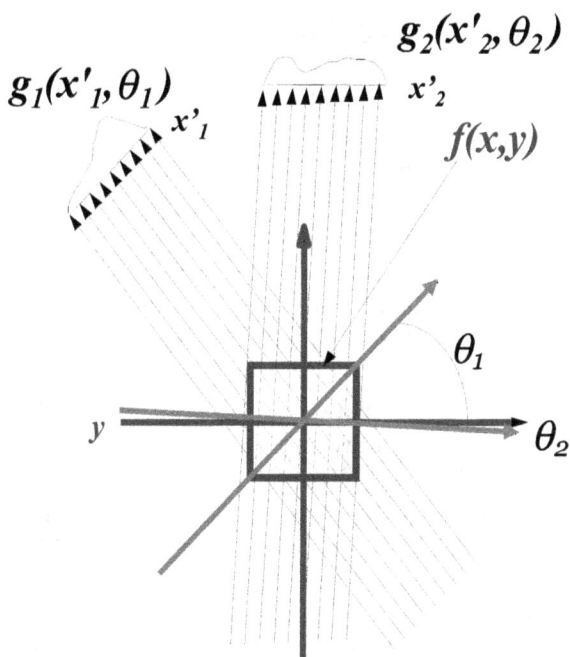

Fig. 1.17.1 Radon transform: parallel projections for two angles θ_1 and θ_2. $f(x,y)$ represents an image.

integrals, $g(s,\theta)$ along lines inclined at angles θ from the y-axis and at a distance s from the origin. The line integrals are space-limited in s and are periodic in θ with period 2π (Fig. 1.17.1).

Referring to Fig. 1.17.1, if the Radon transform at angle θ, or projection g_θ of the image $f(x,y)$ is:

$$g_\theta(\theta, x') = \int f(x' \cos\theta - y' \sin\theta, x' \sin\theta - y' \cos\theta)dy'$$

and if $f(x,y)$ is a WVD transform of a 1D signal, then:

$$g_\theta(\theta, x')[WVD] = |FRFT_\alpha|^2$$

or, in words, the 1D Radon transform or projection of the 2D WVD of a 1D signal onto an axis x'_i making an angle $\theta = a\pi/2$ (radians) with the x-axis, is equal to the squared modulus of the a'th order fractional Fourier transform of the signal. More simply: the Radon-Wigner transform is the squared magnitude of the fractional Fourier transform (Wood & Barry, 1994a, b, c).

One motivation for using Radon transforms in signal processing is to achieve omni-directional filtering (0–180°) of 2D images. 1D signals are given a 2D image representation by the WVD and AF transforms and the enhancement of features requires the application of filters omni-directionally. However, conventional signal/image representation methods do not provide algorithms that filter omni-directionally, i.e., unbiased in any direction. For example, the 2D generalization of quadrature mirror (wavelet) filters, addressing the analysis of trends (averages) and fluctuations (differences), decomposes an image by computing: (1) trends along rows followed by trends along columns; (2) trends along rows followed by fluctuations along columns, (3) fluctuations along rows followed by trends along columns; and (4) fluctuations along both rows and columns. In a pyramidal filtering scheme, 4 arrays of coefficients (of decreasing size) are produced at each level and the filtering is performed only in the vertical and horizontal directions. This choice of procedure is biased to the right angles of rows and columns; and conventional generalization of a 1D filter to 2D form provides no detection capability for image features that lie at specific angles. However, it is possible to analyze trends and fluctuations omni-directionally (0–180°) *by applying 1D filters to the 2D projection space representation of the image*, i.e., the image's Radon transform (Barrett, 2008). Thus in projection space the image is a 2D omni-directional representation with the image distributed along the x-axis as a function of the Cartesian angle of the image; and can be filtered through all angles 0–180° — yet with 1D filters.

The same observation applies to image compression for transmission. In the case of 2D wavelet compression of 2D images by conventional methods, significant coefficients resulting from a pyramidal analysis are transmitted, together with a significance map. If an omni-directional representation is required and using these conventional methods, the same procedure must be applied n times to compress the same 2D images at all n angles. However, using Radon transform methods, it is possible to provide methods for compressing 2D images oriented omni-directionally (0–180°) with 1D filters, e.g., wavelets, but using only one procedure. The resulting coefficients are then transmitted with the significance map.

This observation also applies to image enhancement methods. In the case of 2D wavelet image enhancement of 2D images by conventional methods, a 2D, but still specifically oriented, e.g., wavelet, is used to process an image. To provide omni-directional processing of an image, the orientation of the wavelet must be progressively changed with the image being processed by the same methods at all the orientations adopted by the wavelet. Thus to process an image at n angles requires n processing sequences. However, applying the detecting filter in projection space, it is possible to enhance 2D images in any orientation by applying 1D filtering of the rows of the Radon transformed image and then by inverse Radon transforming the resultant of that filtering, all in one processing sequence.

Thus transforming to projection space permits the analysis of trends and fluctuations omni-directionally by applying 1D filters to the 2D projection space representation of the image, i.e., the image's Radon transform. In projection space the image is in a linearized omni-directional representational form and can be filtered through 0–180° in a series of but one 1D procedure. Briefly, these results are achieved by (i) a Radon transform of the image(s) or array(s); (ii) a convolution of the chosen 1D filter with e.g., a 1D Ram-Lak, or other band-limited filter; (iii) a convolution of the resultant 1D filters with each of the 1D columns of the 2D Radon transform or projection space version of the image; and (iv) an inverse Radon transform of the linearly filtered projection space image back to a Cartesian space omni-directionally filtered form.

The inverse Radon transform is obtained by means of a back-projection operator. The back-projection operator at (r, ϕ) is the integration of the line integrals $g(x', \theta)$ along the sinusoid $x' = r \cos(\theta - \phi)$ in the (x', θ) plane. Thus the back-projection operator maps a function of (x', θ) coordinates into a function of spatial coordinates, (x, y) or (r, ϕ), and integrates into every image pixel over all angles θ. A drawback is that the resulting image is blurred by the point spread function; but the remedy lies in the fact that the back-projection is the adjoint of the inverse Radon transform, which can be obtained by a filtering operation that has a variety of approximations, e.g.: Ram-Lak, Shepp-Logan, etc., Ram-Lak being the most common (Ramm & Katsevich, 1996; Kak & Slaney, 2001).

Figure 1.17.2 illustrates the relation of objects in Cartesian space and Radon transform projection space using two test images in 1.17.2A and B. Peaks in Cartesian space become lines in projection space; and lines in projection space become peaks. In Fig. 1.17.2C an image corrupted with Gaussian noise is restored.

The aforementioned relationship between the Radon-Wigner transform is the squared magnitude of the fractional Fourier transform is illustrated in Figs. 1.17.3 & 1.17.4. Slices through the Radon-Wigner spectra at $\theta = 90°$

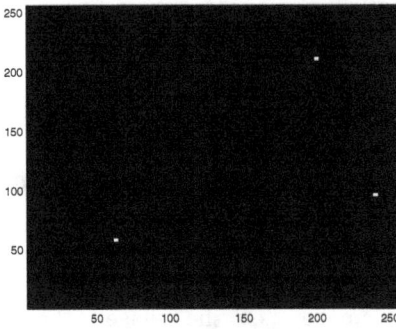

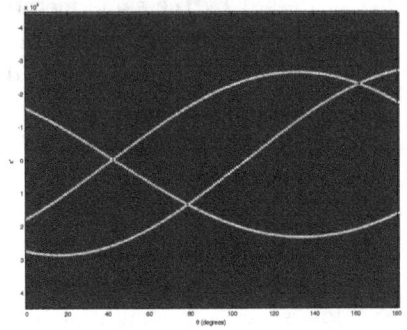

A1   A2

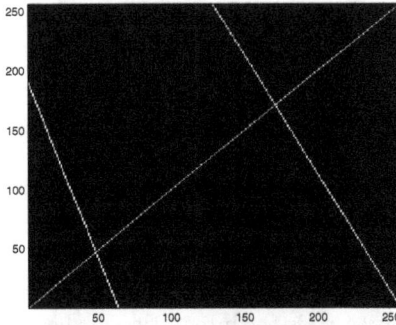

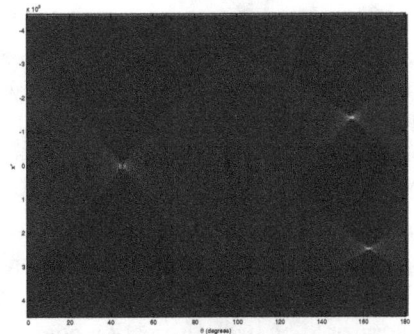

B1   B2

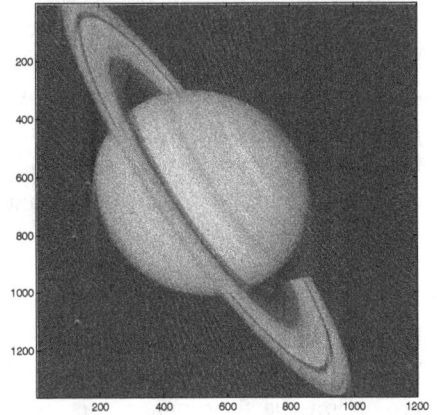

C1   C2

Fig. 1.17.2   A1 Image with 3 points or peaks (square).
A2 Radon transform of image with 3 points or peaks. The 3 points (square) appear as
three curved lines.
B1 Images with 3 lines (straight).

should provide an estimate of the power spectra of the analyzed signal. Figure 1.17.4 shows that this is the case, with the slight difference that the Radon-Wigner transforms, or modulus FRFT, procedure amplified the major band and/or minimized the minor bands.

1.18. Frequency Decomposition: Independent Component Analysis, Matching Pursuit, Complexity Pursuit, Blind Source Separation

In previous sections the WVD transform time-frequency approach was considered as a form of analysis that decomposed a target's RX into a time-frequency spectrum that makes manifest the local, or transient, features of signals, as opposed to the classical Fourier-based approaches which make manifest the global, or constant wavelength. The WVD is based on instantaneous autocorrelations and provides an energy picture of signals. For example, (a) the sum over frequency at a specific time, or (b) the sum over time at a specific frequency, provides the signal energy (a) at that time, or (b) at that frequency. As indicated previously, WVD has major advantages, one being that it is related to the AF, FRFT and other measures known in optics. However, a drawback of the WVD is the presence of cross-terms generated in the transform, that although physically meaningful in optics, are yet a hindrance when the WVD transform is used to describe and interpret non-optical signals. The application of filters can mitigate the presence of cross-terms (Choi & Williams, 1989),

---

Fig. 1.17.2 (*Continued*)

B2 Radon transform of image with 3 lines (straight). The 3 lines appear as 3 peaks in projection space. The peaks in projection space are related to the lines in Cartesian space as follows:

(i) A peak at angle θ on the x-axis in projection space represents a line at an inclination θ to the vertical in Cartesian space.

(ii) A peak at x' on the y-axis in projection space represents a line located at a distance x' from the center of the Cartesian space image.

C1 Image of Saturn with Gaussian noise.
C2 Image of Saturn with noise removed by:

(i) Radon transform to projection space or domain.

(ii) Application of one-dimensional low-pass filter at all angles in projection space, i.e., equivalent to omnidirectional filtering in the in the Cartesian space or domain.

(iii) Inverse Radon transform to Cartesian space.

After Barrett (2008).

MRX TRUCK ASPECT AV SQUARE ABS FRAC FOURIER TRANSFORM

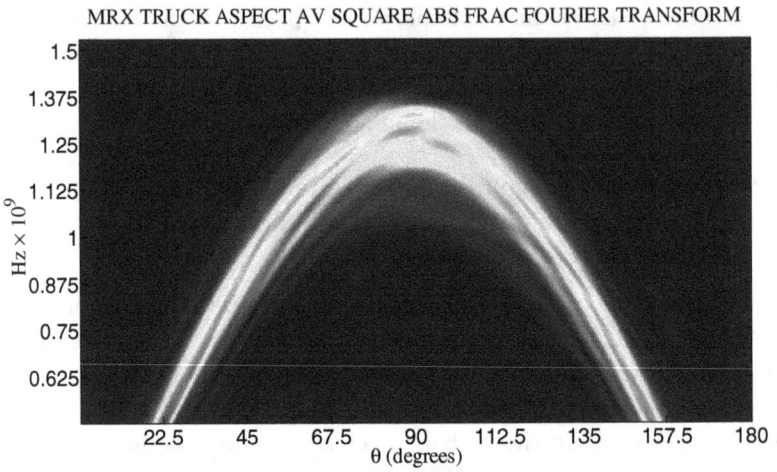

(a)

MRX HUMVEE ASPECT AV SQUARE ABS FRAC FOURIER TRANSFORM

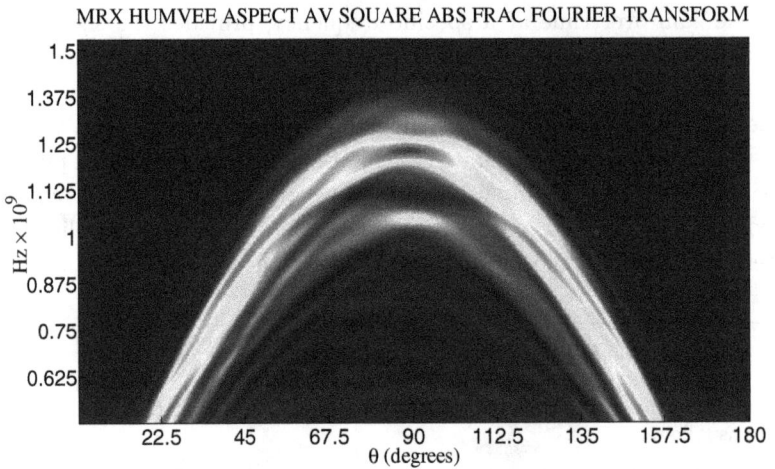

(b)

Fig. 1.17.3   Radon-Wigner transforms of (a): truck target; and (b): humvee target, for average MRXs over aspect angles 000°, 045°, 090° and 180°. The Radon-Wigner transforms were calculated from the squared modulus of the fractional Fourier transform of the signal.

but the application of *optimum* filters requires *a priori* knowledge of the frequency content of the signals being analyzed. In contrast to the WVD, all classical approaches based on the Fourier transform have deficiencies that are addressed in Marple (1987).

MRXs - AVERAGE ASPECT ANGLES 000/045/090/180 deg

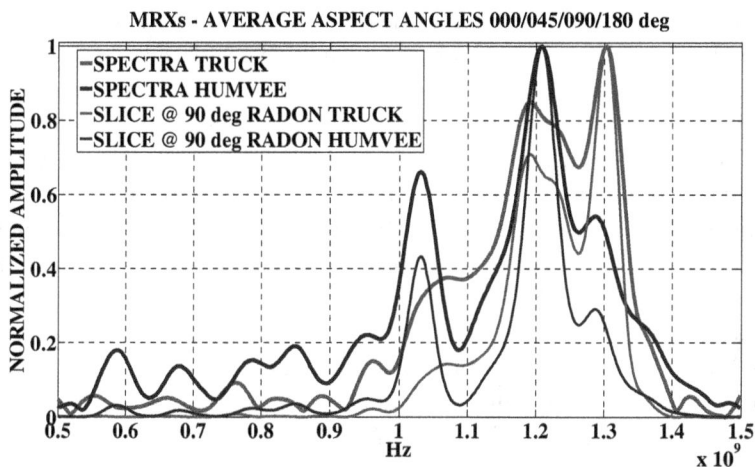

Fig. 1.17.4   Slices through the Radon-Wigner spectra of Fig. 1.16.3 a & b at 90° provide
an estimate of the power spectra. Here it is seen that the major spectral bands coincide
but the Radon-Wigner transforms, or modulus FRFT procedure, amplified the major
band and/or minimized the minor bands.

Other spectral estimation and analysis methods also have assumptions
which, may, or may not apply in a particular instance. For example,
Multiple Signal Classification Algorithm (MUSIC) is a non-parametric
eigen-analysis frequency estimation procedure (Bienvenu & Kopp, 1983;
Schmidt, 1986). MUSIC has a better resolution and better frequency
estimation characteristic than classical methods, especially at high white
noise levels. However, its performance is worse in the presence of col-
ored noise. The eigen-decomposition produces eigenvalues of decreasing
order, and orthonormal eigenvectors. The resulting spectra are not considered true power spectra estimates as signal power is not preserved.
Therefore MUSIC spectra are considered pseudospectra and frequency
estimators.

It is instructive to examine the assumptions of other signal decomposition methods. For example, Principle Component Analysis (PCA) eigen-
decomposition, related to Factor Analysis[3] (FA), is a multivariate analysis
or analysis of multiple variables treated as a single entity which makes
manifest latent information in the original data and both are based on
correlation techniques for source signal separation. This process can result

---

[3]FA is a form of PCA with the addition of extra terms for modeling the sensor noise
associated with each signal mixture.

in data reduction. Information in the data is divided into two subspaces — a signal subspace and a noise subspace. The signal subspace should be twice the number of sinusoids in the data if known. PCA operates by transforming to a new set of uncorrelated variables — the principle components (PCs). The PCs are also orthogonal and ordered in terms of variability. However, only if the original variables are Gaussian are the uncorrelated PCs also independent, because PCA uses only 2nd order statistics, e.g., variances. (Variables with Gaussian distributions have zero statistics above 2nd order.) A general form of PCA is Singular Value Decomposition (SVD).

In comparison, Independent Component Analysis (ICA) assumes that signals are the product of instantaneous linear combinations of independent sources. There are two major assumptions: the sources are independent and are non-Gaussian. Unlike PCA, ICA uses higher-order statistics. Correlation is a measure of the amount of covariation between two signals, e.g., x and y, and depends only on the first moment of the probability density function of x and y: pdf_{xy}. Independence — a stronger condition for signal separation — is a measure of the covariation of all the moments of the pdf_{xy}.

Yet another signal decomposition method, Complexity Pursuit, is based on the assumption that a mixture of signals is usually more complex than the simplest (least complex) of its constituent source signals (the complexity conjecture). Thus Complexity Pursuit is a blind source separation (BSS) method and seeks a weight vector which provides an orthogonal projection of a set of signal mixtures such that each extracted signal is minimally complex. Complexity can be defined in different ways, but in Complexity Pursuit it is defined using criteria related to Kolmogoroff complexity (Cover & Thomas, 1991), and the least complex search strategy is based on gradient ascent.

There are other approaches to signal decomposition including: Autoregressive (AR) parametric modeling in general, the Modified Covariance Method — an AR method, and IIR parametric modeling, all of which assume linearity and a correct model order.

Now, the underlying assumption of a MAP radar is that there is no multiplication of input and TX components, and RX components are statistically independent. RXs are assumed to be mixtures of independent components and the temporal complexity of a mixture is assumed to be greater than that of its simplest (least complex) source signal.

In the case of ICA, the assumptions likewise require that the sources of extracted signals are statistically independent. (It does not follow that uncorrelated sources are also statistically independent, which is a stronger requirement than correlation as noted previously.) Therefore, under this assumption, as the variance of a target's MRX is provided by changes in aspect angle, it might be supposed that sources, or ICA components, extracted from a mixture of several of MRXs from a responding target at various aspect angles, should roughly match. In fact, this approximate match is shown in Fig. 1.18.1 for the case of MRXs from truck and

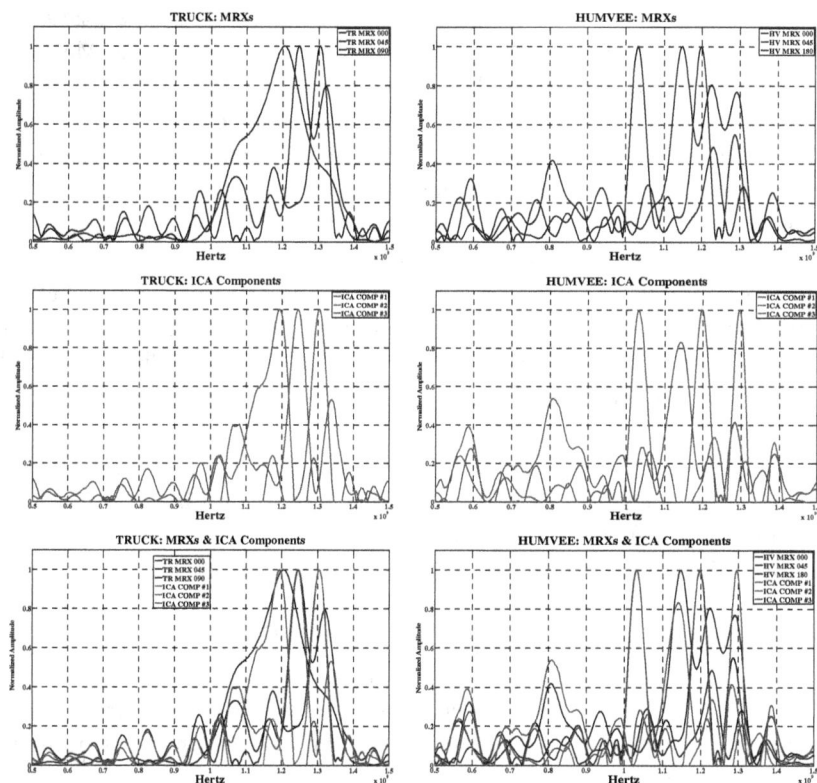

Fig. 1.18.1   Power Spectra, ICA Component Spectra & Their Overlap: truck (Left Column) vs. humvee (Right Column).
Top Row: The power spectra for the MRXs of the two targets at aspect angles 00°, 45° & 90°.
Middle Row: ICA component spectra for a mixture of the same MRXs.
Bottom Row: The two upper rows overlapped.
The good fit of the power spectra and the spectra of the ICA components supports the assumption that the MRXs are (a) the product of instantaneous linear combinations of independent sources (located on the target) and the sources are (b) independent and (c) non-Gaussian, as (a), (b) and (c) are the assumptions of ICA analysis. However, extrapolation of such atrributes to all targets under all conditions is not justified.

humvee targets. The good fit of the power spectra and the spectra of ICA components supports the assumption that the MRXs are (a) the product of instantaneous linear combinations of independent sources (located on the target) and the sources are (b) independent and (c) non-Gaussian — as (a), (b) and (c) are also the assumptions of ICA analysis.

Of course, extrapolation of such attributes, (a), (b) and (c) to all targets under all conditions would not be warranted, but extraolation is warranted, to a degree, in the case of these tested targets and similar targets. One can then ask of these cases: what is the optimum basis representation of target RXs? A technique designed to answer this question is Matching Pursuit, according to which efficient decomposition can only be achieved in a dictionary containing functions reflecting the structure of the analyzed signal (Mallat & Zhang, 1993). An advantage of this approach is that both transients and constant wavelength signals can be captured in an analyzed signal, but a disadvantage is that the choice of filter dictionaries presupposes knowledge of the types of signal being processed. Furthermore, efficient decomposition can only be achieved in a dictionary containing functions reflecting the structure of the analyzed signal, as the Matching Pursuit procedure is: find in the dictionary a function that best fits the signal; subtract its contribution from the signal; then repeat on the remaining residuals.

There is much leeway in the choice of dictionary, or basis, in Matching Pursuit decomposition as shown in Fig. 1.18.2. In Fig. 1.18.2 a time domain MRX signal is analyzed into a dictionary, or basis, of sinusoid vectors. The reconstruction from the basis or dictionary representation of sinusoid vectors is seen to be an accurate representation of the original MRX signal. However, a decomposition into WH filters/wavelets of increasing time-bandwidth product provides an equally accurate reconstruction (Fig. 1.18.3); but, as noted in Section 16, whereas the Fourier transform is optimum for constant wavelength signals, the WH transform is optimum for transient signals.

The criteria used in these blind source separations (BSSs) shown in Figs. 1.18.1–1.18.3 have involved the moments of a joint *pdf* or the fit to an arbitrary dictionary or basis. Another criterion is that of maximizing entropy. Entropy is, in one sense, a measure of the uniformity of the distribution of a bounded set of values such that complete uniformity corresponds to maximum entropy (Cover & Thomas, 1991). The critical observation is then that variables with maximum entropy distributions are statistically independent (Stone, 2004, 2005).

MATCHING PURSUIT

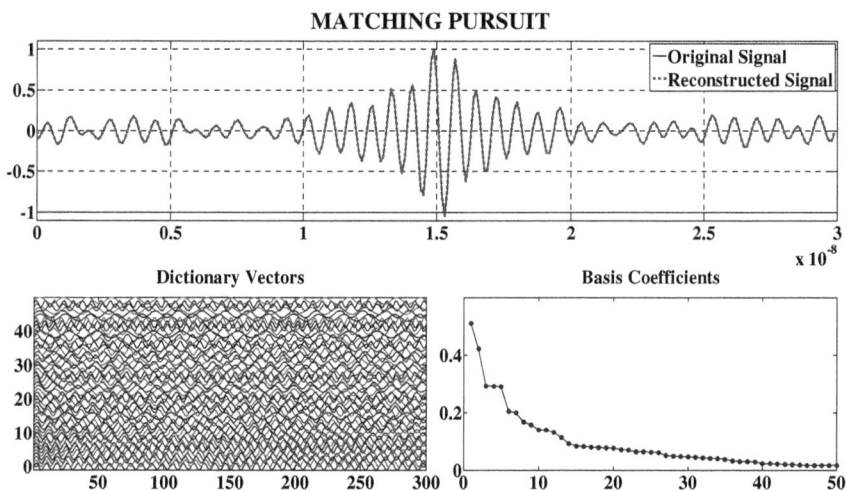

Fig. 1.18.2   A time domain MRX signal, upper, is analyzed into a dictionary or basis of sinusoid vectors, lower left, with coefficients shown in lower right. The reconstruction from the basis or dictionary representation is an accurate representation of the original MRX signal (upper).

WH Transform

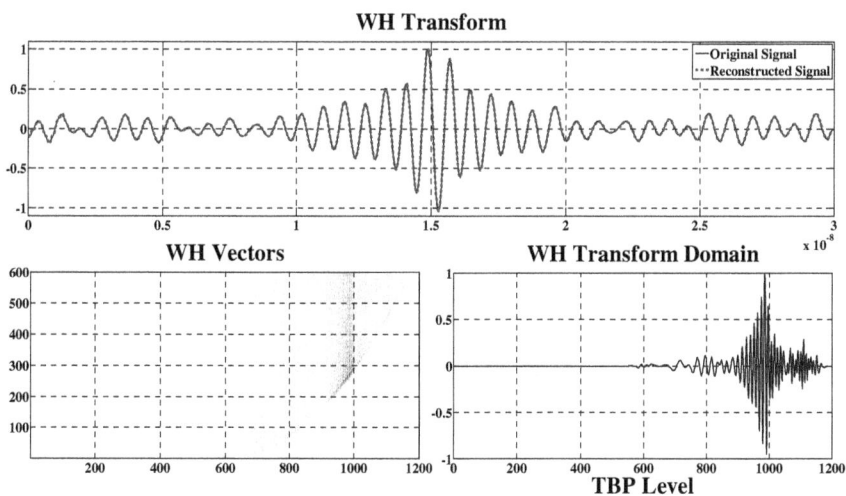

Fig. 1.18.3   The same time domain MRX signal as shown in Fig. 1.18.2, upper, is WH transformed, lower left, with a time-bandwidth product WH vector representation shown in lower right. The reconstruction from the WH transform representation is also an accurate representation of the original MRX signal (upper).

COMPLEXITY PURSUIT, SOURCE SIGNALS: TRUCK MRX

(a)

COMPLEXITY PURSUIT, SOURCE SIGNALS: HUMVEE MRX

(b)

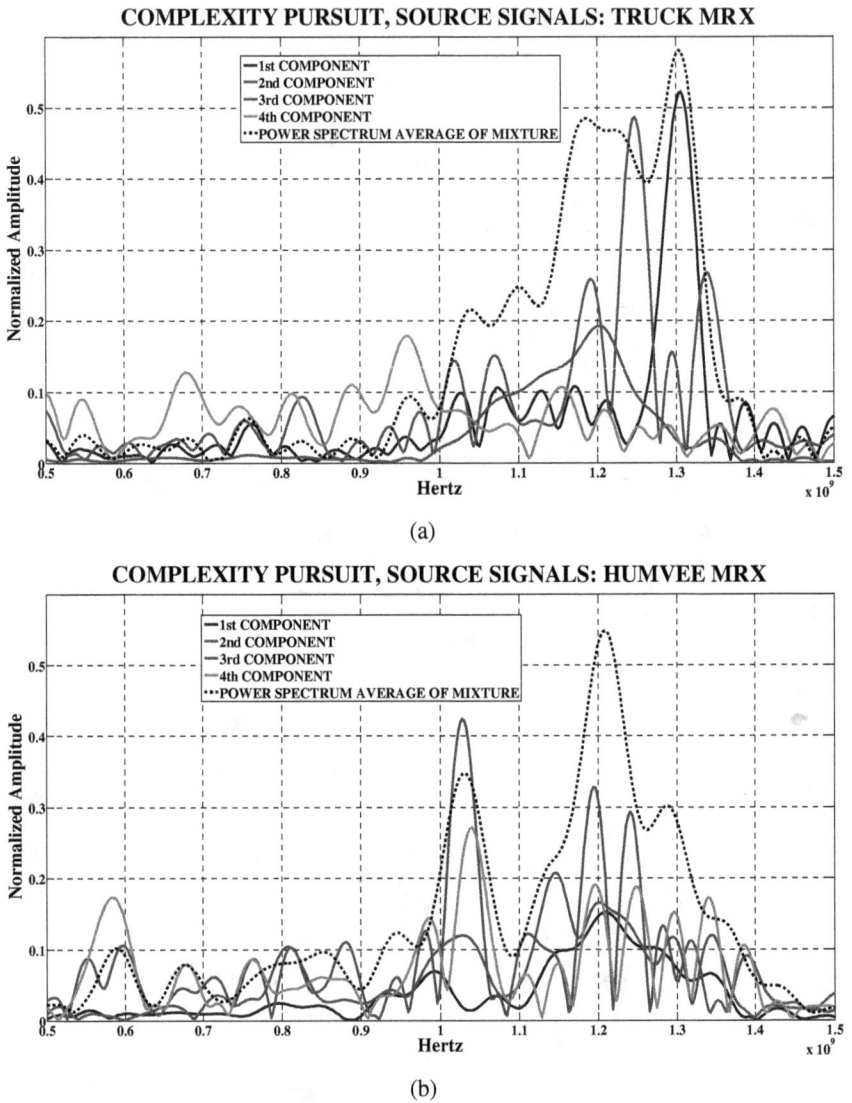

Fig. 1.18.4   Source signals separated from a mixture of MRX signals by a Complexity Pursuit algorithm. Upper: Truck target; Lower: Humvee target. The 4 source components separated in each case are compared with the conventional power spectrum of the average of the mixture. It can be seen that the separated components largely agree with the average power spectrum.

COMPLEXITY PURSUIT, SOURCE SIGNALS: TRUCK MR X

COMPLEXITY PURSUIT, SOURCE SIGNALS: TRUCK HUMVEE

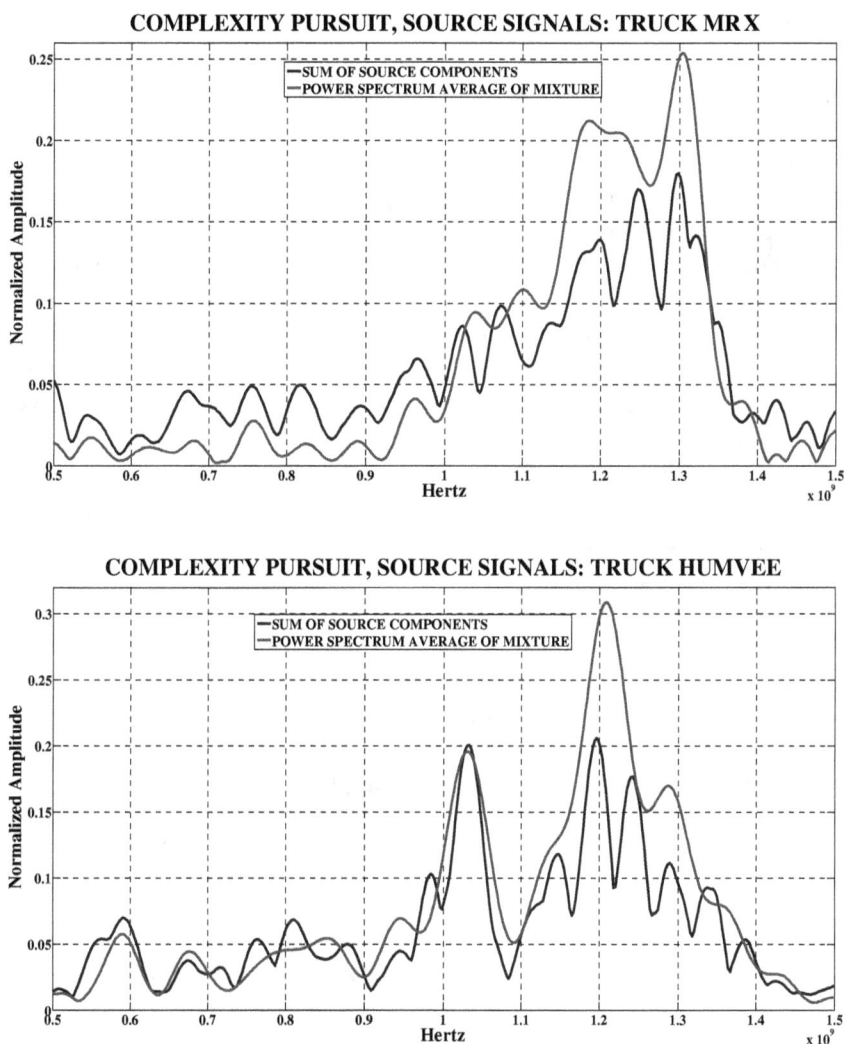

Fig. 1.18.5   The same data as shown in Fig. 1.18.4 with the 4 source components summed and compared with the conventional power spectrum of the average of the mixture. It can be seen that the separated components largely agree with the average power spectrum.

Finding independent signals by maximizing entropy is known as info-max (Bell & Sejnowski, 1995) which is equivalent to Maximum Likelihood Estimation (MLE). The term Complexity Pursuit was coined (Hyvärinen, 2001) to describe the process of extracting from mixtures the minimally complex source signals. The supposition in this case is that maximum predictability is equal to minimal complexity (Xie *et al.*, 2005). Thus unlike PCA and ICA which presuppose power density function models of the signal, Complexity Pursuit depends only on the complexity of the signal.

An example of BSS extraction of Complexity Pursuit source components from a mixture of MRX signals is shown in Figs. 1.18.4 & 1.18.5. Here 4 extracted source components are compared with the conventional power spectrum of the average of the mixture. It can be seen that the separated components largely agree with the average power spectrum. However, once again extrapolation of the validity of the assumptions of infomax to all targets under all conditions is not warranted.

In this section the application of several approaches to blind source separation (BSS) using a variety of assumptions was reviewed. It was shown that some of these techniques and assumptions apply to vehicle targets. However, as it is (statistically) impossible to prove the null hypothesis, one cannot extrapolate the claim to all targets under all conditions. The lesson is probably that a variety of signal processing techniques using a variety of assumptions should be applied in the blind source separation of mixtures of MRXs from a newly encountered class of targets. Nonetheless, for the targets tested, it appears that the assumption that there is no multiplication of input and TX components, and the RX components are statistically independent, appears supported. The RXs can be assumed to be mixtures of independent components and the temporal complexity of a mixture is greater than that of its simplest (least complex) source signal.

PART 2 — UHF-BAND MAP PROTOTYPE

2.0.0 UHF MAP System

The UHF MAP system was designed to penetrate and detect targets under forest canopy. This is a difficult task in that both the envelope modulation and the carrier must be of low enough frequency to penetrate twice — entering and exiting — through the forest canopy, but yet of high enough frequency to detect designated targets by their resonance excitation. This means that such a system must be designed to "thread the eye of the needle" offered by common frequencies possessed by both target (reflecting/scattering) resonances and foliage (non-absorbing/dispersing) transparencies. This requires a double spectral overlap of transparencies and reflectancies. The task is yet further complicated by government regulations requiring emissions to be off-limits for transmissions at certain FCC-dedicated frequencies. In the case of the United States and at the time of data collection addressed here, the regulation limitations permitted UHF emissions only in the bandwidth 215–730 MHz, and the system (Fig. 2.0.0.1) was designed to utilize this, and only this, specific bandwidth. As with the case of the Ka-Band system, it is noteworthy that a high sample rate scope was used both as an essential part of the receiver, and also to view the RX signals. Sampling rate was 20 GSs. Systems that transmit an impulse and receive and process the RX signal using a sampling receiver are considered to operate in the time domain. Systems that transmit individual frequencies in a sequential manner or as a swept frequency, i.e., as in LFM, and receive and process the RX signal using a frequency conversion receiver are considered to operate in the frequency domain. Therefore, the MAP receiver is a time-domain receiver.

The antenna is a critical system MAP component — more so at UHF than at Ka band. Not only is a wide instantaneous bandwidth — as opposed to wide sequential bandwidth — required, but the closer the

UHF MAP System

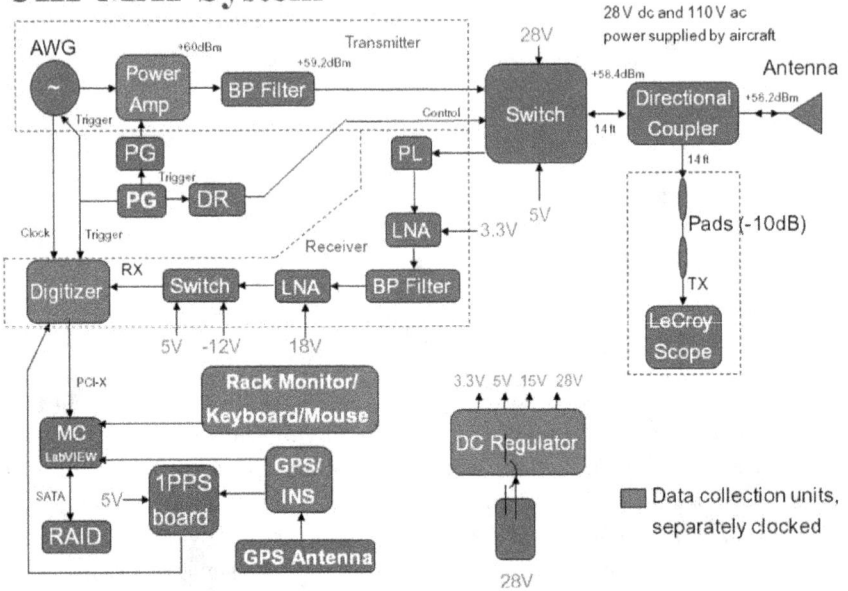

Fig. 2.0.0.1   UHF MAP System for 215–730 MHz bandwidth. MAP reception requires the preservation of all frequency components in the RX signal and thus a high sample rate. A LeCroy high sample rate scope was used both as an essential part of the receiver, and also to view the RX signals. Sampling rate was 10 GSs. Systems that transmit an impulse and receive and process the RX signal using a sampling receiver are considered to operate in the time domain. Systems that transmit individual frequencies in a sequential manner or as a swept frequency, i.e., as in LFM, and receive and process the RX signal using a frequency conversion receiver are considered to operate in the frequency domain. Therefore, the MAP receiver is a time-domain receiver. The AWG — Arbitrary Wave Generator — provided the capability of matching the TX signal to the target on the basis of *a priori* information by amplitude modulation. The UHF System transmits at relatively low frequencies (215–730 MHz) so that, unlike the Ka Band MAP System in which only TX envelope, and not the carrier, is matched to the target, in the case of this UHF system the entire TX signal is matched to the target with no distinction of envelope and carrier.

antenna achieves a focused beam, the less background clutter is obtained. The intention was to test an airborne system at 10,000 feet, and the higher the system, the larger the ground footprint and the higher the probability of exciting ground clutter. (In the actual tests, the ground was actually about 5,000 feet above sea level, and the air platform was about 5,000 feet above the ground, so the ground footprint, although large, was not as large as an airborne system 10,000 feet above ground at sea level.)

The transmit array used in the data presented was composed of Vivaldi horns, which sufficiently preserve small pulse characteristics across the bandwidth of interest and beyond (cf. Anderson *et al.*, 2003). The test data obtained with the Vivaldi array provided information concerning the target's response to a wide instantaneous bandwidth, in contrast to a target's continuous wave, or single frequency, response across a wide sequential bandwidth. The identification of a target's frequency response by a wide TX instantaneous bandwidth signal, instead of by a wide TX sequential bandwidth FM signal, is important. If a target were to act as a nonlinear transducer, then the target response to a certain TX single frequency (i) applied as a narrowband cw signal, or (ii) even applied as part of an LFM signal, within an FM band, may not be the same as the target response elicited by that same single frequency component applied within a band of instantaneous frequencies. In other words, a target resonance response at a particular frequency identified by application of a broad band of instantaneous frequencies, i.e., a PTX, can be the result of the interaction of the multiple, simultaneously-applied TX frequencies within that band, and not due to the linear summation of the elicited effects of narrowband cw signals within the band. The emphasis, here, is on the possibility rather than the actuality, for, as discussed in the INTRODUCTION, no nonlinear target impulse/frequency responses were detected in these tests.

The tests conducted were of three kinds: (a) Ground tests transmitting through foliage (Section 2.1, below); (b) Anechoic Chamber tests (Section 2.2, below) to obtain *a priori* information of targets in a clutter-free environment; and, using that *a priori* information, (c) flight tests with targets under forest canopy (Section 2.3, below).

To compare the total energy of PTX, MTX and DTX pulses, an energy equalization protocol was followed using a scaling factor. The scaling factor was calculated based on the total pulse relative energy, or the integration of the square of the voltage samples of each individual TX. The voltage scaling factor is then the square root of the ratio of total energy of one TX compared with another. Each RX sample was then divided by its corresponding TX voltage scaling factor.

2.1.0 Ground Tests Through Foliage

The ground tests were conducted with foliage, or plants, placed before targets. The density of the cover could be varied by adding more plants. The signal attenuation varied with the type of foliage, its density and layering, and time lapse after watering. Therefore it is impossible to give an accurate quantitative analysis of the foliage target cover. However, plant-covered corner reflector tests indicated that the one-way foliage signal attenuation varied over the range 4–7 dB. Apart from plastic barrels, all targets tested had metal surfaces. The poorly-conducting plastic surface barrels were more difficult to detect in comparison with metal surface barrels.

2.1.1 Target: Barrels

The major barrel resonances were in the ∼(525–575) MHz range (Figs. 2.1.1.1–2.1.1.3). The spectral profile for barrel targets was unaltered with the imposition of foliage between the transmitter and receiver (Fig. 2.1.1.3). A clear SNR advantage is obtained by using MTX and DTX signals over PTX signals of equal TX energy (Figs. 2.1.1.4–2.1.1.5).

2.1.2 Target: Roof Panels

As expected, the roof panel major resonance was of lower frequency than that of barrels (Fig. 2.1.2.1). Only minor differences between the aluminum and steel panels, and between the no cover and 31 plant cover conditions, were found (Fig. 2.1.2.3). Figure 2.1.2.3 shows the RXs elicited by a PTX (UWB) and a series of DTXs. The indication is that the roof panel target acts as a linear LTV system, as the frequency composition of the DRXs agrees with that of the DTXs.

2.1.3 Target: Microwave Oven

The microwave oven used as a target was smaller than conventional ovens. The top surface aspect was 8.5 × 15.5 inches. As anticipated for a target smaller than both barrels and roof panels, a microwave oven provides a major resonance at ∼600 MHz (Fig. 2.1.3.1). This resonance was the dominant feature of the microwave oven target return and a WH transform revealed no major difference between the PRX and the DRX (for DTX = 600 MHz) — Fig. 2.1.3.2. Using a DTX signal, up to 21 dB of signal-to-clutter ratio (SCR) enhancement could be obtained.

2.1.4 Targets: Trucks

The two trucks used as targets in these tests were: a Dodge truck and a Humvee truck (Fig. 2.1.4.1). There is a clear difference in the PRX spectra for the two targets at ∼600 MHz (Fig. 2.1.4.2). DRX

PRX 1 BARREL
90 minutes after watering

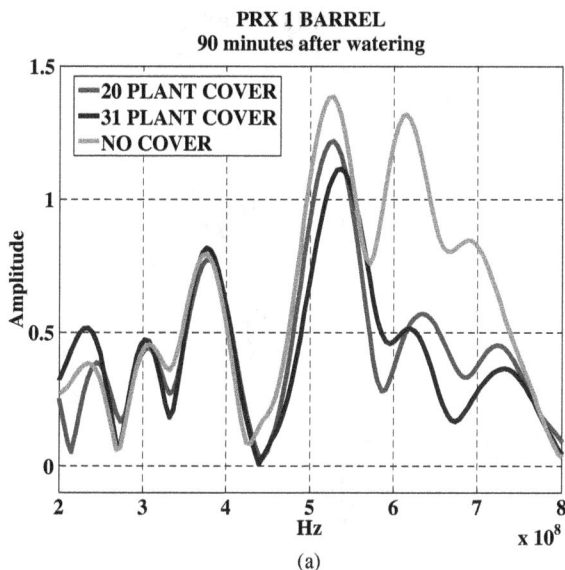

(a)

PRX 2 BARRELS
90 minutes after watering

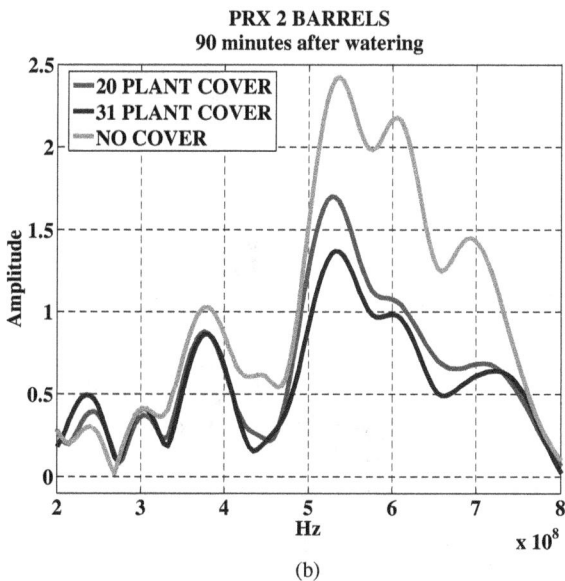

(b)

Fig. 2.1.1.1   PRX spectra — A: Target: 1 Barrel, with (a) 20 plant cover; (b) 31 plant cover; and (c), no cover. B: Targets: 2 Barrels, with (a) 20 plant cover; (b) 31 plant cover; and (c), no plant cover; C: Targets: 3 Barrels, with (a) 20 plant cover; (b) 31 plant cover; and (c), no plant cover. The main resonance is approximately at the same frequency for 1, 2 and 3 barrels, and the plant cover reduces RX signal amplitude.

PRX 3 BARRELS
1hr after watering

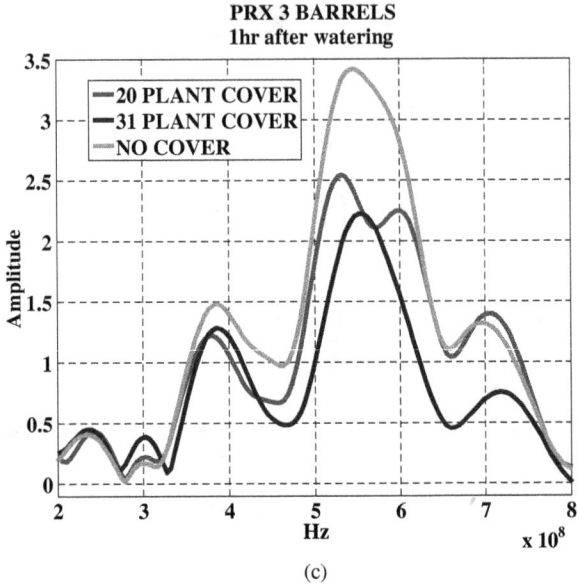

(c)

Fig. 2.1.1.1 (*Continued*)

PRX BARRELS

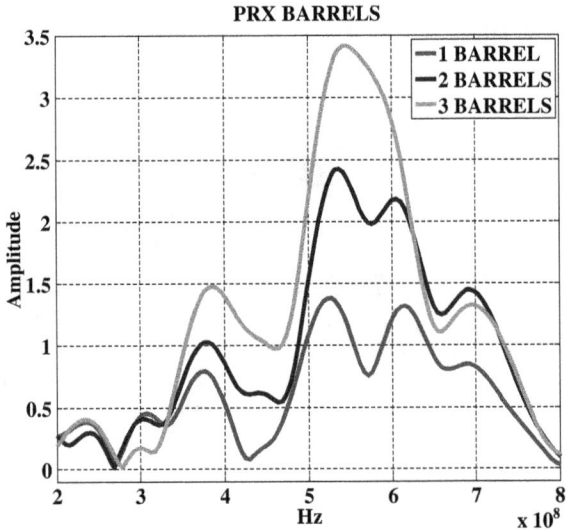

Fig. 2.1.1.2   PRX spectra for targets: 1, 2 and 3 Barrels, and for the no foliage cover condition. The resonance at ∼530 MHz is present in the case of 1, 2 and 3 barrel targets, but the resonances above 600 MHz shift downward as the number of barrel targets increase.

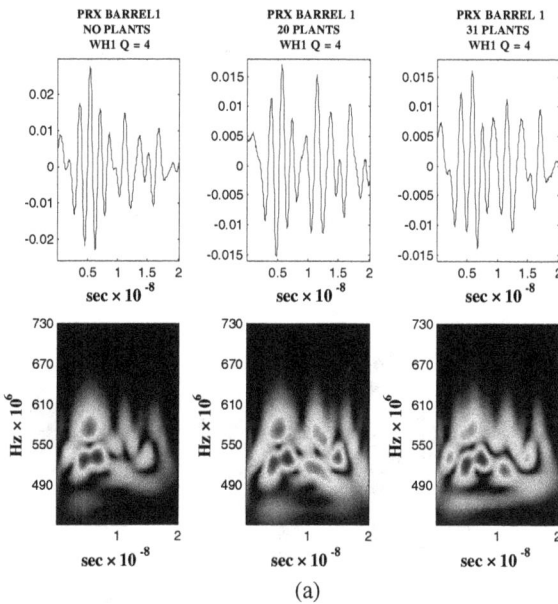

(a)

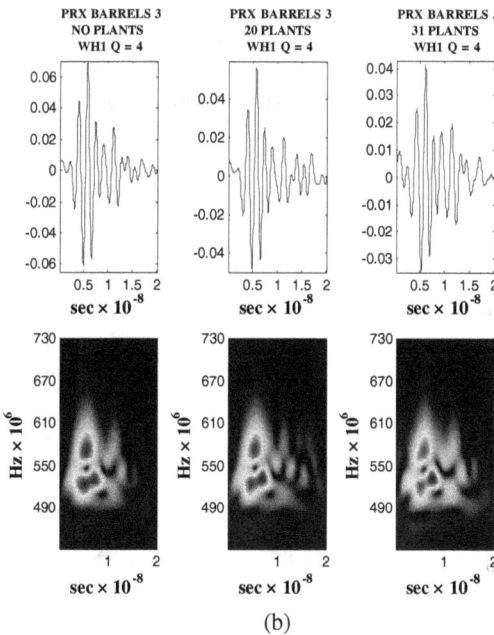

(b)

Fig. 2.1.1.3 Time-frequency PRX spectra with 1 Barrel target and over increasing plant coverage. Upper: PRX time domain. Lower: time-frequency spectra. The no plant cover spectrum on the left is similar to the 20 and 31 plant cover spectra, but with additional plant-derived spectral components.

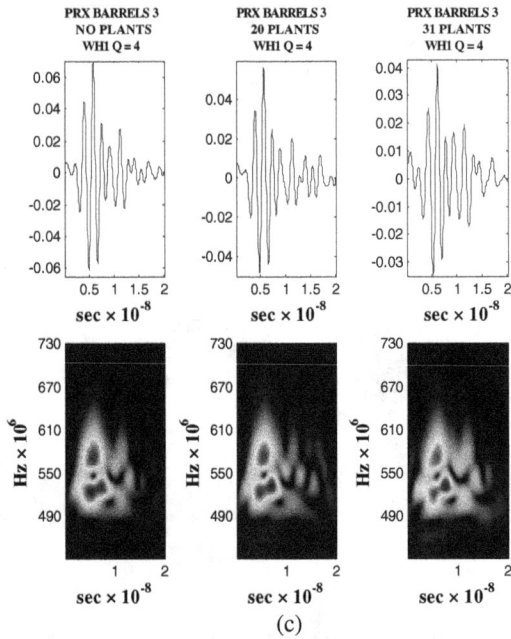

Fig. 2.1.1.3 (*Continued*)

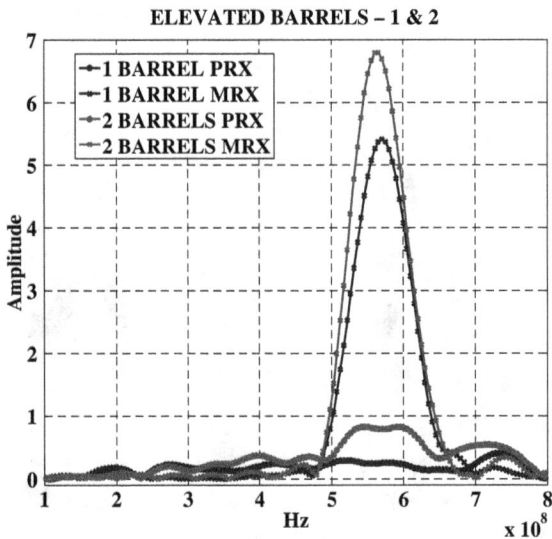

ELEVATED BARRELS – 1 & 2

Fig. 2.1.1.4 A comparison of PRX and MRX spectra, target: 1 and 2 barrels. There is a clear SNR advantage in the case of MRX at equal TX energy to the PTX (UWB) — in this case an advantage of ∼8 dB.

DRX versus PRX

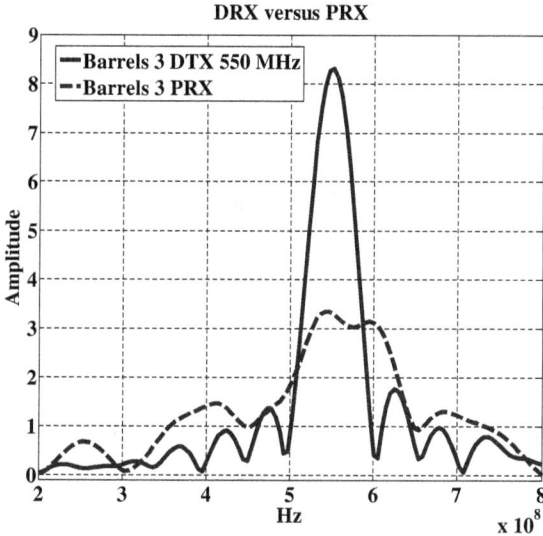

Fig. 2.1.1.5 A comparison of PRX and DRX spectra, target: 3 barrels. The DTX was 550 MHz. There is a clear SNR advantage in the case of DRX (UWB) at equal energy of PTX and DTX.

TWO ROOF PANELS LOW ANGLE NORM

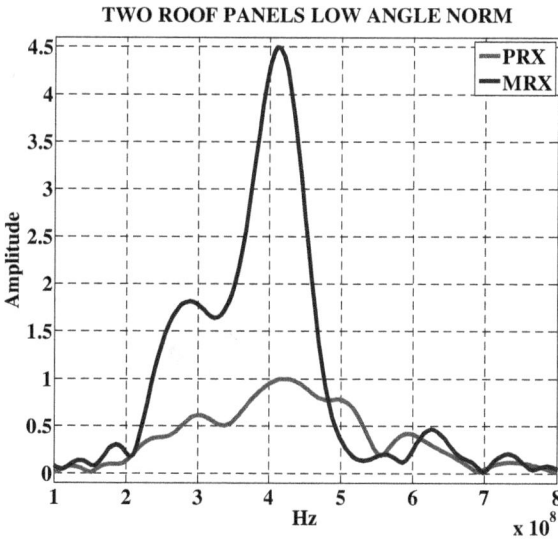

Fig. 2.1.2.1 PRX and MRX spectra; Target: 2 Roof Panels. The MRX spectral peak is ~6.5 dB higher in amplitude than that of the PRX (UWB) spectral peak at equal TX energy for MTX and PTX.

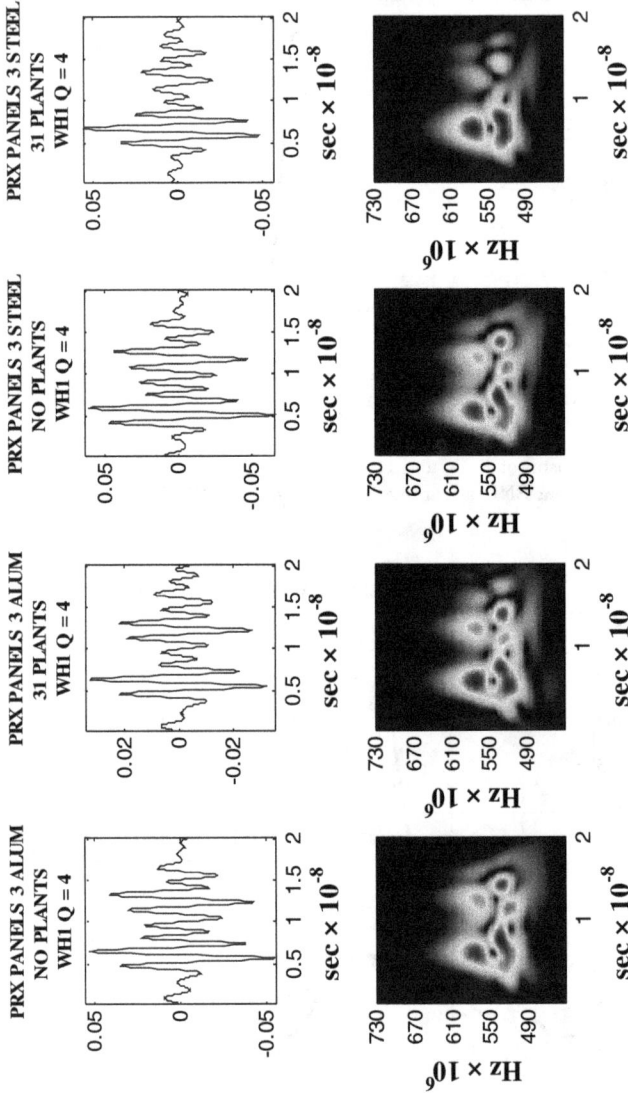

Fig. 2.1.2.2   PRX Time-frequency spectra; Target 3 Roof Panels. Left pair: Aluminum Panels; Right Pair: Steel Panels. First of each pair: No plant cover. Second of each pair: 31 plant cover. There are only minor differences between the aluminum and steel panels, and between the no cover and 31 plant cover conditions. Time-frequency filter: WH1, Q = 4.

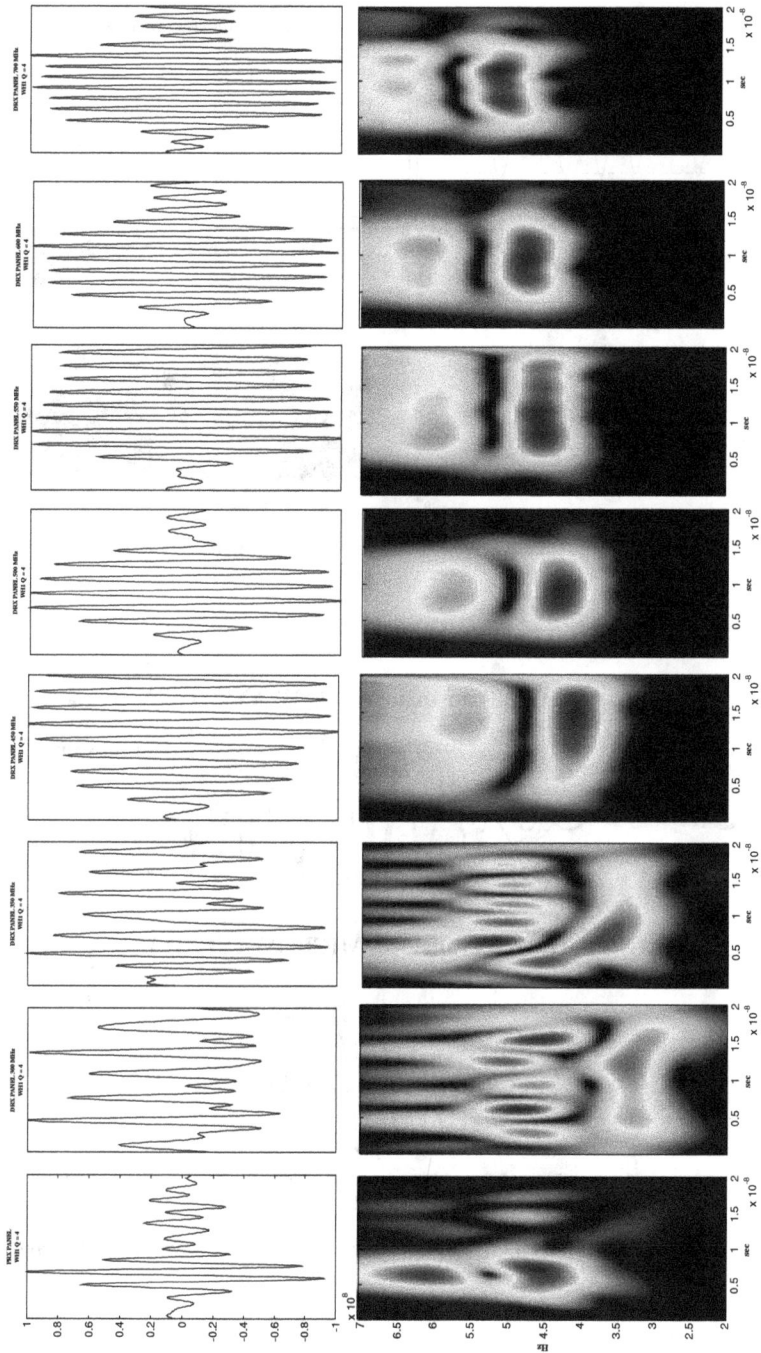

Fig. 2.1.2.3 PRX & DRX Time-frequency spectra; Target 1 Roof Panel. Leftmost: PRX; followed by 7 DRXs, elicited by DTXs, 300, 350, 450, 500, 550 and 600 MHz. Time-frequency filter: WH1, $Q = 4$. The DRXs follow the spectral composition of the DTXs indicating linear transduction.

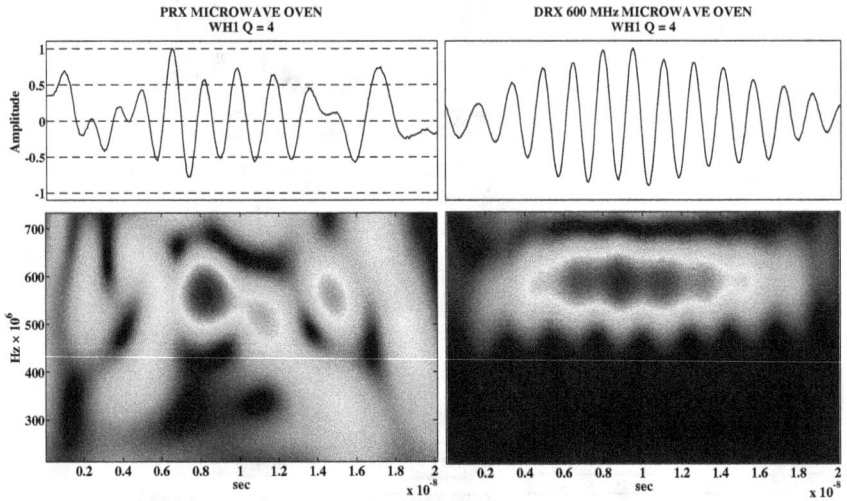

Fig. 2.1.3.1   Time-frequency spectra, target: microwave oven. Left: PRX; Right: DRX (DTX = 600 MHz). Filter WH1, Q = 4.

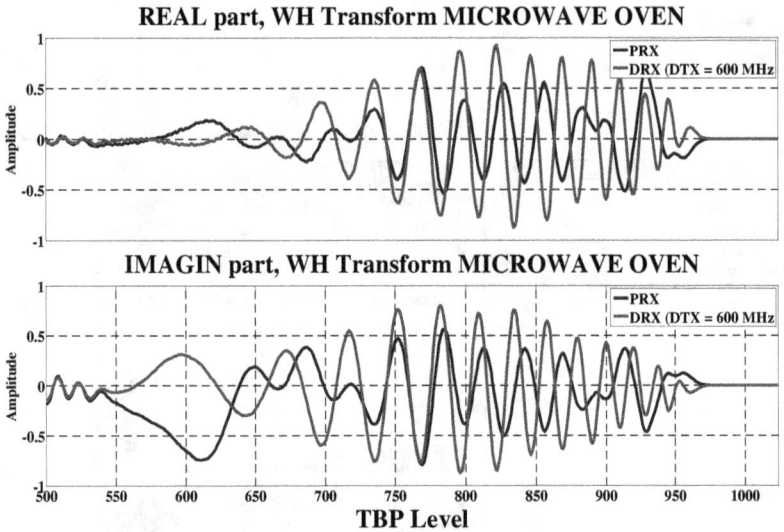

Fig. 2.1.3.2   Real and Imaginary Parts of WH Transform of microwave oven PRX and DRX (DTX = 600 MHz).

Microwave Oven

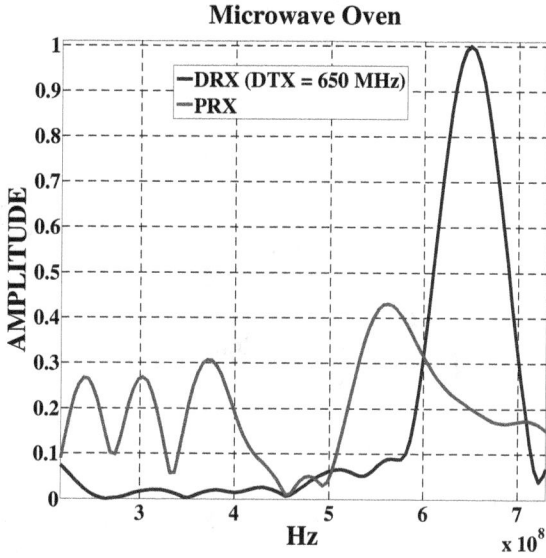

Fig. 2.1.3.3   DRX (DTX = 650 MHz) and PRX (UWB) spectra. Target: microwave oven. A DTX of 650 MHz converts to a resonance for an object of length 0.4615 meters, which is approximately 1.5 feet. The length of the microwave oven was 1.3 feet.

0° Aspect   180° Aspect   0° Aspect   180° Aspect

90° Aspect Humvee   90° Aspect Dodge Ram

Fig. 2.1.4.1   Truck targets: Humvee and Dodge Ram.

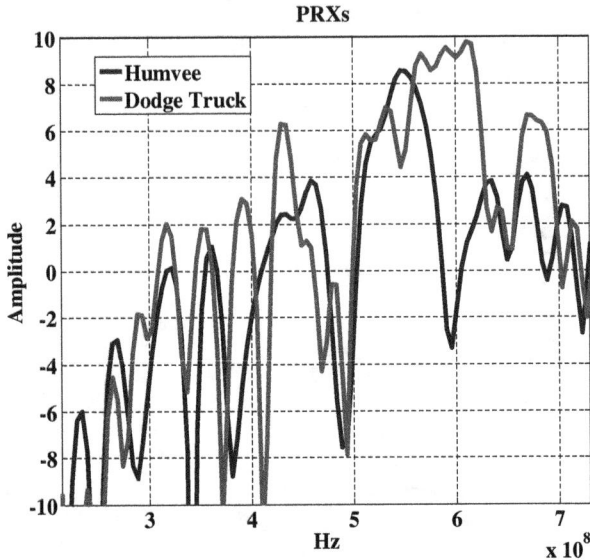

Fig. 2.1.4.2   PRX spectra of Humvee and Dodge Trucks. A spectral difference around 600 MHz is clearly indicated.

(DTX = 600 MHz) spectra reveal this major difference, ~23.7 dB, even more clearly (Fig. 2.1.4.3). Significantly, a WH transform of the target PRXs accentuated the difference to 31.2 dB at a filter time-bandwidth product (TBP) of around 850 (Fig. 2.1.4.4).

At the lower frequencies resonance aspect independence is exhibited (Figs. 2.1.4.5–2.1.4.6), with aspect dependence in the time of arrival of those resonances, as expected for an LTV system.

2.1.5 Target: Artillery Shell

An artillery shell was tested (Fig. 2.1.5.1), and a time-frequency analysis — but, significantly, not a Fourier spectral analysis — revealed a resonance ~600 MHz (Fig. 2.1.5.2). Other minor resonances occur separated at 1.8 MHz separations probably due to surface harmonics (Fig. 2.1.5.3). These resonances were unaffected by the target's aspect — horizontal or vertical — (Fig. 2.1.5.4).

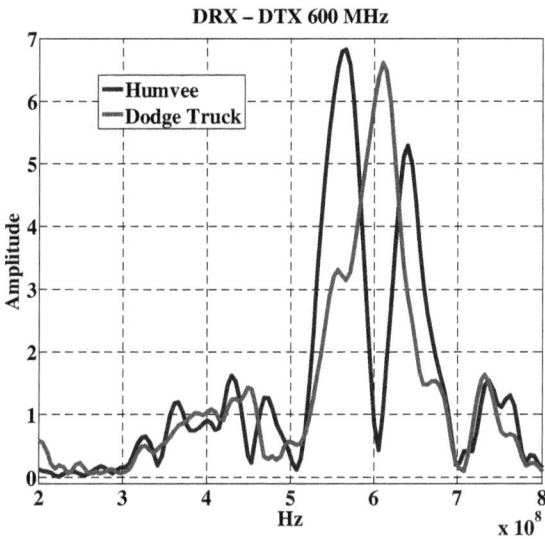

DRX – DTX 600 MHz

(a)

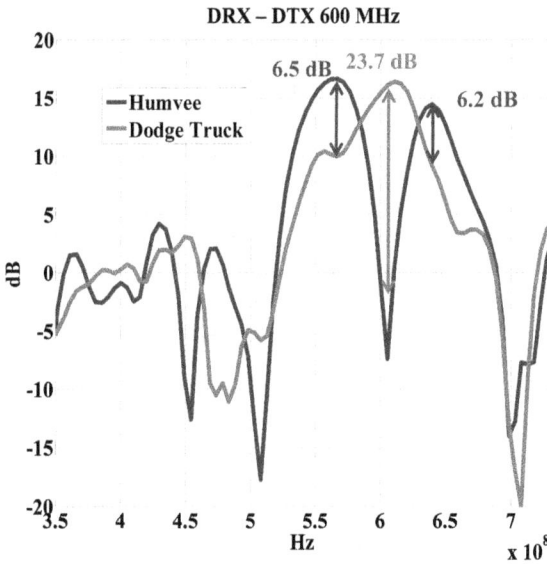

DRX – DTX 600 MHz

(b)

Fig. 2.1.4.3   DRX spectra (DTX = 600 MHz); Targets: Humvee and Dodge Trucks. (a) Amplitude in linear units; (b) Amplitude in dB units. Although both trucks have maximum resonances in the 500–715 MHz band, the Humvee's DRX in this range is bimodal, while the Dodge Truck's DRX is unimodal and with lesser side bands. At its maximum at 600+ MHz, the difference between the two spectra is 23.7 dB.

WH Transform PRX @ 345 feet

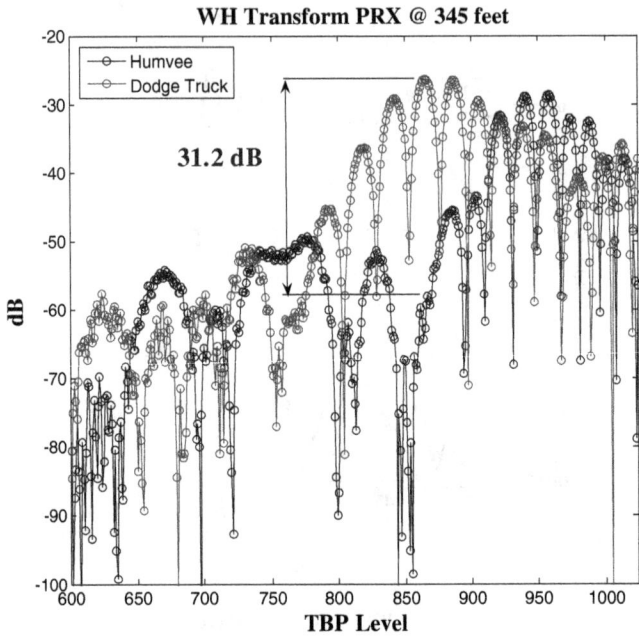

Fig. 2.1.4.4 PRX WH Transform spectra of Humvee and Dodge trucks. As in Fig. 2.1.4.2, but with a RX difference of 31.2 dB at a filter time-bandwidth product (TBP) of around 850.

HUMVEE PRX ASPECT ANGLES

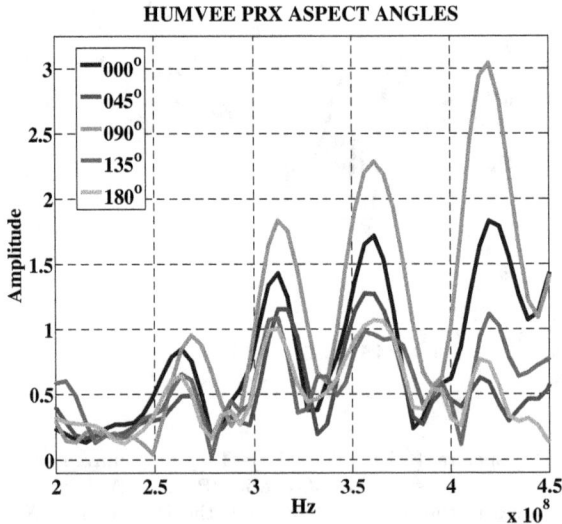

Fig. 2.1.4.5 Lower frequency PRX spectra, target: Humvee truck. These lower frequency resonances exhibit aspect independence.

(a)

(b)

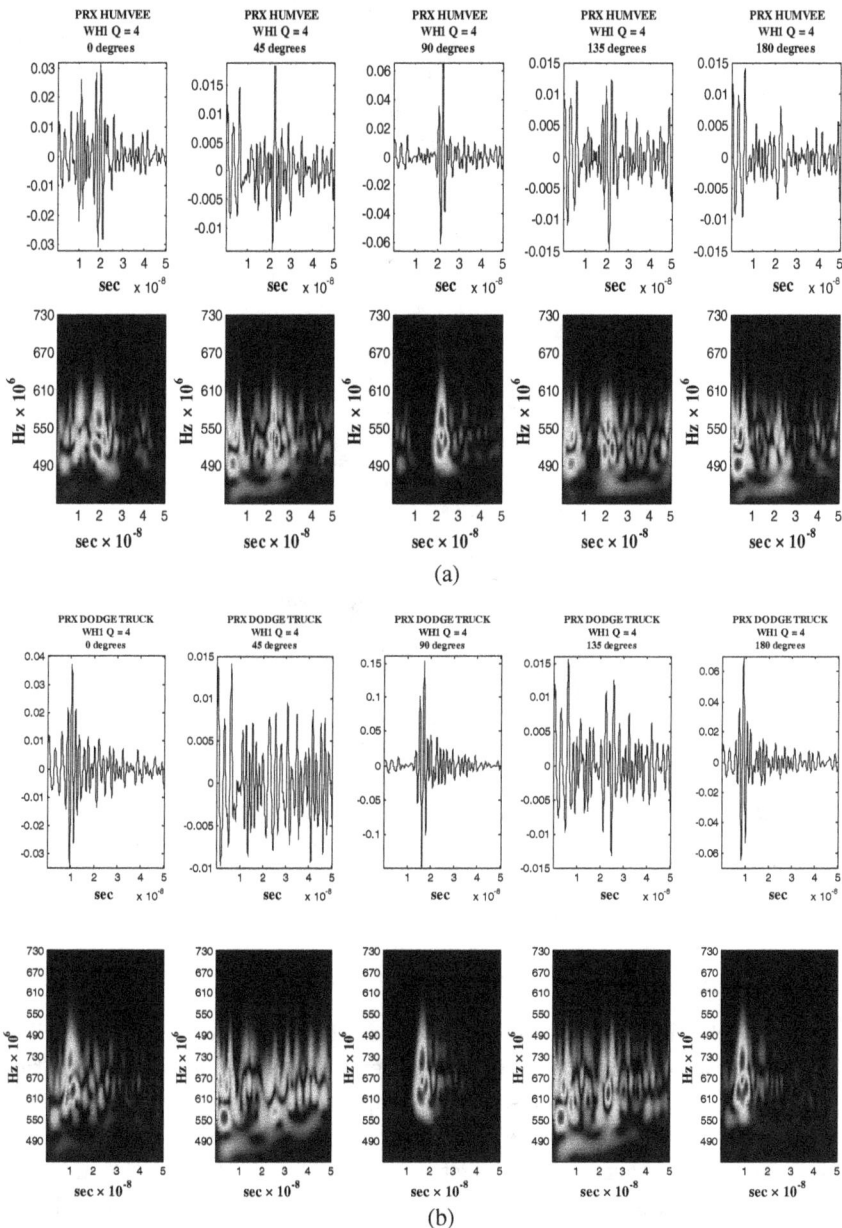

Fig. 2.1.4.6 Time-frequency PRX spectra for (a) Humvee truck; (b) Dodge truck at aspect angles 0° (head on), 45°, 90°, 135° and 180°. Differential filter: WH1, $Q = 4$. There is aspect-dependence exhibited in the time domain (x-axis), but aspect-independence in the frequency domain (y-axis).

Fig. 2.1.5.1   Artillery Shell 155×667 mm. Expected resonance is 3e8/(667e-3) = 4.4978e8 or approximately 450 MHz — neglecting the effect of shell's taper. Empirical testing showed the main resonance at around 600 MHz.

2.2.0 Anechoic Chamber Tests

Anechoic chamber tests were conducted at the Benefield Anechoic Facility (BAF), Edwards Air Force Base in order to obtain *a priori* information concerning the PTX response — the impulse response — of targets in a clutter-free environment. The test arrangement is shown in Fig. 2.2.0.1. In the case of light targets, the target was hoisted over the TX-RX antenna. In the case of a heavy target — a truck — the antenna was hoisted over the target.

In order to obtain equipment-free estimates of the clutter-free PRX of the targets tested, TX, antenna and RX compensations were applied to the raw RX data (Fig. 2.2.0.2).

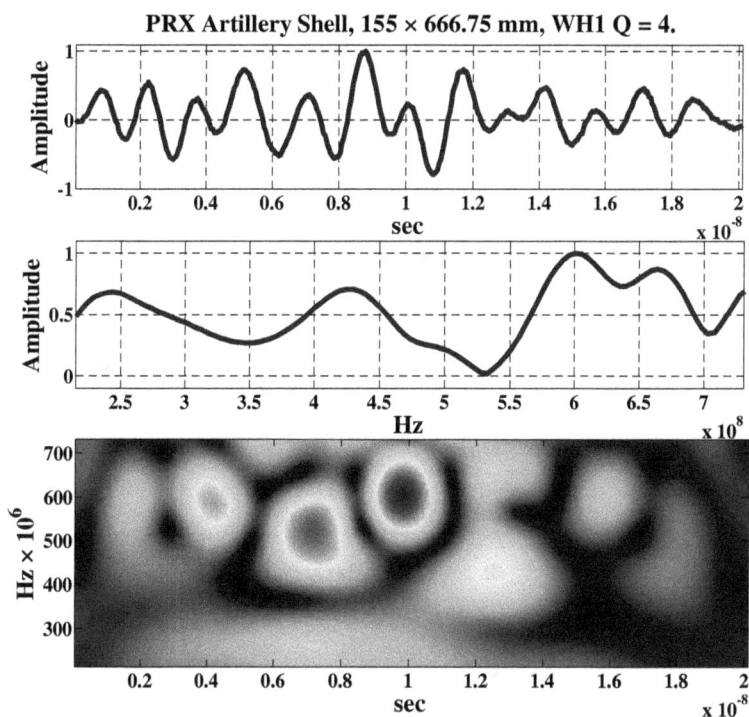

Fig. 2.1.5.2   PRX for artillery shell. Top: time domain. Middle: Fourier spectrum. Bottom: time-frequency spectrum. Whereas the Fourier spectrum (sufficient for an LTI system response description) indicates multiple resonant peaks, the time-frequency spectrum (appropriate for LTV system response description) clearly reveals the major resonance at 600 MHz.

2.2.1 Target: Barrel Aspect Up: PRX

The PTX (UWB) response of a Barrel target — positioned up — was obtained for two TX orthogonal linear polarizations, and three orientations. The RX reception was at both polarizations. The time domain PRXs are shown in Fig. 2.2.1.1(a), and the PRX spectra in Fig. 2.2.1.1(b). As can be seen from inspection, the effect of TX polarization on resonance features and target orientation is inconsequential. However, positioning the barrel up or sideways is consequential — compare Fig. 2.2.2.1, below, for results for the barrel target positioned sideways. The difference is because positioning the barrel up or sideways exposes or hides from the TX signal significant target resonant features of this target.

PRX ARTILLERY SHELL : 155 mm × 666.75 mm

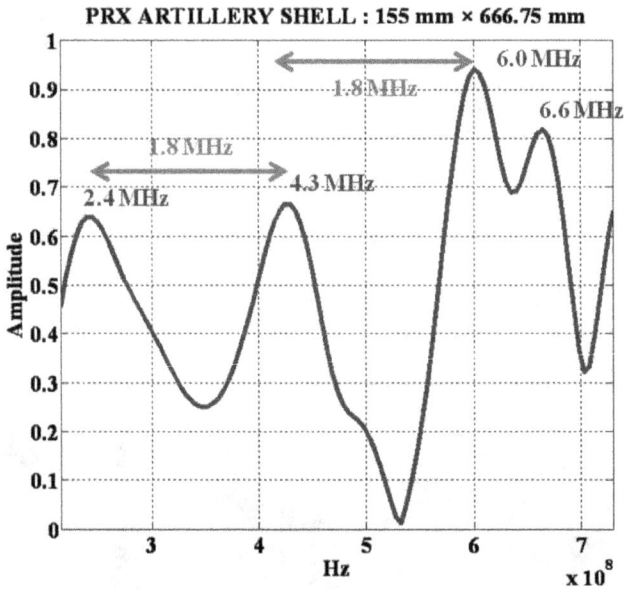

Fig. 2.1.5.3 PRX artillery shell spectrum exhibits two other resonances in the 215–730 MHz bandwidth separated by 1.8 MHz.

2.2.2 Target: Barrel Aspect Side: PRX

The PTX response of a Barrel target — positioned sideways — was obtained for two TX orthogonal linear polarizations, and three orientations. The time domain PRXs are shown in Fig. 2.2.2.1(a), and the PRX spectra in Fig. 2.2.2.1(b). As can be seen from inspection, the effect of TX polarization on resonance features and target orientation is inconsequential. However, positioning the barrel up or sideways is consequential — compare Fig. 2.2.1.1, above, for results for the barrel target positioned up. The difference is because positioning the barrel up or sideways exposes or hides from the TX signal significant target resonant features of this target.

2.2.3 Target: Generator: PRX

The PTX response of an electric generator target was obtained for two TX orthogonal linear polarizations, and three orientations. The time domain PRXs are shown in Fig. 2.2.3.1(a), and the PRX spectra in Fig. 2.2.3.1(b). As can be seen from inspection, the effect of TX polarization on resonance features and target orientation is inconsequential.

ARTILLERY SHELL ELEVATED PRX & DRX 400-500 MHz, 500 MHz.
HORIZONTAL vs VERTICAL

PRX   DRX=400-500 500 MHz   PRX   DRX=400-500 500 MHz

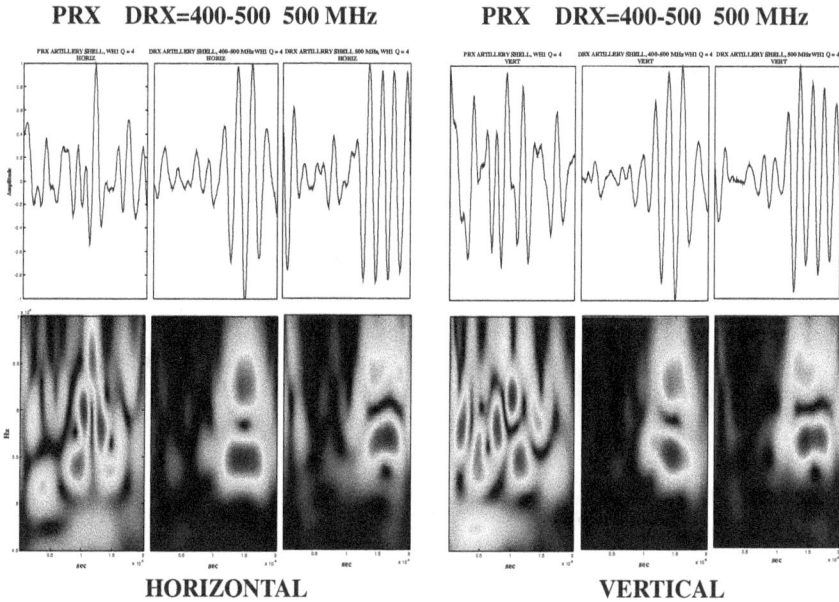

HORIZONTAL   VERTICAL

Fig. 2.1.5.4   PRX and DRX artillery shell time-frequency spectra, vertical vs. horizontal target aspect. Inspection shows that target aspect — horizontal or vertical — does not influence the resonances viewed from the side frequency axis — the y-axis.

2.2.4 Target: Microwave Oven: PRX

The top surface aspect of the microwave oven was 8.5×15.5 inches. The PTX response of a microwave oven target was obtained for two TX orthogonal linear polarizations, and three orientations. The time domain PRXs are shown in Fig. 2.2.4.1(a), and the PRX spectra in Fig. 2.2.4.1(b). As can be seen from inspection, the effect of TX polarization on resonance features and target orientation is inconsequential.

2.2.5 Target: Roof Panel: PRX

The PTX response of a Roof Panel target was obtained for two TX orthogonal linear polarizations, and three orientations. The time domain PRXs are shown in Fig. 2.2.5.1(a), and the PRX spectra in Fig. 2.2.5.1(b). As can be seen from inspection, the effect of TX polarization on resonance features and target orientation is inconsequential.

- 55 gallon drum, metal panel, diesel generator, microwave oven:
  - Hoist overhead with antenna below
- Truck
  - Hoist antenna overhead with truck below
- Data at 0°, ±6° incidence for H and V polarizations

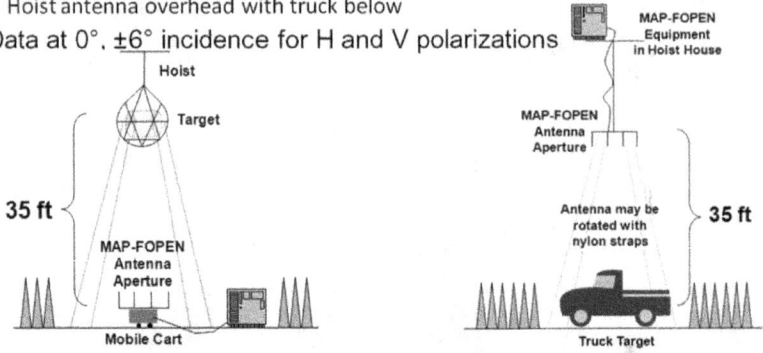

Fig. 2.2.0.1   BAF Edwards AFB test arrangement. In the case of light targets, the target was hoisted over the TX-RX antenna. In the case of a heavy target — a truck in this case — the antenna was hoisted over the target.

COMPENSATIONS – ANECHOIC CHAMBER TESTS

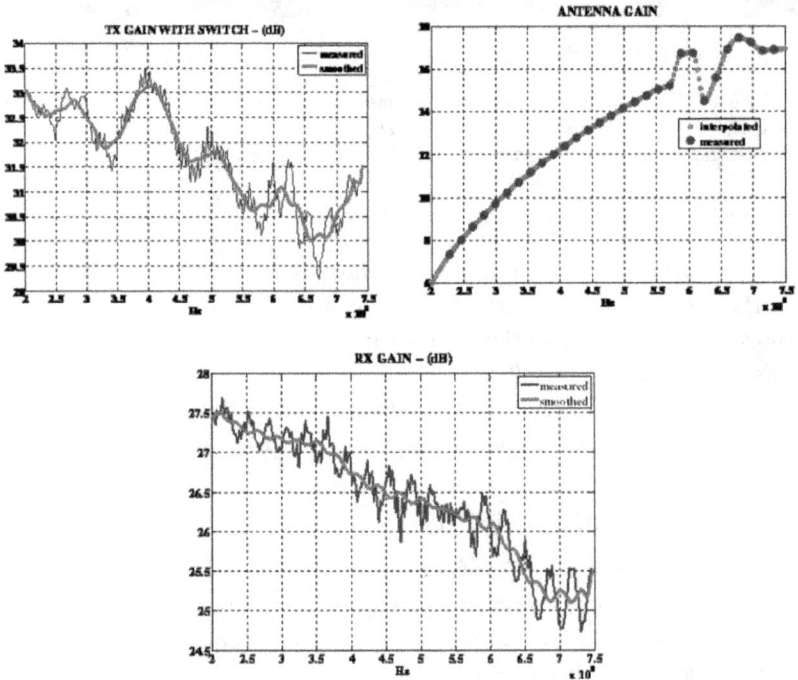

Fig. 2.2.0.2   TX, antenna and RX compensations applied to the raw RX data.

PRX BARREL UP

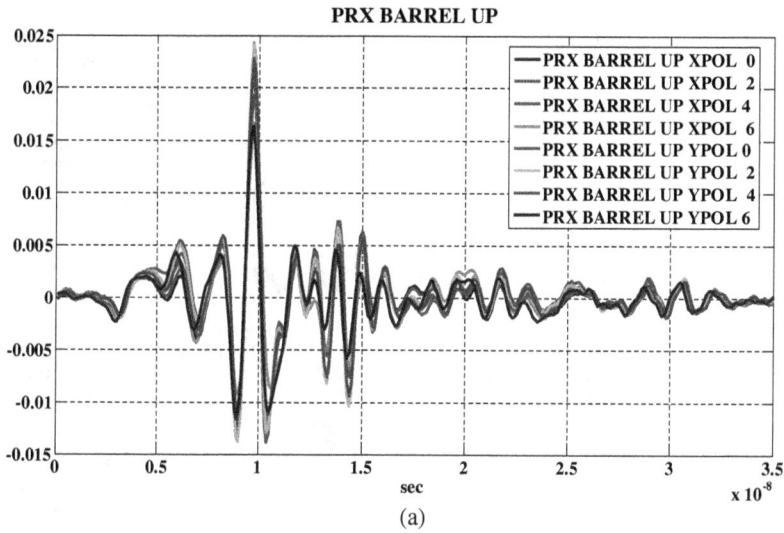

(a)

PRX BARREL UP

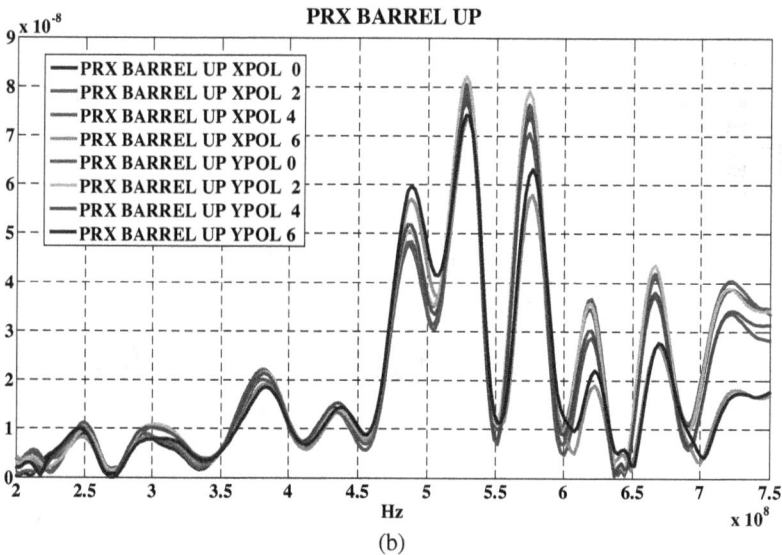

(b)

Fig. 2.2.1.1   The PTX response of a Barrel target — positioned up — was obtained for 2 TX orthogonal linear polarizations, and 3 orientations. The time domain PRXs are shown in (a), and the PRX spectra in (b). The effect of TX polarization on resonance features and target orientation is inconsequential. But positioning the barrel up or sideways is consequential — see Fig. 2.2.2.1, below. Positioning the barrel up or sideways exposes or hides from the TX signal significant target features.

PRX BARREL SIDEWAYS

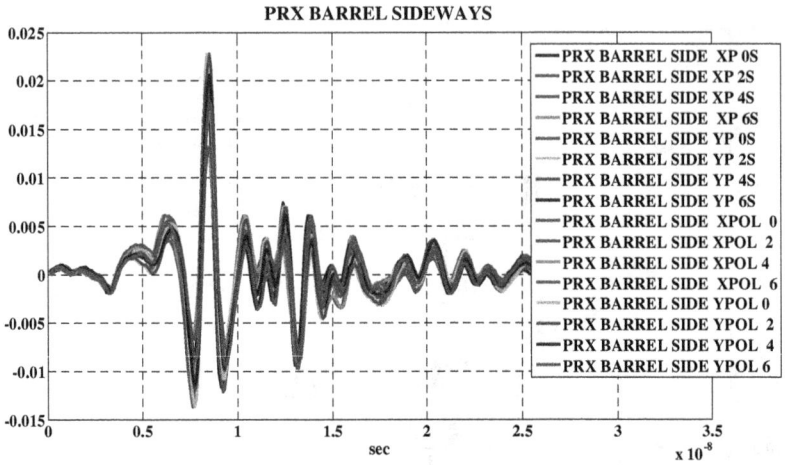

(a)

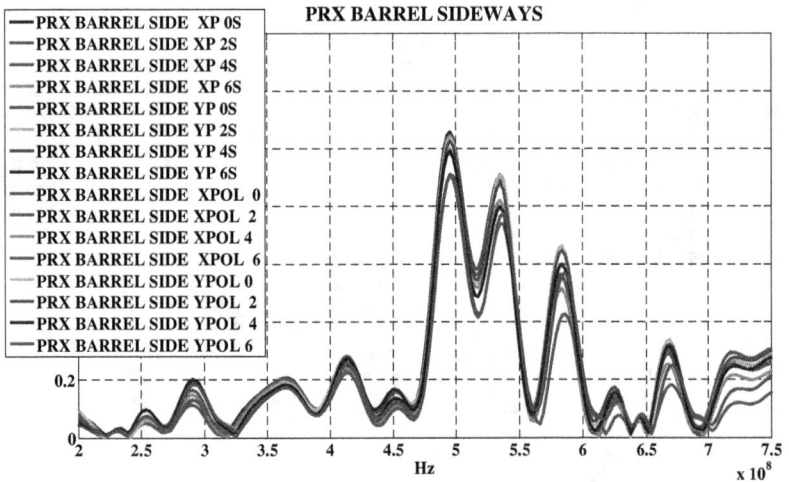

(b)

Fig. 2.2.2.1   The PTX response of a Barrel target — positioned sideways — was obtained for 2 TX orthogonal linear polarizations, and 3 orientations. The time domain PRXs are shown in (a), and the PRX spectra in (b). The effect of TX polarization on resonance features and target orientation is inconsequential. Positioning the barrel up or sideways is consequential — see Fig. 2.2.1.1, above. Positioning the barrel up or sideways exposes or hides from the TX signal significant target features.

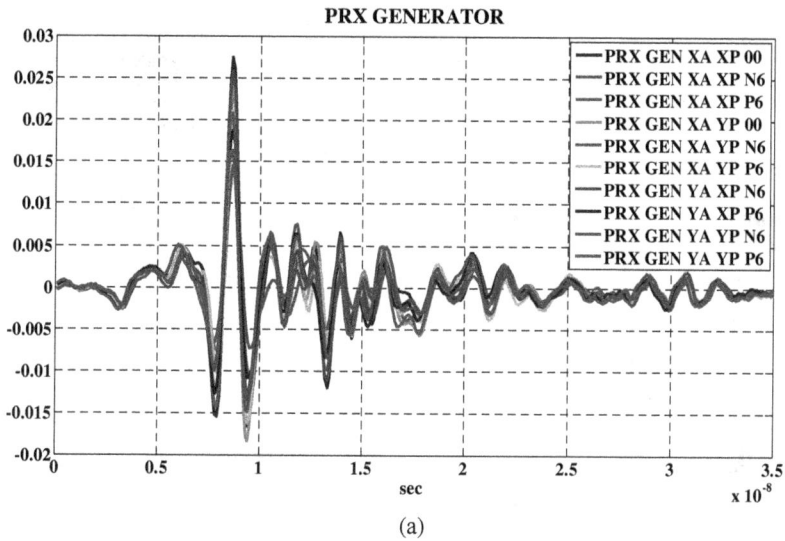

PRX GENERATOR

(a)

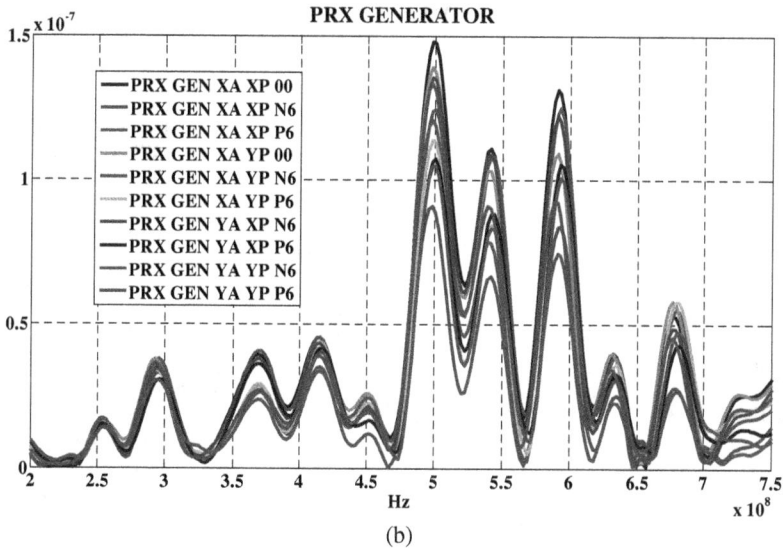

PRX GENERATOR

(b)

Fig. 2.2.3.1 The PTX response of an electric generator target was obtained for 2 TX orthogonal linear polarizations, and 3 orientations. The time domain PRXs are shown in (a), and the PRX spectra in (b). The effect of TX polarization on resonance features and target orientation is inconsequential for this target.

PRX MICROWAVE OVEN

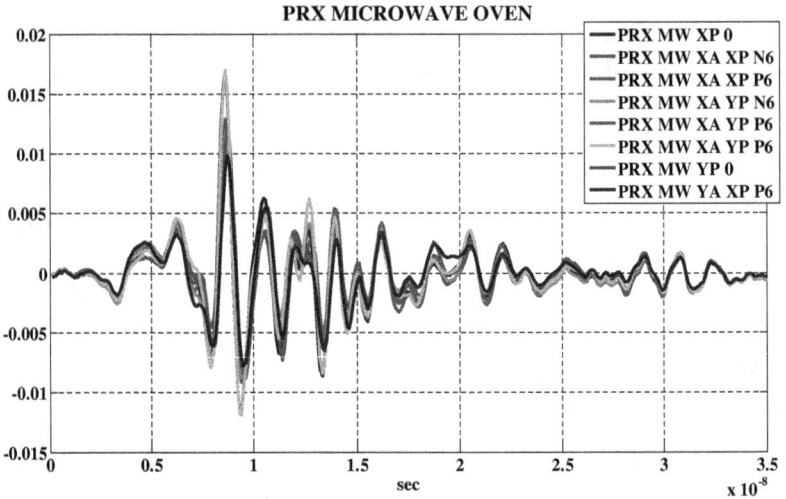

(a)

AUTOSPECTRUM PRX MICROWAVE OVEN

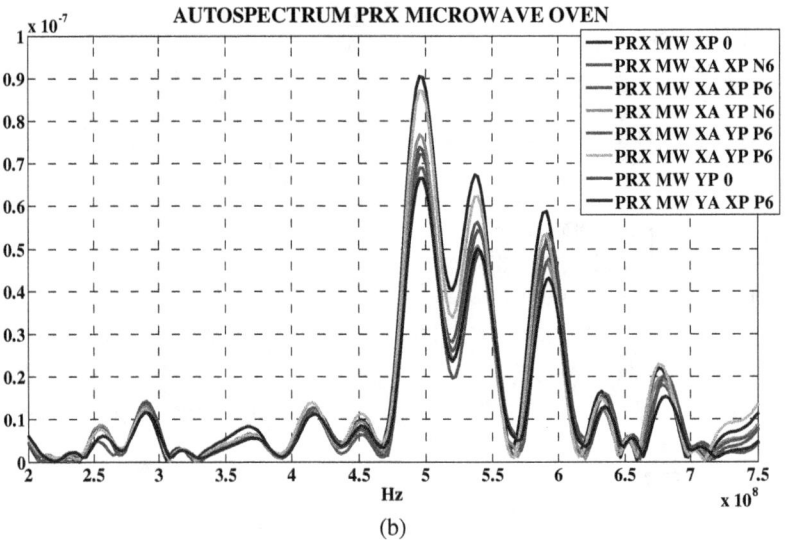

(b)

Fig. 2.2.4.1   The PTX response of a microwave oven target was obtained for 2 TX orthogonal linear polarizations, and 3 orientations. The time domain PRXs are shown in (a), and the PRX spectra in (b). The effect of TX polarization on resonance features and target orientation is inconsequential for this target.

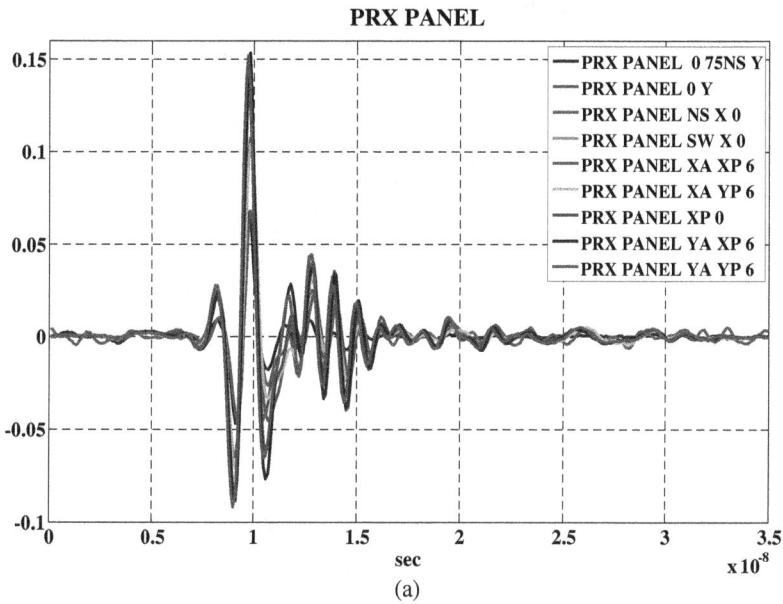

(a)

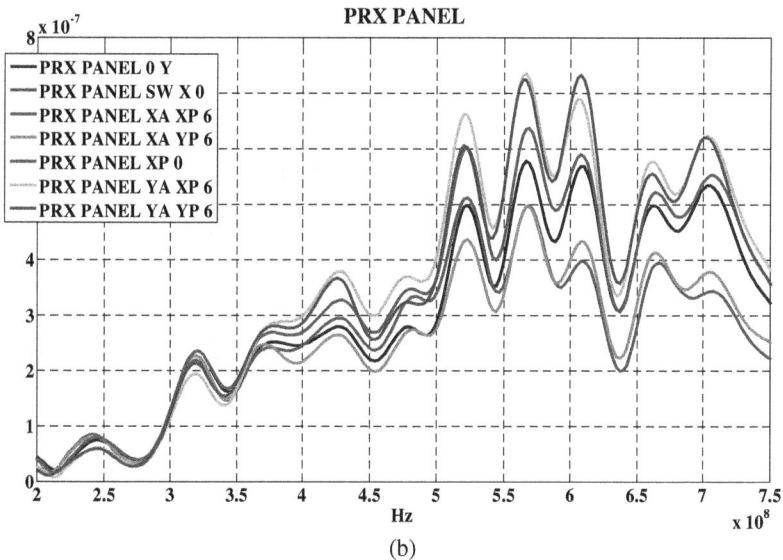

(b)

Fig. 2.2.5.1 The PTX response of a Roof Panel target was obtained for 2 TX orthogonal linear polarizations, and 3 orientations. The time domain PRXs are shown in (a), and the PRX spectra in (b). The effect of TX polarization on resonance features and target orientation is inconsequential for this target.

2.2.6 Target: Truck: PRX

The PTX response of a Truck target was obtained for two TX orthogonal linear polarizations, and three orientations. The time domain PRXs are shown in Fig. 2.2.6.1(a), and the PRX spectra in Fig. 2.2.6.1(b). Positioning this large target at a distance relative to the TX/RX antenna is consequential, because positioning exposes or hides from the TX signal significant target features. Corresponding to the three positions of the target and with the TX/RX antenna 35 ft removed from target, there resulted three responses differing in certain spectral aspects. The effect of TX polarization on RX resonance features (with reception at both polarizations), however, is inconsequential for this target, as for the other targets.

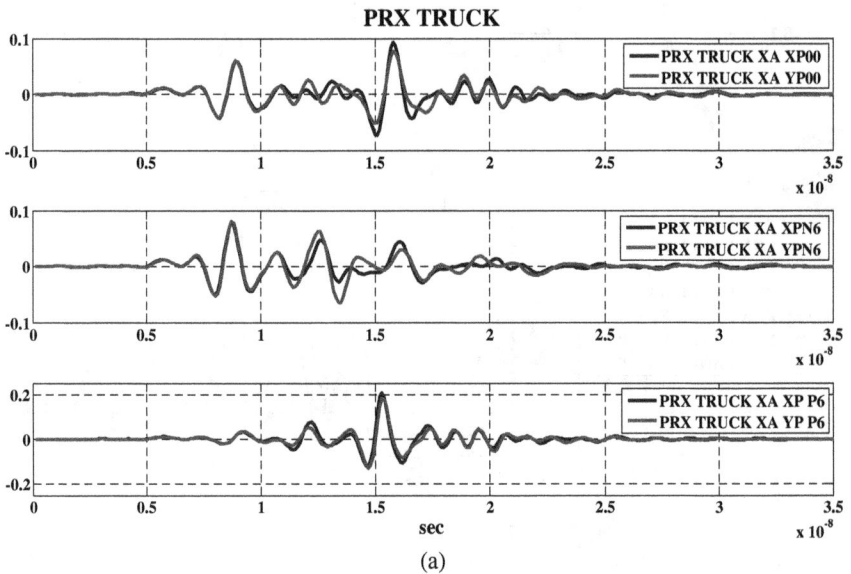

Fig. 2.2.6.1   The PTX response of a Truck target was obtained for 2 TX orthogonal linear polarizations, and 3 position orientations. The time domain PRXs are shown in (a), and the PRX spectra in (b). Positioning this relatively large target is consequential, because in the case of large targets close the TX/RX antenna, positioning exposes or hides from the TX signal significant target features. Corresponding to the three positions of the target and with the TX/RX antenna 35 ft removed from target, three RX were obtained, differing in certain spectral aspects. The effect of TX polarization on RX resonance features, however, is inconsequential for this target.

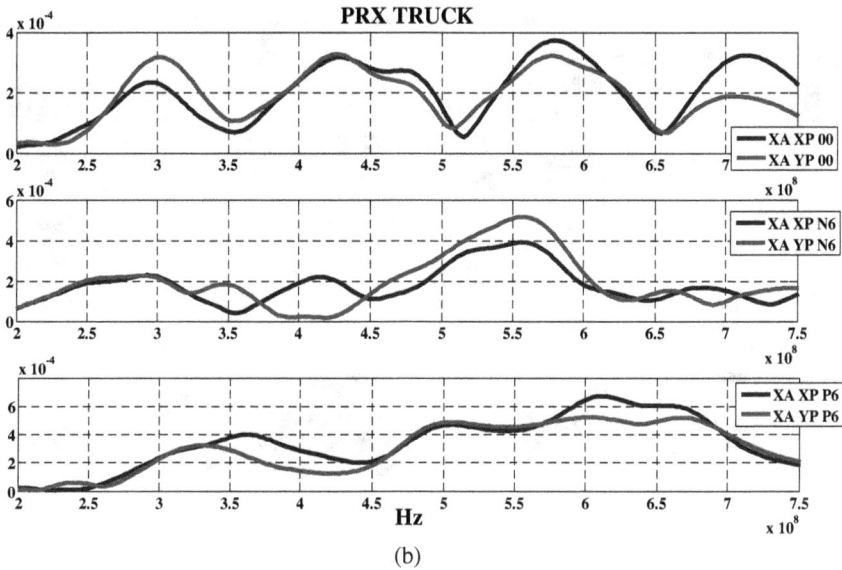

(b)

Fig. 2.2.6.1 (*Continued*)

2.2.7 Multiple-Window Spectra

The averaging of LTV radar signals is not straightforward due the problem of nonstationarity. The multiple-window spectral method is designed to address random, nonstationary signals and to provide a low bias and variance spectral estimator (Thomson, 1982; Martin & Flandrin, 1985; Frazer & Boashash, 1994; Xu *et al.*, 1999). In the following, the Weber-Hermite (WH) wave functions (Section 1.16) were used to obtain Multiple-Window spectra.

Transforms of RX signals were calculated using the WH series, that are pairs of scaling functions (averaging filters) and wavelets (differentiating filters). For example, all even numbered filters are scaling functions, or averagers, e.g., WH0, WH2, WH4, WH6, etc., and all odd numbered filters are wavelets, or differentiators, e.g., WH1, WH3, WH5, WH7, etc. Arbitrarily, only the eight filters, WH0-7, were used. Scaling these filters provided 8 localized time-frequency spectra as shown in Fig. 2.2.7.1. Averaging these spectra provided the aggregate localized magnitude multiple-window time-frequency spectra shown in Fig. 2.2.7.2.

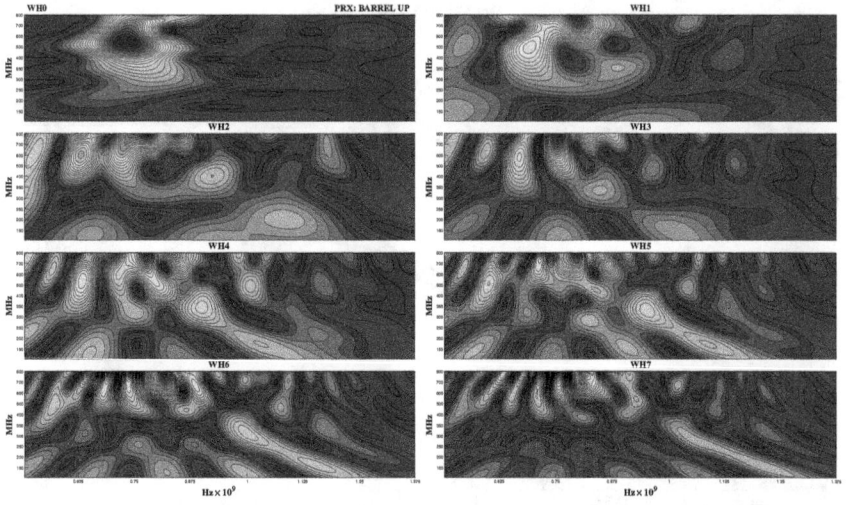

(a)

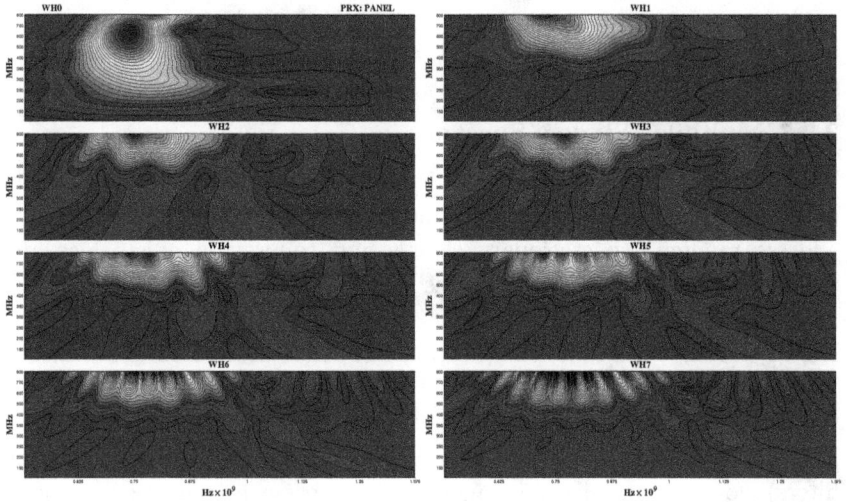

(b)

Fig. 2.2.7.1   WH transforms of RX signals were calculated using the WH series, that are pairs of scaling functions (averaging filters) and wavelets (differentiating filters). For example, all even numbered filters are scaling functions, or averagers, e.g., WH0, WH2, WH4, WH6, etc., and all odd numbered filters are wavelets, or differentiators, e.g., WH1, WH3, WH5, WH7, etc. Arbitrarily, only the eight filters, WH0-7, were used here. Scaling these filters provided 8 localized time-frequency spectra. In (a) is shown 8 spectra (WH0-7) for the barrel target PRX, and in (b) for the roof panel target PRX.

The multiple-window time-frequency spectra shown in Fig. 2.2.7.2 for the barrel target, positioned up, are SRX spectra. The TX signals were STXs, or bundles of DTXs with the following frequencies.

A: STX: $330 + 515$ MHz.
B: STX: $380 + 515$ MHz.
C: STX: $330 + 515 + 570$ MHz.
D: STX: $380 + 515 + 570$ MHz.
E: STX: $380 + 515 + 570 + 675$ MHz.
F: STX: $330 + 380 + 515 + 570 + 675$ MHz.

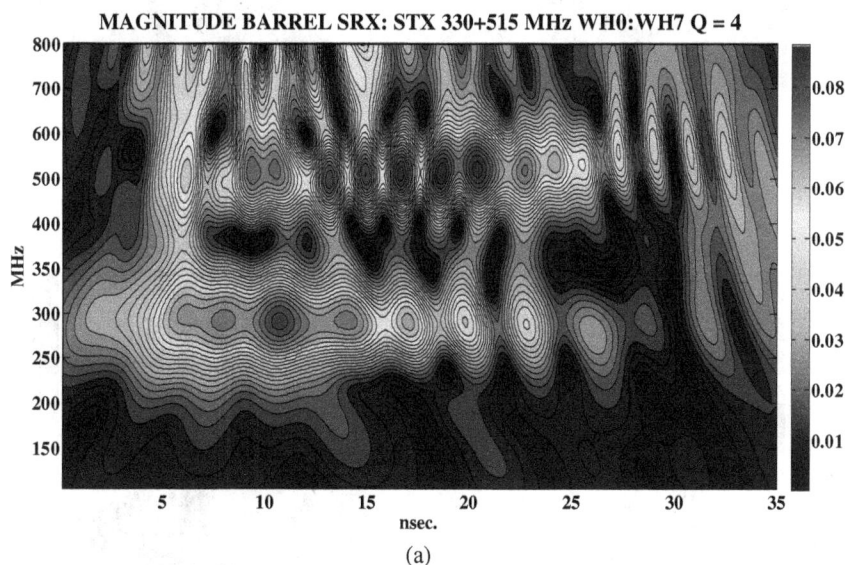

(a)

Fig. 2.2.7.2 Multiple-Window time-frequency spectra, based on the average of 8 WHWFs: WH0-7. These filters provided 8 localized time-frequency spectra. Summing these plots provided an *aggregate localized magnitude time-frequency spectrum*. Target: Barrel positioned up. TX signals are STXs, or bundles of DTXs at the following frequencies.
(a): STX: $330 + 515$ MHz.
(b): STX: $380 + 515$ MHz.
(c): STX: $330 + 515 + 570$ MHz.
(d): STX: $380 + 515 + 570$ MHz.
(e): STX: $380 + 515 + 570 + 675$ MHz.
(f): STX: $330 + 380 + 515 + 570 + 675$ MHz.
The frequencies in the transmitted STX signals are returned in the SRX signals indicating an LTV target. There is also an amplitude modulation in time with a harmonic component that is addressed Section 2.2.9.

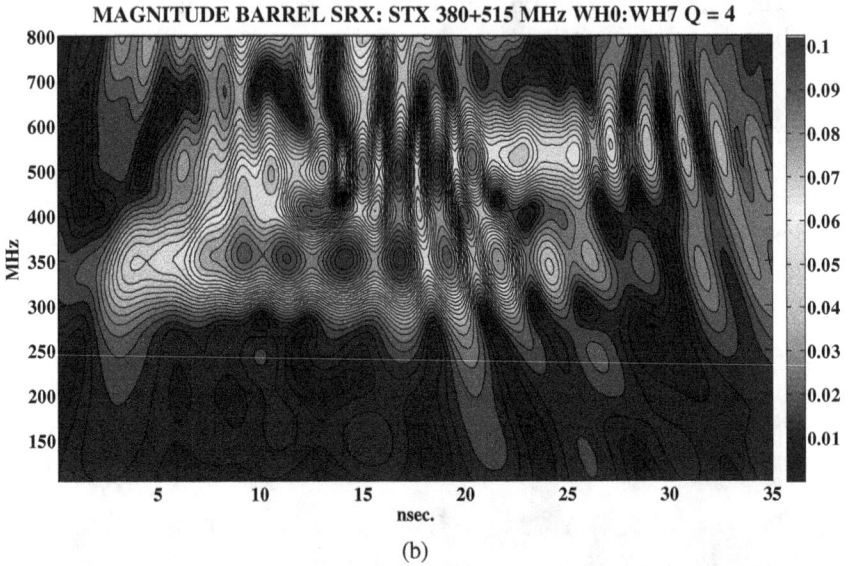

(b)

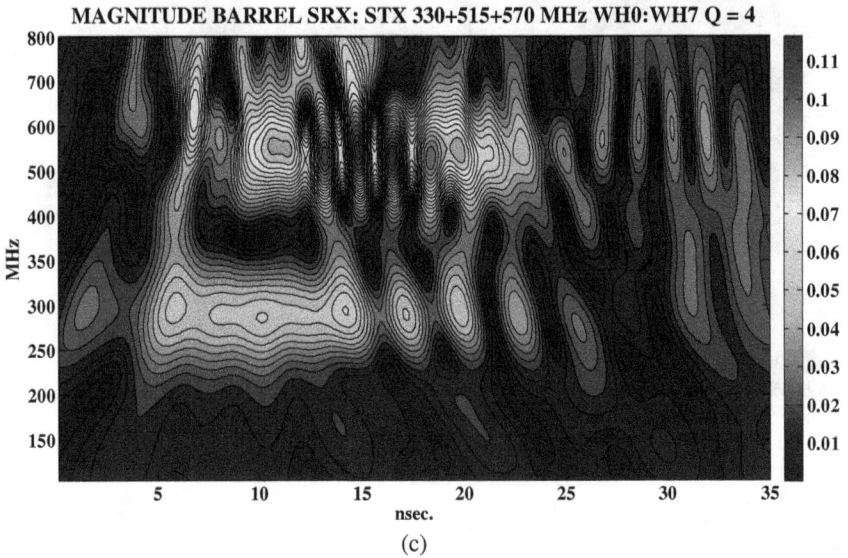

(c)

Fig. 2.2.7.2   (*Continued*)

MAGNITUDE BARREL SRX: STX 380+515+570 MHz WH0:WH7 Q = 4

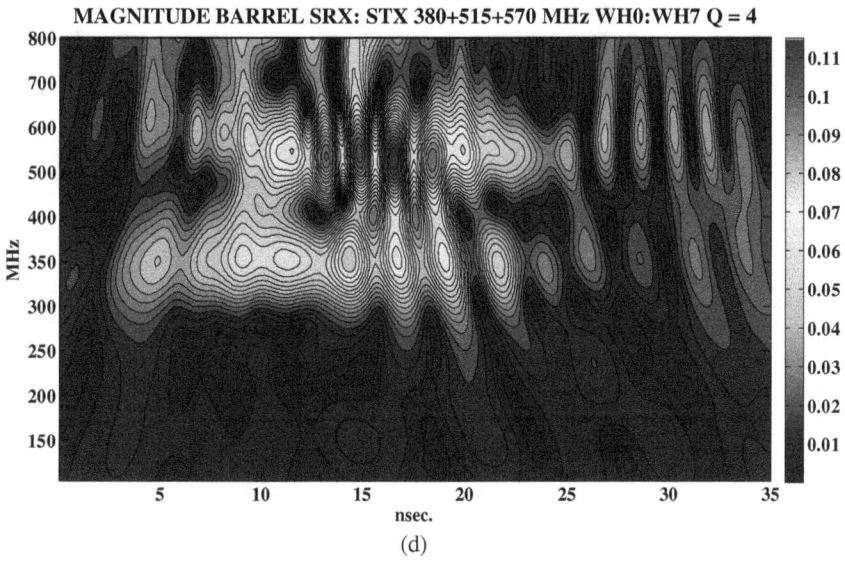

(d)

MAGNITUDE BARREL SRX: STX 380+515+570+675 MHz WH0:WH7 Q = 4

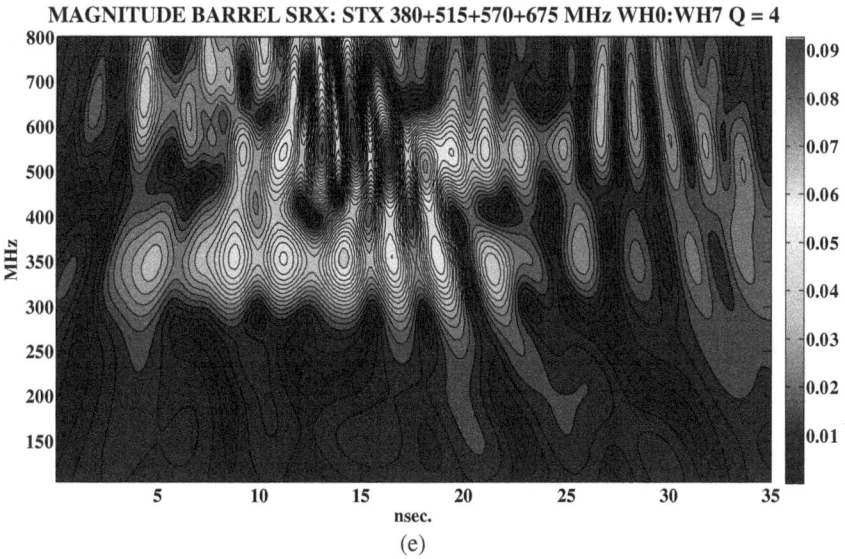

(e)

Fig. 2.2.7.2 (*Continued*)

MAGNITUDE BARREL SRX: STX 330+380+515+570+675 MHz WH0:WH7 Q = 4

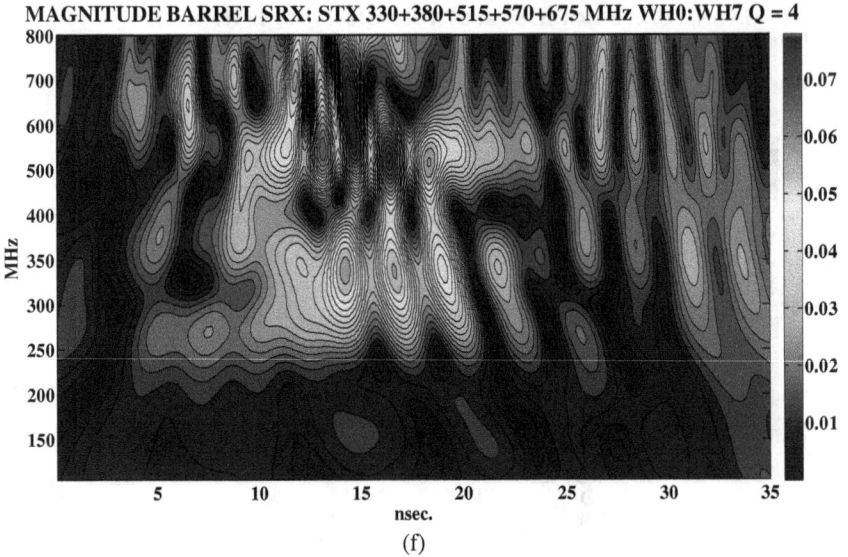

(f)

Fig. 2.2.7.2　(*Continued*)

From inspection, Fig. 2.2.7.2 indicates that the frequencies in the transmitted STX signals were returned in the SRX signals indicating an LTV target. There is also an amplitude modulation in time with a harmonic component that is further addressed in Section 2.2.9.

2.2.8 Target Linear Frequency Response Functions

There are three major approaches to calculating the *linear* frequency response function (FRF). The three methods are described by the following equations:

$$H(f)_1 = \frac{S(f)_{xy}}{S(f)_{xx}}$$

$$H(f)_2 = \frac{S(f)_{yy}}{S(f)_{yx}}$$

$$H(f)_3 = \frac{S(f)_{yy} - S(f)_{xx} + \sqrt{(s(f)_{xx} - S(f)_{yy})^2 + 4|S(f)_{xy}|^2}}{2S(f)_{yx}}$$

where

$S(f)_{xx}$ is the autospectrum or spectral density function of TX;

$S(f)_{yy}$ is the autospectrum or spectral density function of RX;

$S(f)_{xy}$ and $S(f)_{yx}$ are the cross-spectra of TX and RX;

and these FRFs, of $H(f)_1, H(f)_2$ and $H(f)_3$, are unbiased with respect
to the output noise, to the input noise, and to both the input and output
noise, respectively.

Calculation of the FRF is dependent on calculating the input to the
system (i.e., the transmitted signal directly before, and exciting, the target)
and the output from the system (i.e., the return signal directly after, and
radiating from, the target). Therefore the following compensations were
applied to the transmitted signals at the transmitter and the received
signals at the receiver:

(1) switching parasitics (on the TX path);
(2) transmit magnitude compensation function (on the TX path);
(3) inverse of receive magnitude compensaton function (on the RX path);
(4) inverse of cable loss (on the TX path).

The empirical TX and RX BAF Edwards AFB collected signals were
compensated in this way to provide $S(f)_{xx}, S(f)_{yy}, S(f)_{xy}$ and $S(f)_{yx}$.
These permitted the calculation of the $H(f)_1, H(f)_2$ and $H(f)_3$, that are
shown in Figs. 2.2.8.1–2.2.8.2.

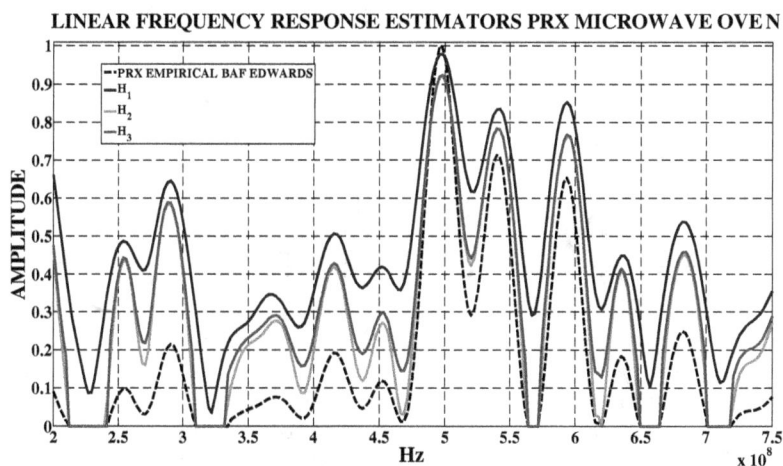

Fig. 2.2.8.1   The frequency response function for the microwave oven estimated by the
method
Autospectrum: Empirical obtained RX.
TF1 — $H(f)_1$ unbiased with respect to the output noise.
TF2 — $H(f)_2$ unbiased with respect to the input noise.
TF3 — $H(f)_3$ unbiased with respect to both the input and output noise.

BARREL UP SRX
STX CONVOLVED WITH TRANSFER FUNCTION OBTAINED FROM PRX
STX = 330 & 515 MHz

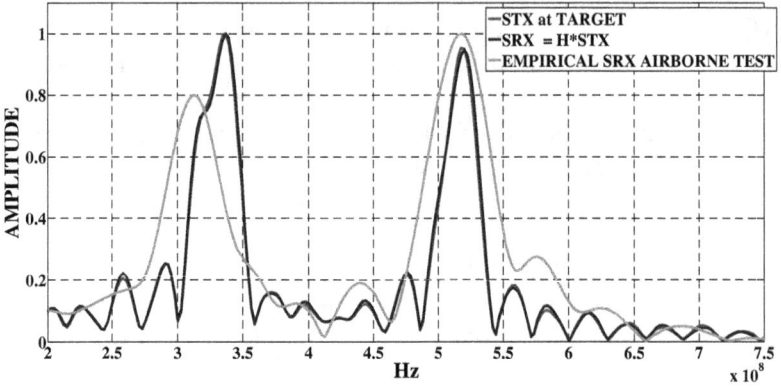

(a)

BARREL UP SRX
STX CONVOLVED WITH TRANSFER FUNCTION OBTAINED FROM PRX
STX = 380 & 515 MHz

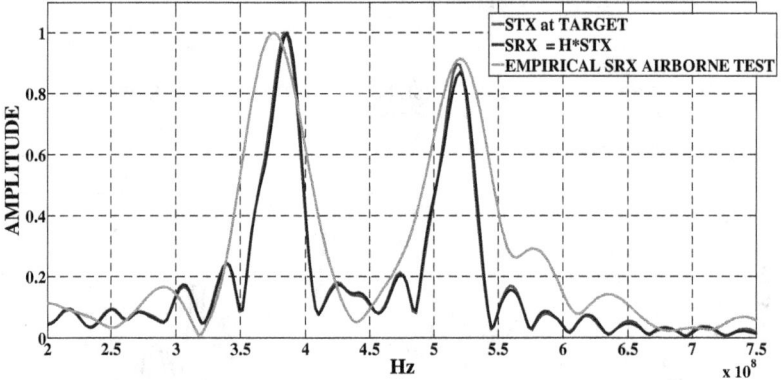

(b)

Fig. 2.2.8.2 In (a) and (b): (1) is the calculated and compensated STX *at the target*, commencing with the empirical STX at the transmitter (2) is the SRX calculated from the target transfer function, H, that was itself calculated using the STX and SRX *at the target*; (3) are empirical SRX spectra obtained from airborne tests — see Section 4.0 below.

The STXs are:
A: STX = 330 & 515 MHz
B: STX = 380 & 515 MHz
The STXs and SRXs almost completely overlap.

2.2.9 Carrier Frequency-Envelope Frequency (CFEF) Spectra

Pre-processing of real signals into their WH transforms (Section 1.16), provides a signal decomposition in complex number form. The RX signal is decomposed into a hierarchical time-bandwidth product (TBP) basis more suited to LTV signal analysis, rather than into constant frequency sinusoidal basis functions as in the case of Fourier decomposition.

As described in Section 2.2.7, the WH series is a hierarchy of pairs of scaling functions and wavelets. For example, all even numbered filters are scaling functions, or averagers, e.g., WH0, WH2, WH4, WH6, etc., and all numbered filters are wavelets, or differentiators, e.g., WH1, WH3, WH5, WH7, etc. Arbitrarily, only the eight filters, WH0-7, were used in this analysis. Scaling these filters provided 8 localized time-frequency spectra. Summing these spectra provided the Multi-Window aggregate localized magnitude time-frequency spectra as shown in Fig. 2.2.7.2 and in Fig. 2.2.9.1(a), below. The real part of these spectra was used in the following additional analysis.

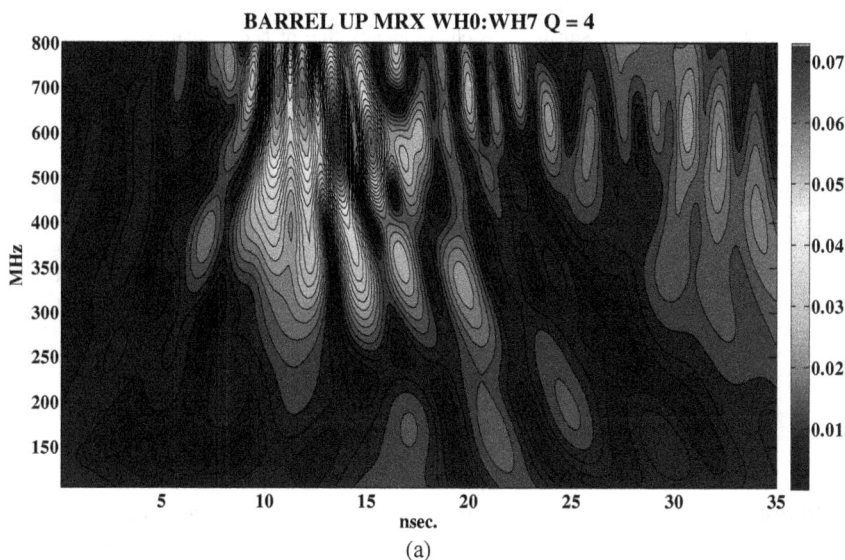

(a)

Fig. 2.2.9.1(a)   A Multiple Window time-frequency spectrum for MRX, target Barrel Up, calculated by the average of 8 time-frequency spectra using filters WH0-7, $Q = 4$. See Section 2.2.7.

CARRIER-FREQUENCY/ENVELOPE-FREQUENCY TRANSFORM
BARREL UP MRX WH0:WH7 Q = 4

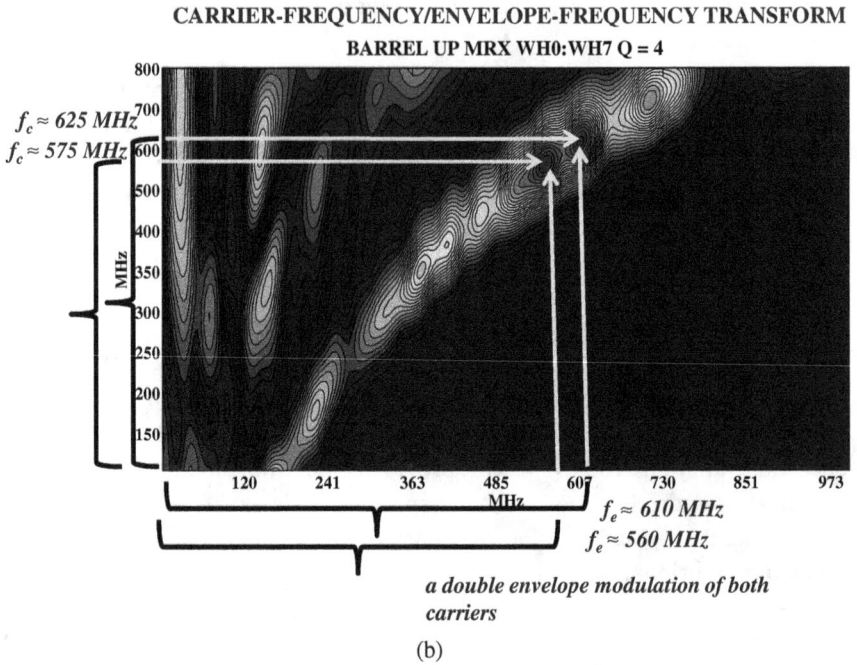

a double envelope modulation of both carriers

(b)

Fig. 2.2.9.1(b) A *Carrier* Frequency-*Envelope* Frequency (CFEF) spectrum. These plots are the Fourier transforms of each frequency line of the Multiple Window time-frequency spectrum, providing the spectra, 100 MHz–1 GHz, of the modulating envelopes of the carrier signal frequencies of the multiple-window spectrum. There are peaks in these spectra at "carrier frequencies" and "envelope" modulations at specific frequencies. There are also lower envelope frequency peaks indicating steady bursts cross all carrier frequencies, i.e., a spike, that is indicated on the left as a vertical band.

The motivation for such CFEF spectra is to circumvent the difficulty that a comparison of different time-frequency spectra requires that the compared spectra represent signals with receiver arrivals at precisely the same time, i.e., time frequency spectra are only comparable if the time axis is aligned, which requires that the different targets must be at precisely the same distance from the transmitter. By Fourier transforming at each frequency line of the time-frequency spectrum shown in (a), the link to a specific RX receiver time of signal arrival is broken. There is, of course, the cost that all time-of-RX-signal arrival information is lost.

The Fourier transform was taken of each frequency line of a Multi-Window spectrum, resulting in a frequency × frequency spectrum, or a *Carrier* Frequency-*Envelope* Frequency (CFEF) spectrum (e.g., Fig. 2.2.9.1(b)) identifying both the frequencies of the modulating envelopes of the carrier signal frequencies of, e.g., Fig. 2.2.9.1(a). There are peaks in Fig. 2.2.9.1(b) at approximately the RX Fig. 2.2.9.1(a) "carrier frequencies" indicating amplitude envelope modulations of the carrier frequency components of the RX signal at the same frequency as the RX carrier frequency components — *a double modulation of both carrier and envelope at the same frequency.* There are also lower envelope frequency peaks indicating steady bursts cross all carrier frequencies, i.e., a spike, that are indicated on the left as vertical bands.

The rationale for these plots is to remedy the fact that a time comparison (or a time average) of time-frequency plots requires that the compared plots represent signal arrival at the receiver at precisely the same time, i.e., the same position in all time-frequency spectra. By Fourier transforming at each frequency line of, e.g., Fig. 2.2.9.1(a), the impediment of different signal arrival times is remedied, and provides additional information concerning the envelope modulation. There is, of course, the cost that all time-of-RX-signal arrival information is lost.

Figure 2.2.9.2 shows PRX CFEF spectra for the targets: microwave oven, barrel up, barrel sideways, and a roof panel; and Fig. 2.2.9.3 for the three truck PRXs.

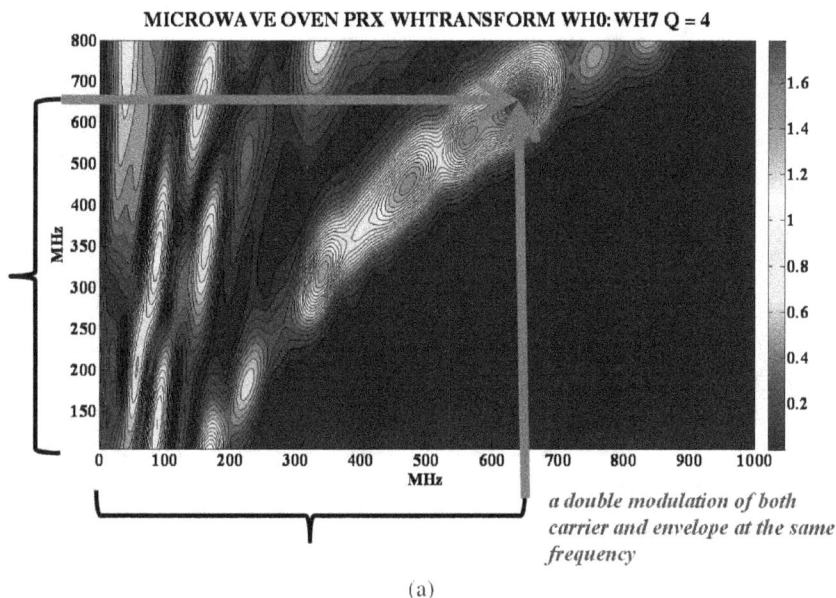

(a)

Fig. 2.2.9.2   CFEF PRX spectra for targets: (a) microwave oven; (b) barrel up; (c) barrel sideways; (d) roof panel.

BARREL UP PRX WHTRANSFORM WH0:WH7 Q = 4

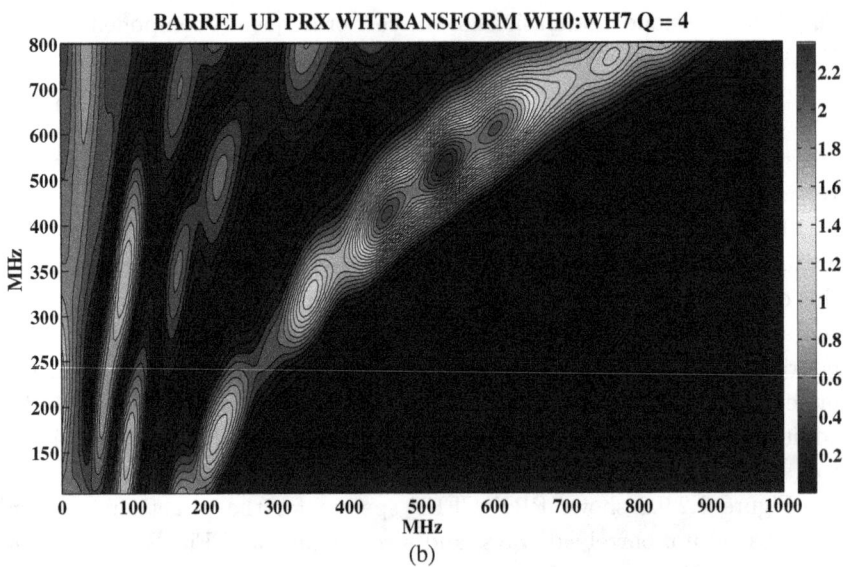

(b)

BARREL SIDE PRX WHTRANSFORM WH0:WH7 Q = 4

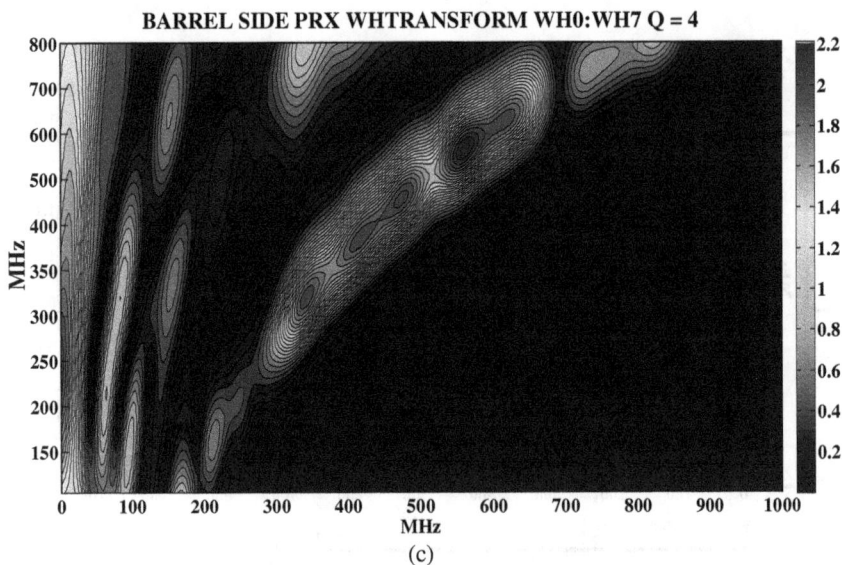

(c)

Fig. 2.2.9.2   (*Continued*)

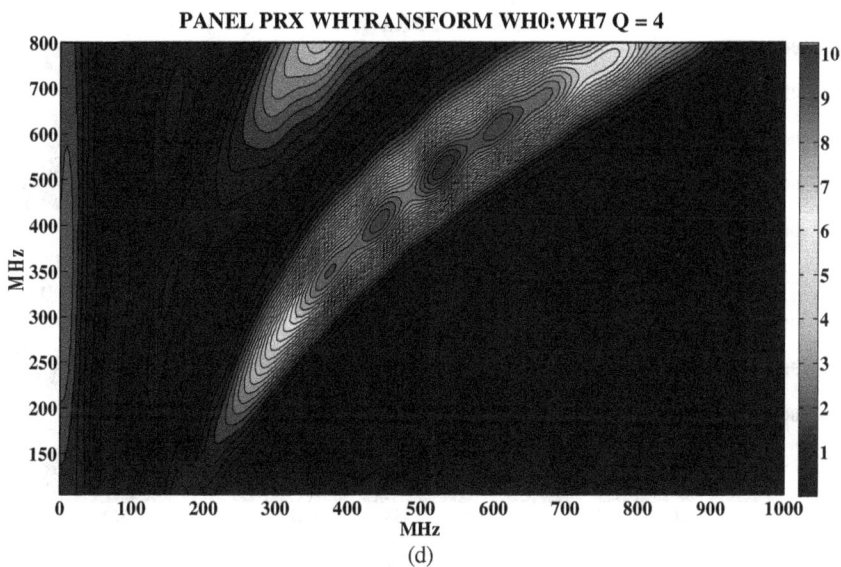

Fig. 2.2.9.2 (*Continued*)

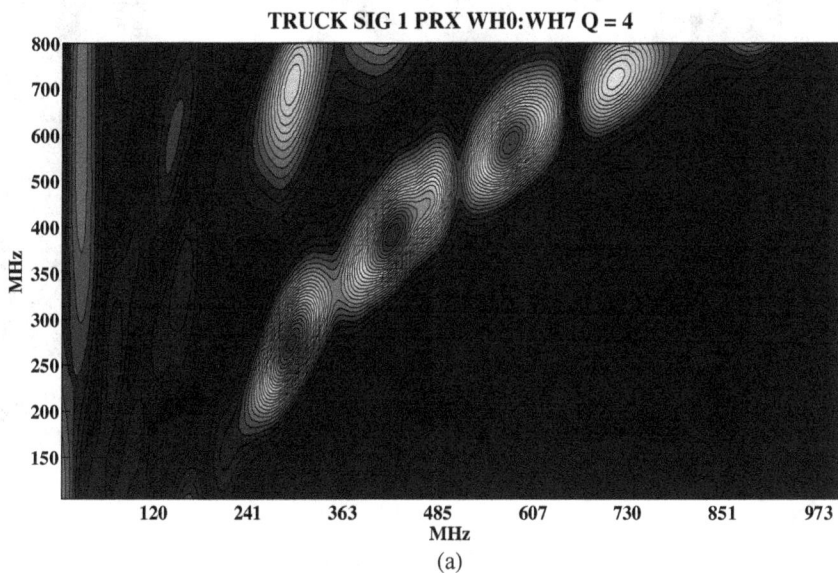

Fig. 2.2.9.3   CFEF PRX spectra for the truck target. (a) Aspect 1; (b) Aspect 2;
(c) Aspect 3.

TRUCK SIG 2 PRX WH0:WH7 Q = 4

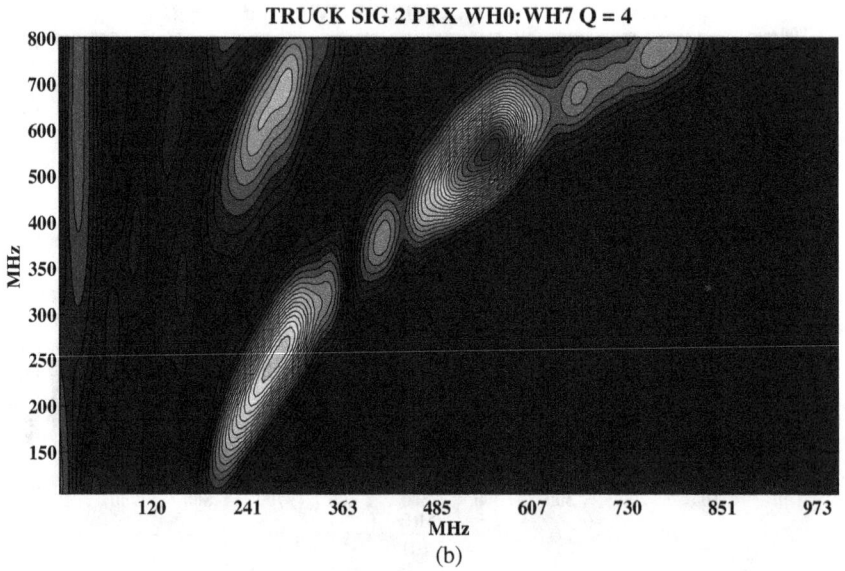

(b)

TRUCK SIG 3 PRX WH0:WH7 Q = 4

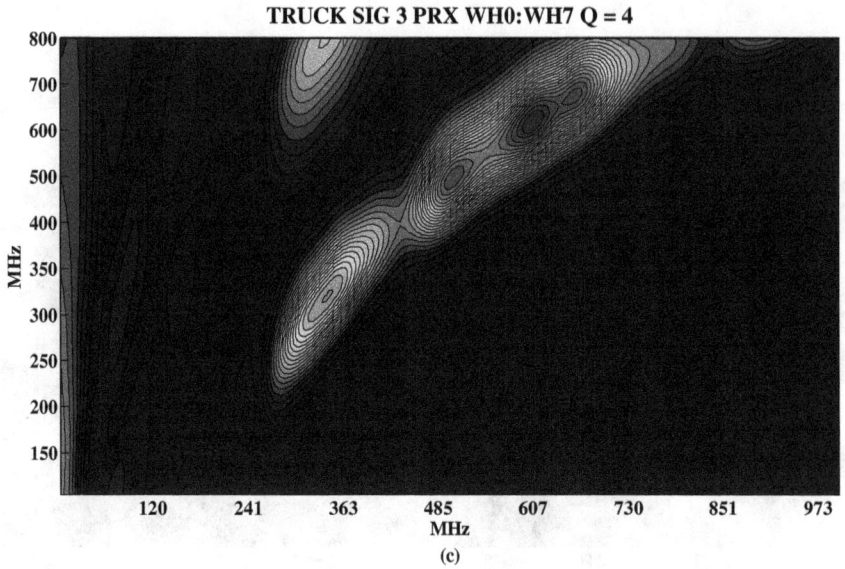

(c)

Fig. 2.2.9.3   (*Continued*)

2.3.0 Flight Tests of FOPEN RAMAR

Originally, the flight test project goal was to test a MAP UHF foliage penetrating (FOPEN) radar operating from an airplatform at 10,000 ft above ground level. However, the ground in Huachuca Canyon, Arizona, was actually about 5,000 feet above sea level, and the air platform was flown about 5,000 feet above the ground, so the ground footprint, although large, was not as large as an airborne system 10,000 feet above ground at sea level. The transmitted signals were chosen from those used in the BAF anechoic chamber tests (Section 2.2 above). Targets addressed were a reduced set: barrels, metal panels and trucks.

The antenna mounting and equipment installation in the Twin Otter air platform are shown in Fig. 2.3.0.1. A typical flight path over Huachuca Canyon and the type of forest cover are shown in Figs. 2.3.0.2 and 2.3.0.3, respectively.

The transmitted signals used in the flight tests were STX signals — bundles of DTX signals addressing the target resonances identified in the BAF anechoic chamber tests. MTX signals were not used in the flight tests because to obtain sufficient backscattered energy at relatively long range requires high transmit repetition rates, and the equipment available was not

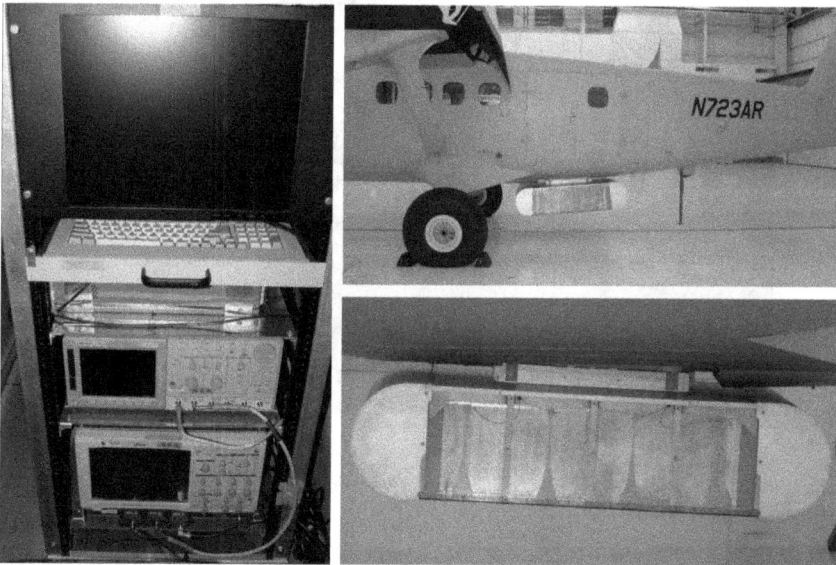

Fig. 2.3.0.1   Twin Otter installation of equipment and antenna. Fixed mount antenna, and two stacks of electronics.

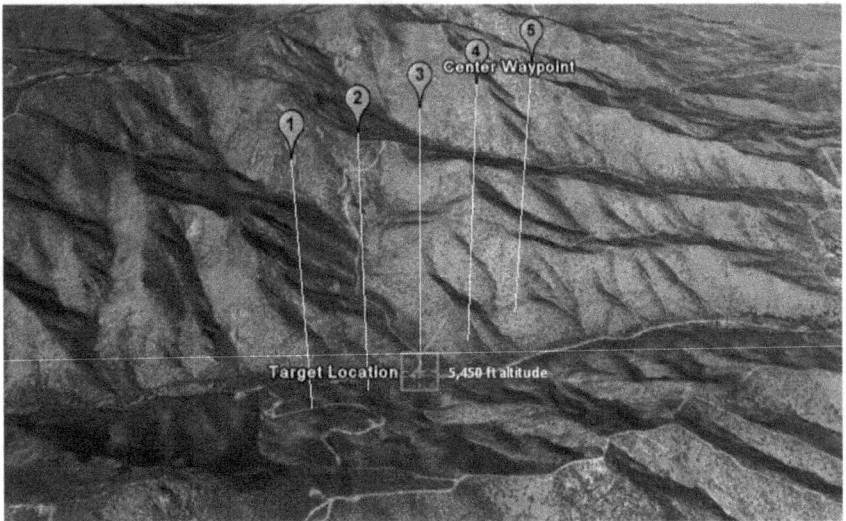

Fig. 2.3.0.2   Flight path over Huachuca Canyon, Arizona, flying at 100 knots.

(a)

Fig. 2.3.0.3   Forest cover, Huachuca Canyon. Ground view of covered barrels (a), roof panels (b); and trucks (c).

(b)

(c)

Fig. 2.3.0.3 (*Continued*)

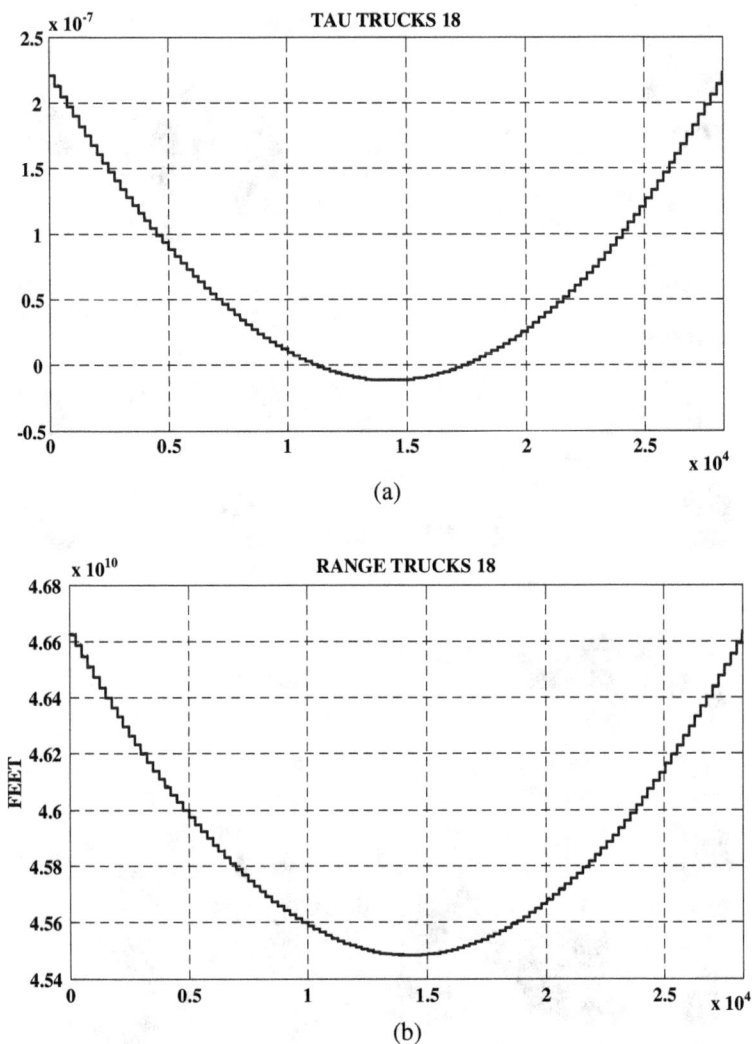

Fig. 2.3.0.4 (a) Representative calculated time, τ, compensation (in seconds) along the transmit ground footprint (in meters) with respect to the target reference point. (b) Representative calculated estimated range from target (feet $\times 10^7$).

capable of high repetition rate transmission. The STX signals' duration was 1.1 microseconds.

Motion compensation was applied to the collection of the RX signals for the ground footprint over the target. Figure 2.3.0.4 shows a representative calculation for time and range correction; and Fig. 2.3.0.5 shows representative flight course latitude, longitude, altitude, roll and pitch. The STX signal bundles used were constructed from the following frequency components (MHz):

$300 + 425$

$300 + 580$

$500 + 540$

$510 + 610$

$510 + 560 + 610$

$510 + 560 + 610 + 660$

$500 + 540 + 585 + 670$

$300 + 425 + 580 + 660$

These STX signals were transmitted under two flight conditions: with Target Present (TP) under forest canopy and Target Absent (TA); and

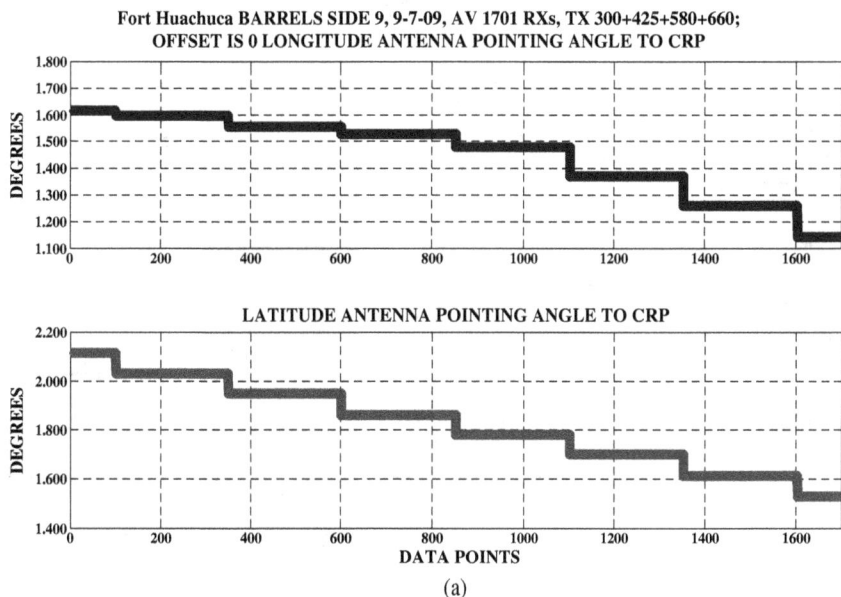

Fig. 2.3.0.5   Representative collecting flight over Huachuca Canyon, Arizona.
(a) Latitude and Longitude.
(b) Pitch and Roll.
(c) Ground boresight trajectory near target location (represented by green dot).
(d) Ground boresight (red) and airplatform (blue) trajectories. The corner reflector point (CRP) is at 0,0 and marked in green.

ALTITUDE: Fort Huachuca, PANELS13, 9-7-09, NO TARGET 16, 9-5-09, AV 100 RXs, TX 500+540+585+670

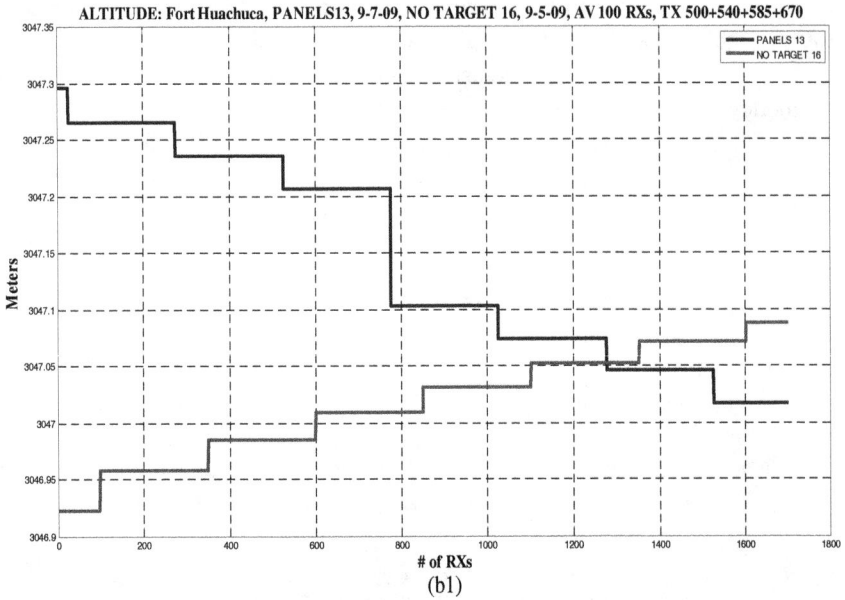

(b1)

PITCH & ROLL DIFFERENCES: Fort Huachuca, PANELS 11, 9-7-09,
NO TARGET 15, 9-5-09, AV 1700 RXs, TX510+560+610+660

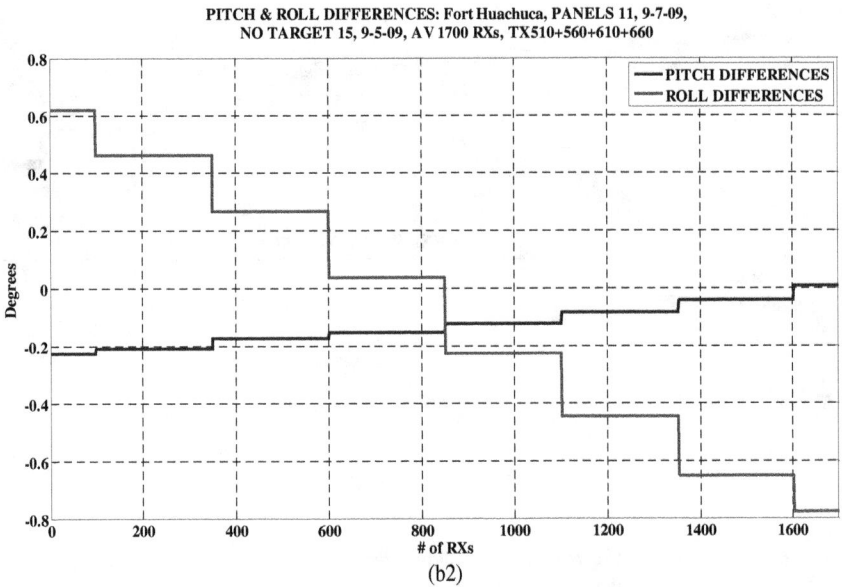

(b2)

Fig. 2.3.0.5   (*Continued*)

**Fort Huachuca, GROUND BORESIGHT BARRELS SIDE 8, 9-7-09,
AV 1701 RXs, TX 500+540+585+670**

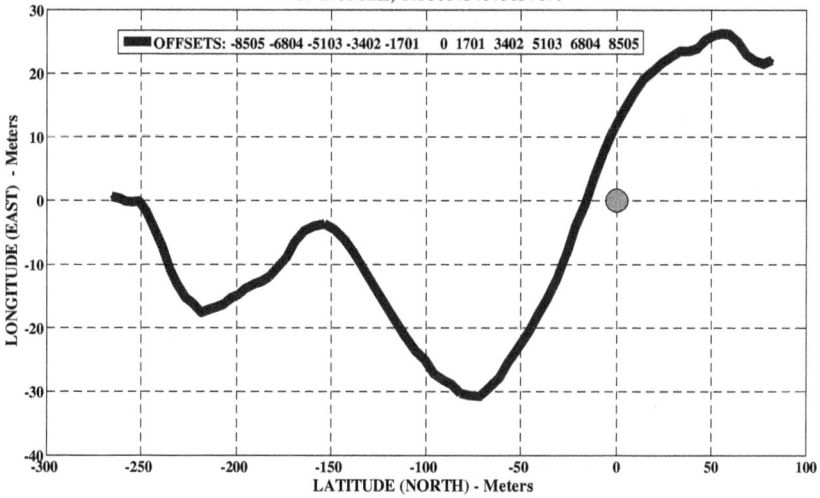

**Fort Huachuca, GROUND BORESIGHT NO TARGET 16, 9-5-09,
AV 1701 RXs, TX 500+540+585+670**

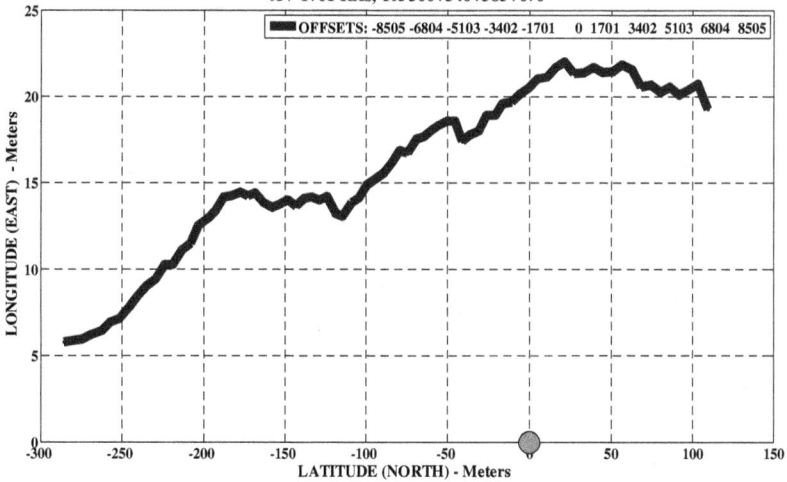

(c)

Fig. 2.3.0.5   (*Continued*)

Fort Huachuca, BARRELS SIDE 8, 9-7-09, AV 1701 RXs, TX 500+540+585+670

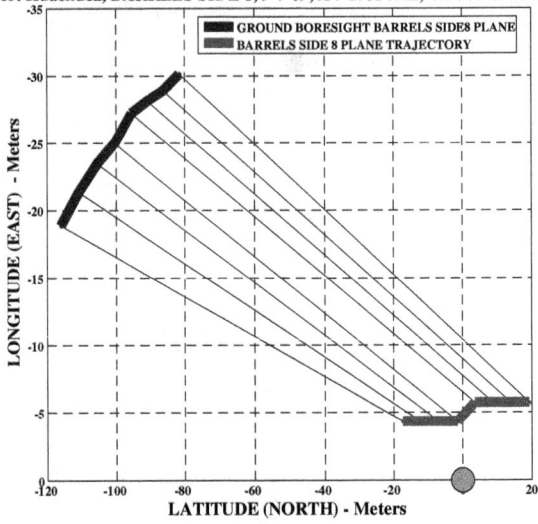

Fort Huachuca, NO TARGET 16, 9-5-09, AV 1701 RXs, TX 500+540+585+670

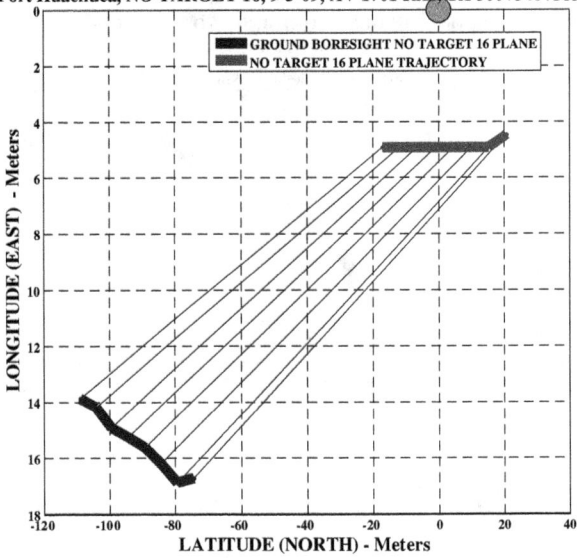

(d)

Fig. 2.3.0.5 (*Continued*)

amplitude comparisons were made between the corresponding RX signals collected under the two conditions. The targets were (1) Barrels laid sideways; (2) Panels; and (3) Trucks — all under forest canopy. The SRXs under the TP condition, or TP-SRXs, were compared with the SRXs under the TA condition, or TA-SRXs, to determine the degree of target detection in the presence of foliage and clutter.

2.3.1 Target Detection Under Foliage

Representative residual (TP-TA) fopen spectra of the averaged return signals (RX) in which Target Absent (TA) condition spectra were subtracted from the Target Present (TP) condition spectra. It should be noted that these RX data were collected with the transmit power at a very low level in order to prevent satiation of the receiver. Use of a more capable receiver will enhance the results. TP-SRX/TA-SRX ratios of 4 major frequency bands for each target are summarized in Table 2.3.2.1, in Section 2.3.2, below.

The protocol for the raw data treatment and motion compensation for all data collection flights was as follows:

(1) The RX signals were averaged over 1701 data points $(-850{:}+850)$ around an offset point. This resulted in 11 of these 1701 data point minor frames. When displayed together, an $11 \times 1701 = 18,711$ data point major frame was obtained.

(2) The altitude, the airplatform position, and the ground boresight trajectory were calculated from GPS data for the major frames — *cf.* Figs. 2.3.0.4–2.3.0.5 above.

(3) Longitude and latitude of the antenna pointing angles to CRP were calculated for the major frames — *cf.* Fig. 2.3.0.5(a), above.

(4) The spectra of the RX signals for each of 11 minor frames were calculated and the maximum value for each spectral band used to obtain the TP-SRX/TA-SRX ratios (Table 2.3.2.1, below).

The maximum was obtained for each band in the case of the data collection flight with target present (TP) and compared with the maximum for those data collection flights with the target absent (TA), to provide signal-to-clutter ratios (SCRs). The STX frequencies used for both the TP and TA conditions were the same, but the TP and the TA data collections were not made on the same flights and sometimes not on the same day — which introduced an unknown variability into the ratio calculations.

Figure 2.3.1.2 shows representative overlaid SRX spectra for target present and absent conditions under forest canopy; and for the targets:

RESIDUAL SPECTRUM
[BARRELS SIDE FLT 9 – NO TARGET FLT 17], AV RXs, STX 300+425+580+660

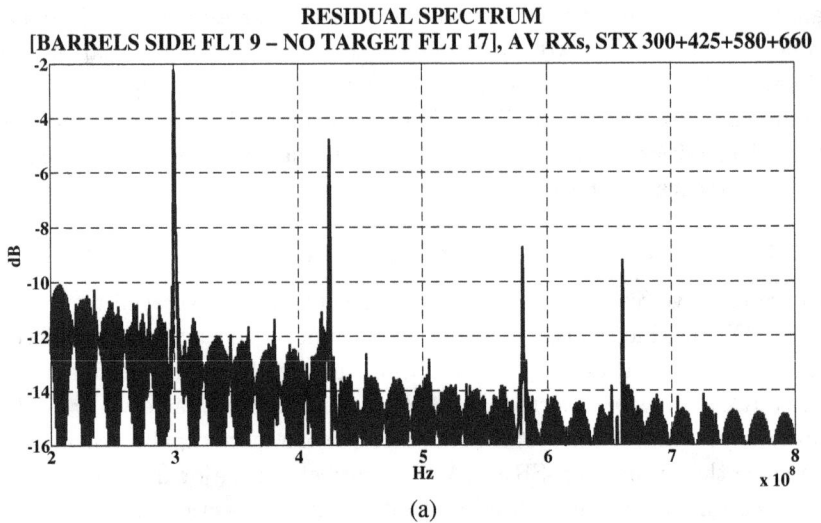

(a)

RESIDUAL SPECTRUM
[BARRELS SIDE FLT 8 – NO TARGET FLT 16], AV RXs, STX 500+540+585+670

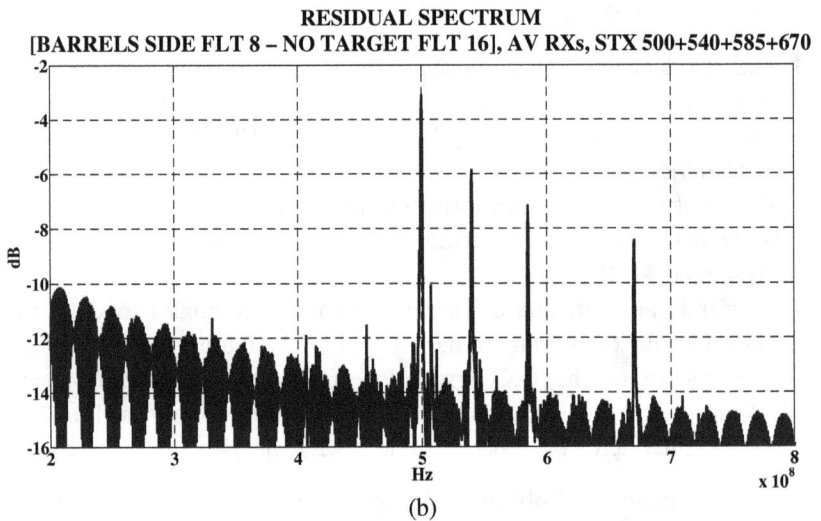

(b)

Fig. 2.3.1.1   Representative residual (TP-TA) fopen spectra of the averaged return signals (RX) in which Target Absent (TA) condition spectra were subtracted from the Target Present (TP) condition spectra. It should be noted that these RX data were collected with the transmit power at a very low level in order to prevent satiation of the receiver. Use of a more capable receiver will enhance the results.

RESIDUAL SPECTRUM
[BARRELS SIDE FLT 10 – NO TARGET FLT 15], AV RXs, STX 510+560+610+660

(c)

RESIDUAL SPECTRUM
[PANELS FLT 12 – NO TARGET FLT 17], AV RXs, STX 300+425+580+660

(d)

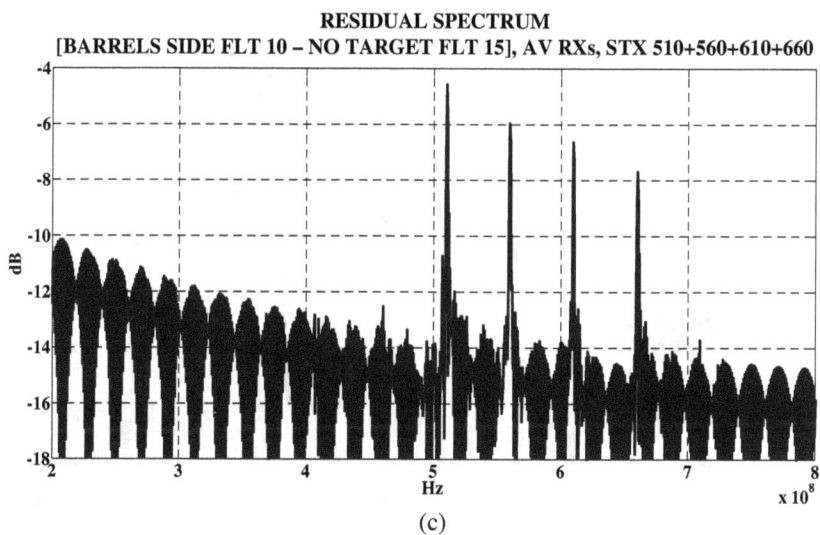

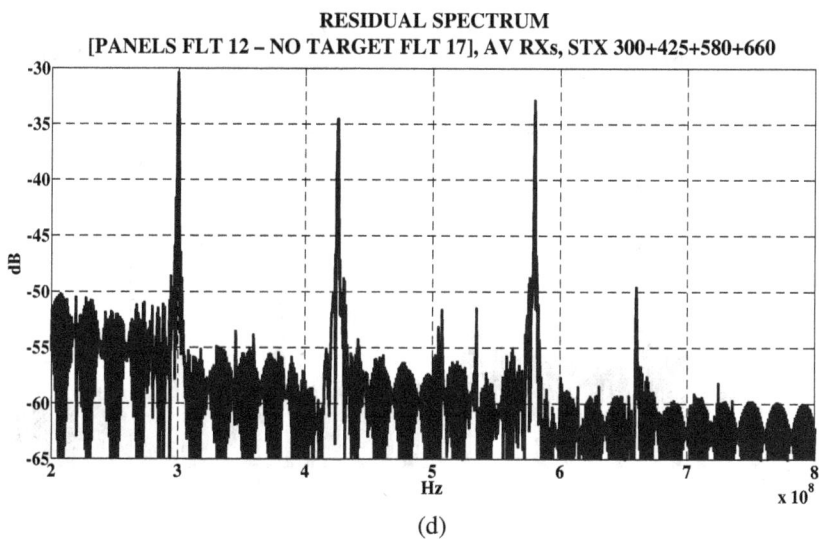

Fig. 2.3.1.1 (*Continued*)

RESIDUAL SPECTRUM
[TRUCKS FLT 16 – NO TARGET FLT 19], AV RXs, STX 300+580

(e)

RESIDUAL SPECTRUM
[TRUCKS FLT 18 – NO TARGET FLT 17], AV RXs, STX 300+425+580+660

(f)

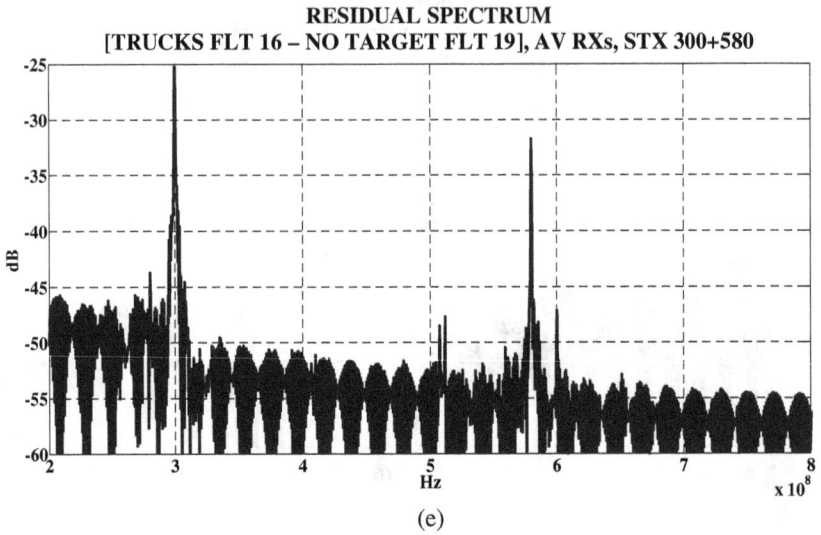

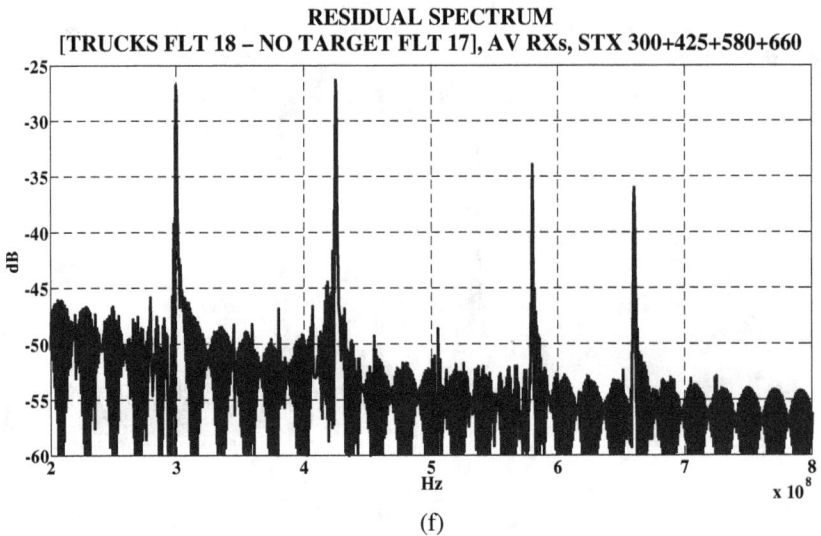

Fig. 2.3.1.1   (*Continued*)

BARRELS SIDE FLT 4 & NO TARGET FLT 16
Average 1700, STX 500+540+585+670

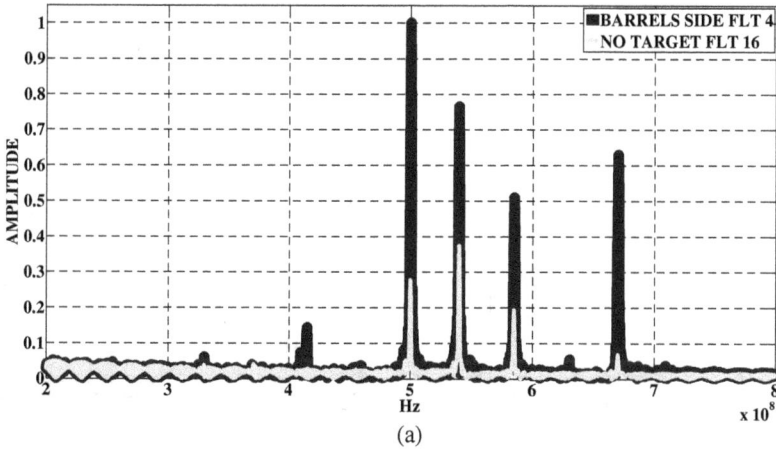

(a)

BARRELS SIDE FLT 4 & NO TARGET FLT 17
Average 1700, STX 300+425+580+660

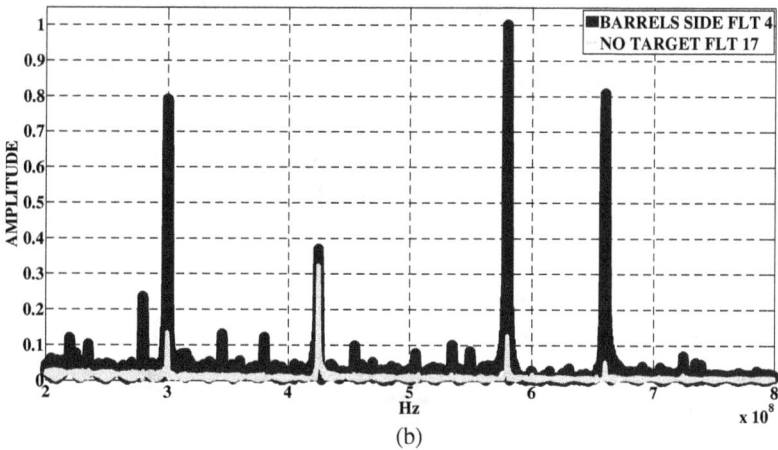

(b)

Fig. 2.3.1.2 Examples of SRX, TP & TA spectra overlaid, Huachuca Canyon. Blue: target present (TP); Yellow: target absent (TA).

(a): Target: Barrels Sideways. STX: 500 + 540 + 585 + 670 MHz.

(b): Target: Barrels Sideways. STX: 300 + 425 + 580 + 660 MHz.

(c): Target: Roof Panels. STX: 510 + 560 + 610 + 660 MHz.

(d): Target: Roof Panels. STX: 500 + 540 + 585 + 670 MHz.

(e): Target: Trucks. STX: 300 + 425 + 580 + 660 MHz.

(f): Target: Trucks. STX: 300 + 425 + 580 + 660 MHz.

Again, it should be noted that these RX data were collected with the transmit power at a very low level in order to prevent satiation of the receiver. Use of a more capable receiver will enhance the results.

PANELS FLT 11 & NO TARGET FLT 15
Average 1700, STX 510+560+610+660

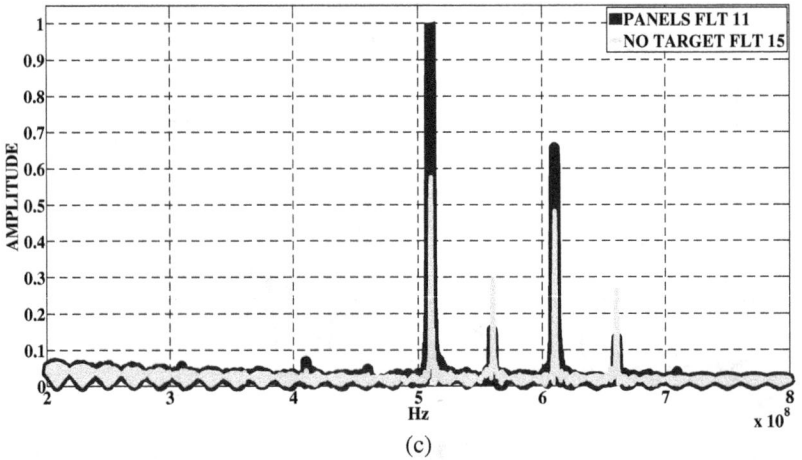

(c)

PANELS FLT 13 & NO TARGET FLT 16
Average 100, STX 500+540+585+670

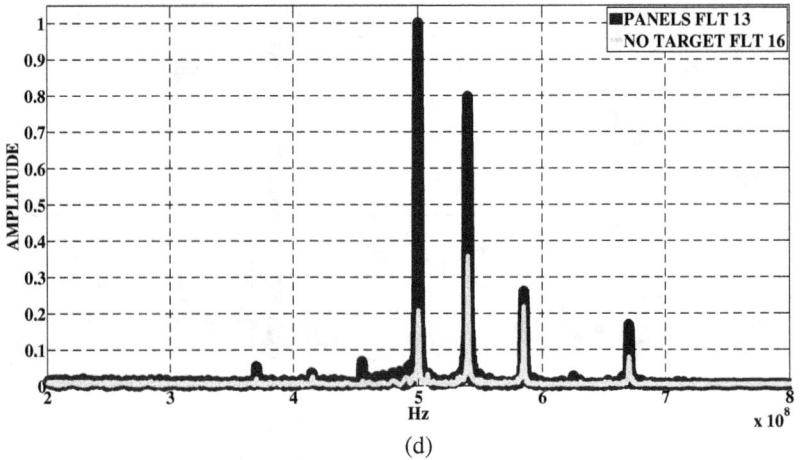

(d)

Fig. 2.3.1.2   (*Continued*)

TRUCKS FLT 9 & NO TARGET FLT 17
Average 1700, STX 300+425+580+660

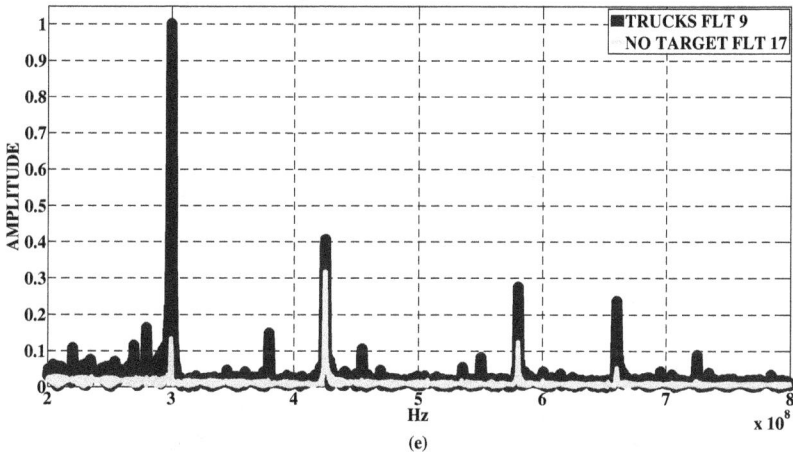

(e)

TRUCKS FLT 18 & NO TARGET FLT 17
Average 1700, STX 300+425+580+660

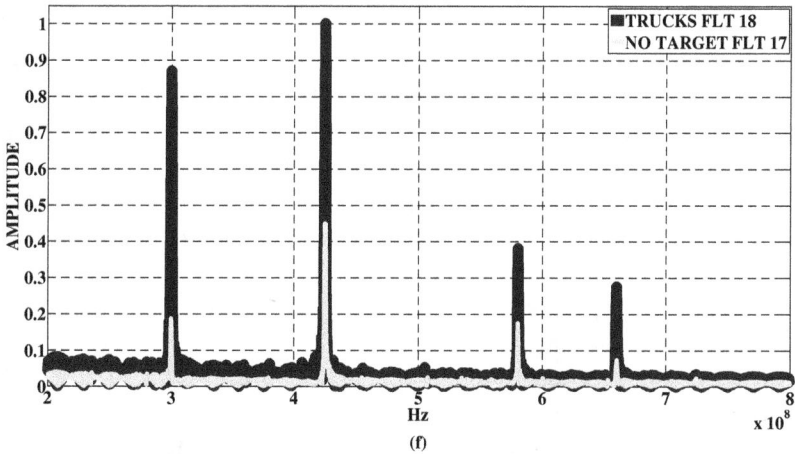

(f)

Fig. 2.3.1.2 (*Continued*)

barrels, panels and trucks. It can be seen that the presence of the targets results in increased amplitude in RX frequency bands — more for some bands than for others. Again, it should be noted that these RX data were collected with the transmit power at a very low level in order to prevent satiation of the receiver. Use of a more capable receiver will enhance the results.

2.3.2 Comparisons of the Results of the Anechoic Chamber Tests and the Flight Tests

The STX signals (i.e., bundles of DTX signals) used in the airborne tests were constructed from the following DTX components (MHz):

300 + 425
300 + 580
500 + 540
510 + 610
510 + 560 + 610
510 + 560 + 610 + 660
500 + 540 + 585 + 670
300 + 425 + 580 + 660

As stated above, these STX signals were transmitted on data collection flights under two TX conditions: with Target Present (TP) and with Target Absent (TA). RX signal amplitude comparisons between the two conditions were then made. The targets used were (1) Barrels Side; (2) Roof Panels; and (3) Trucks — all under forest canopy.

Examples of the TP-SRX (SRX, target present) and TA-SRX (SRX, target absent) spectra are shown in the previous Section 2.3.1. From these spectra TP-SRX/TA-SRX ratios of 4 major frequency bands for each of the three target classes were calculated. Table 2.3.2.1 is a summary of these ratios. As the ground footprint of the transmitted beam was large, the (target) signal-to-clutter ratio was expected to be positive, but not exceptionally high, and the results reflect this expectancy. Furthermore, these results are representative only of the specific data collection methods and equipment used on the flights. The RX data were collected with the transmit power at a very low level in order to prevent satiation of the receiver. Use of a more capable receiver will enhance the results. But above all: with a more focused beam, the expectancy is that the resonance enhancements and SCRs will be higher. This observation is addressed further in Section 2.4.0.

The major differences between the Huachuca Canyon tests and the Ground Tests through Foliage tests (Section 2.1.0) are the comparative criteria. In the case of Ground Tests through Foliage, apart from the demonstration of foliage penetration, MTX pulses were matched against PTX pulses, i.e., the criterion was MAP enhancement of SNR. In the case of the Huachuca Canyon tests, the criterion was SCR.

Table 2.3.2.1 Summary: Huachuca Canyon airborne tests (TP-SRX/TA-SRX§ ratios): 4 major bands per target.

Resonances (MHz)	Enhancement (dB)	F-TP§§	F-TA§§
(A) Target: Panels			
500	6.25	9/7/09	9/5/09
560	2.01	9/7/09	9/5/09
585	5.63	9/7/09	9/5/09
610	2.66	9/7/09	9/5/09
(B) Target: Barrels (side)			
425	7.67	9/7/09	9/5/09
500	4.88	9/7/09	9/5/09
540	6.76	9/7/09	9/5/09
670	10.12	9/7/09	9/5/09
(C) Target: Trucks			
300	8.48	9/5/09	9/5/09
560	9.70	9/5/09	9/5/09
580	7.66	9/5/09	9/5/09
660	9.25	9/5/09	9/5/09

§ TP-SRX = Return Signal (SRX) with Target Present (TP);
TA-SRX = Return Signal (SRX) with Target Absent (TA)
§§ F-TP = Flight date, Target Present; F-FA = Flight date, Target Absent.

Comparing the data obtained at the BAF, Edwards AFB, and the data obtained on flight tests at Huachuca Canyon is difficult and must be tentative, one of the reasons being that whereas in the case of the Huachuca Canyon data collection, the target present (TP-SRX) returns and the target absent (TA-SRX) returns were both obtained in the presence of unknown clutter, clutter was minimized in the case the BAF data collection. Furthermore, the TP-SRXs and the TA-SRXs were collected on different flights over Huachuca Canyon, with slightly different flight paths, and in some instances, not on the same day. Therefore, the ground clutter may not be the same for the TP-SRXs and the TA-SRXs collections, giving an unknown inaccuracy to the TP-SRX/TA-SRX ratios.

Furthermore, the major contamination of the TP-SRX/TA-SRX ratios is due to the large ground footprint transmitted by the UHF antenna, working to lower these SCRs — (target) signature-to-clutter ratios. Nonetheless, despite the non-optimum conditions of data collection at Huachuca Canyon, the comparisons between the BAF spectral results, and the Huachuca Canyon RX spectral results (Table 2.3.2.1) are favorable, and shown in Figs. 2.3.2.1–2.3.2.3. In each case there is a correspondence between the

PRX BARREL SIDE

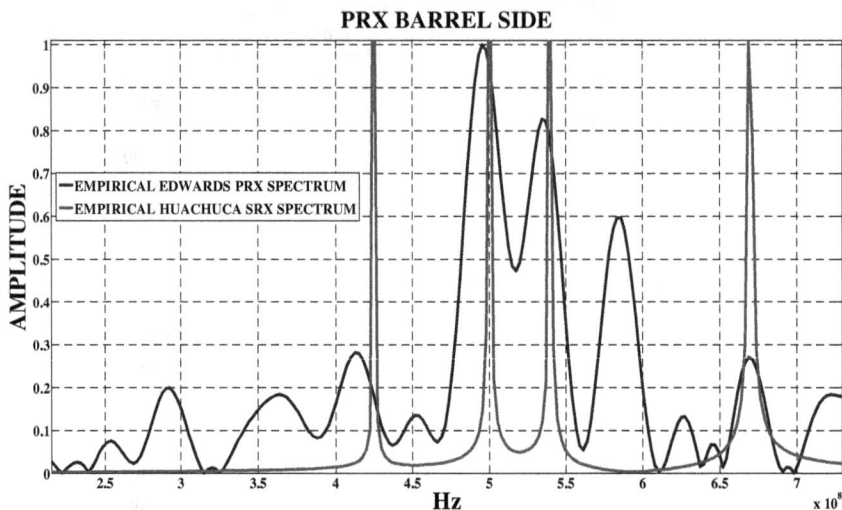

Fig. 2.3.2.1   BARRELS SIDEWAYS: Comparison of PRXs (BAF Edwards AFB) with STXs (Huachuca Canyon). The red spectral lines indicate the SRX bands of 425, 500, 540 and 670 MHz, indicated in Table 2.3.2.1, that provided an increase in TP-SRX/TA-SRX ratios obtained from the Twin Otter air platform. The spectrum is of averaged PRX data obtained at BAF, Edwards AFB.

PRX ROOF PANEL

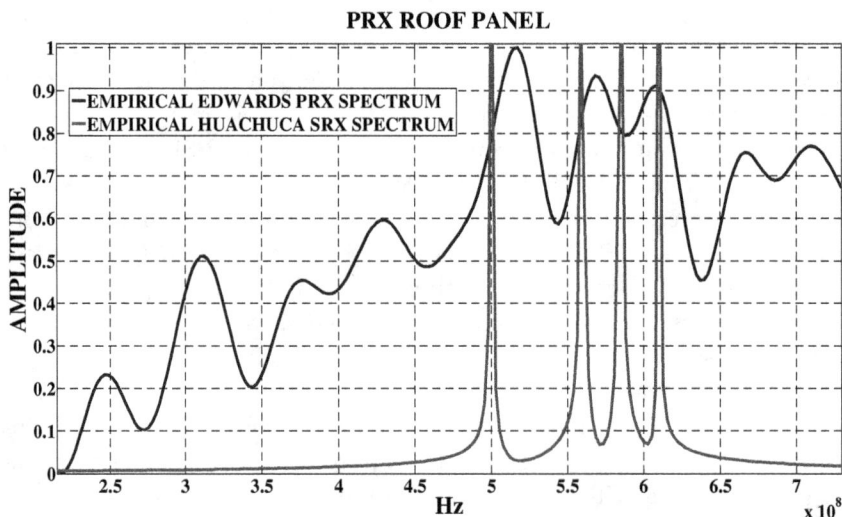

Fig. 2.3.2.2   ROOF PANELS: Comparison of PRXs (BAF Edwards AFB) with STXs (Huachuca Canyon). The red spectral lines indicate the SRX bands of 500, 560, 585 and 610 MHz, indicated in Table 2.3.2.1, that provided an increase in TP-SRX/TA-SRX ratios obtained from the Twin Otter air platform. The spectrum is of averaged PRX data obtained at BAF, Edwards AFB.

TRANSFER FUNCTION PRX TRUCK RX AVERAGE #1

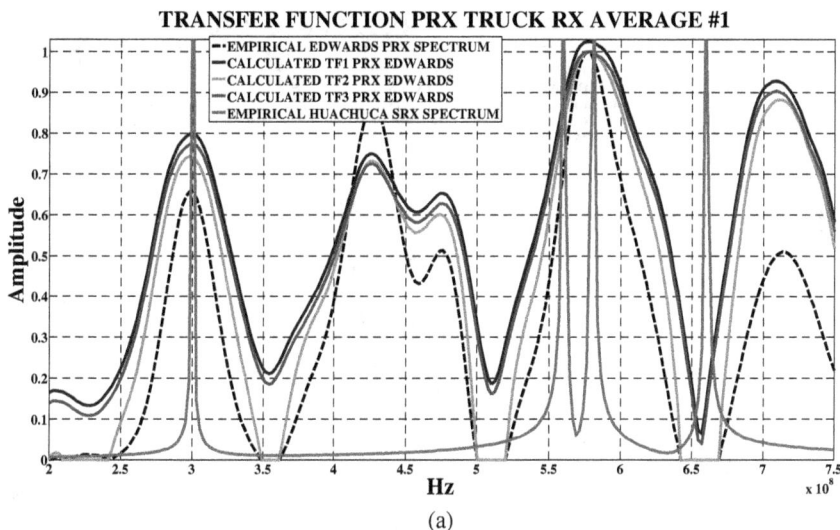

(a)

Fig. 2.3.2.3 TRUCK: Comparison of PRXs (BAF, Edwards AFB) with STXs (Fort Huachuca). Due to the relatively close range of the truck target at BAF and the relatively large target size, the truck target elicited three aspect-dependent PRXs after averaging (with TX polarization irrelevant) — see Section 3.6 above. The three PRX autospectra are shown in Figs. (a)–(c). In each of these, four spectra are shown, which are (1) the autospectra calculated from the PRX, and (2)–(4) three linear system response estimations:

TF1 = the linear system response estimation unbiased with respect to the presence of output noise,
TF2 = the linear system response estimation unbiased with respect to the presence of input noise,
TF3 = the linear system response (total least squares estimator) unbiased with respect to the presence of both input and output noise.
As TF1 = TF2, TF1 is not visible due to plotting overlap.

The red spectral lines indicate the SRX bands of 300, 560, 580 and 660 MHz, (Table 2.3.2.1) that provided an increase in TP-SRX/TA-SRX ratios obtained from the Twin Otter air platform. Figure (d) summarizes the finding that these 4 bands correspond to major peaks in one or more of the three truck PRX spectra.

Huachuca Canyon SRX major spectral bands and bands in the BAF PRX spectra.

In Figs. 2.3.2.1–2.3.2.2 are shown comparisons of PRXs for Barrel Side and Roof Panels (BAF) with STXs (Huachuca Canyon). The SRX spectral bands for each target (Table 2.3.2.1) provided an increase in (TP-SRX/TA-SRX or SCR) ratios detected from the air platform.

TRANSFER FUNCTION PRX TRUCK RX AVERAGE #2

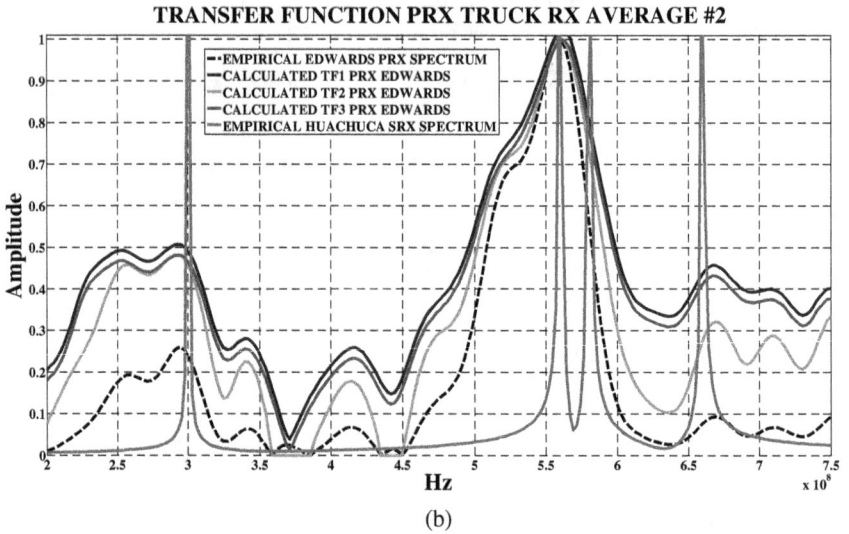

(b)

TRANSFER FUNCTION PRX TRUCK RX AVERAGE #3

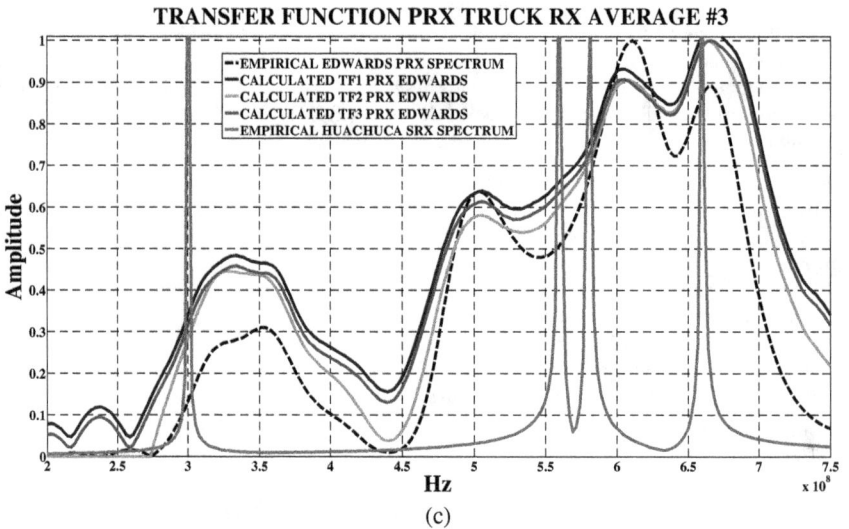

(c)

Fig. 2.3.2.3 (*Continued*)

Similarly, in Fig. 2.3.2.3 are shown comparisons of PRXs for the Truck PRXs (BAF), with STXs (Huachuca Canyon). As noted before, due to the relatively close range of the truck target at BAF and the target's relatively large size, the truck target elicited three aspect-dependent

TRUCK RX AVERAGES #1 #2 & #3

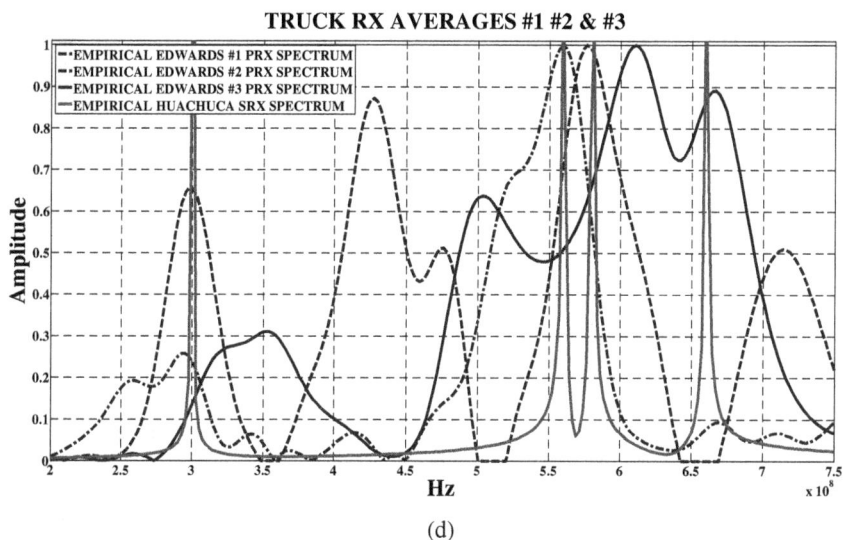

(d)

Fig. 2.3.2.3 (*Continued*)

PRXs after averaging — see Section 2.2.6 above. In each of the (a), (b) and (c) parts of Fig. 2.3.2.3, four spectra are shown, which are the spectra calculated from the PRX, and three linear system response estimations:

TF1 = the linear system response estimation unbiased with respect to the presence of output noise,

TF2 = the linear system response estimation unbiased with respect to the presence of input noise,

TF3 = the linear system response (total least squares estimator) unbiased with respect to the presence of both input and output noise.

The calculations showed that TF1 = TF2, so TF1 is not visible in these figures due to complete spectral overlap.

The Truck target SRX bands of 300, 560, 580 and 660 MHz, (Table 2.3.2.1), provided an increase in (TP-SRX/TA-SRX or SCR) ratios obtained from the air platform; and Fig. 2.3.2.3(d) summarizes the finding that these 4 SRX bands correspond to major peaks in one or more of the three truck PRX spectra.

2.4.0 Summary and System Improvements

The ground tests at BAF Edwards AFB demonstrated that, if a target's features and subcomponents are oriented toward the receive antenna (line-of-sight), then the target radar cross-sections of the whole target, as well as its features and subcomponents are

- frequency-dependent, and
- with resonance characteristics, that are:
- aspect-independent with respect to the location in the spectrum, but aspect-dependent with respect to the amplitude of those resonances;
- aspect-independent with respect to transmit signal polarization — given omni-polarized reception; and
- aspect-dependent with respect to time of arrival at the receiver of returning signal packets.

The limited airplatform tests over Huachuca Canyon demonstrated that a UHF MAP FOPEN system, even one that was not optimized with respect to ground footprint — *cf.* Fig. 2.4.0.1 — or antenna beamwidth, still substantially improves the target signal-to-clutter ratio (SCR) — *cf.* Table 2.3.2.1 above.

The SCRs can be further improved, as the flight tests, equipment used and data collection conditions were not optimum. The separate flight tests in which collection of (i) the RXs — target present (TP) — and (ii) the RXs — target absent (TA) — were not always made on the same day, nor were the exact same flight paths taken. Also, the ground footprint of the

ANTENNA FOOTPRINT

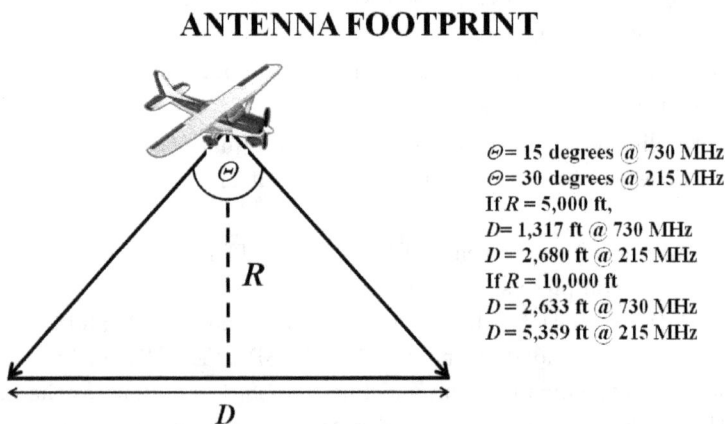

Θ = 15 degrees @ 730 MHz
Θ = 30 degrees @ 215 MHz
If R = 5,000 ft,
D = 1,317 ft @ 730 MHz
D = 2,680 ft @ 215 MHz
If R = 10,000 ft
D = 2,633 ft @ 730 MHz
D = 5,359 ft @ 215 MHz

Fig. 2.4.0.1   The beam ground footprint extent at 5,000 and 10,000 feet.

ANTENNA FOOTPRINT

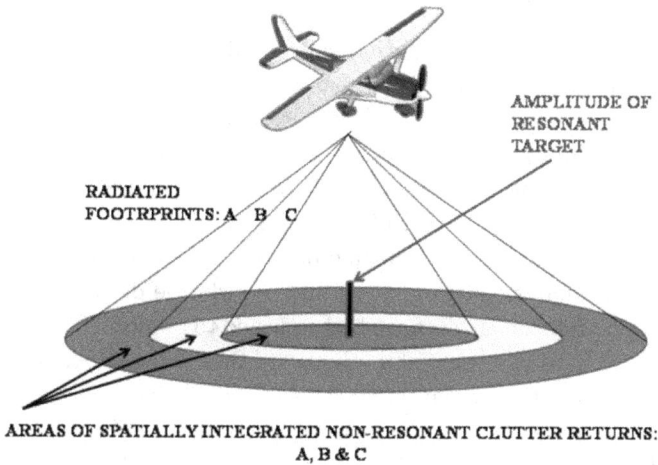

AMPLITUDE OF
RESONANT
TARGET

RADIATED
FOOTRPRINTS: A B C

AREAS OF SPATIALLY INTEGRATED NON-RESONANT CLUTTER RETURNS:
A, B & C

Fig. 2.4.0.2   The relative correspondence of the clutter returns within the ground footprint and the target return.

antenna was large; and while the averaging of the target signal return was compensated for phase and range (*cf.* Fig. 2.3.0.4–2.3.0.5), and the clutter was averaged out-of-phase, it is yet still likely that — due to the relatively long range — the out-of-phase averaging did not maximally reduce the clutter (Fig. 2.4.0.2). Therefore, it is probable that the long range and the size of the footprint acted to increase clutter, reducing the SCRs.

As indicated in Section 2.2.9, a Carrier Frequency-Envelope Frequency (CFEF) spectral estimate removes the difficulty that a comparison of time-frequency spectra requires that different RX signals arrive at the receiver at precisely the same time. Time-frequency spectra are only comparable in time if the time axis is aligned, which means that comparing different target RX time-frequency spectral signatures requires that the different targets must be at precisely the same distance from the transmitter. CFEF comparisons remove the time of returning signal arrival variability between targets, with the acceptable penalty that all time information is lost.

An improved and more capable next generation UHF MAP FOPEN system will require:

- A focused beam antenna with smaller beamwidth and ground footprint.
- An antenna gimbal for precision beam pointing.

- The means for repetitive transmission of MTXs. It should be noted that neither MTXs, nor PTXs were used in the airborne tests, because to elicit sufficient backscattered energy at long range using short duration pulses requires high transmit repetition rates, and the equipment required for high repetition rates was not available.
- The TX power on the test flights was very low in order to avoid overloading the receiver resulting in the "clipping" of the RX signal, or satiation of the receiver. Considerably more power can be used — given a more robust receiver of greater dynamic range.
- The means to implement real-time processing. The compensation, monitoring and analysis methods used in the airborne data collections were performed after the airplatform had returned to ground. The software-, as opposed to hardware-, based methods are time consuming; and therefore cannot provide the system operator with corrective information during flight. The fabrication of the compensation and monitoring algorithms, as well as data treatment algorithms, into dedicated electronic form, is straightforward, and would increase operator efficiency.

The collection of *a priori* information concerning target whole body resonances, target minor resonances, and class-of-target resonances can be carried out "on-the-fly" in real-time, but is addressed optimally, off-line, in a clutter-free environment. In the case of the UHF-Band system, an anechoic chamber was used for such collection. In the case of the Ka-Band system, both targets of real dimensions in a relatively clutter-free environment, as well as dimensionally-corrected model targets in the same environment, were used. The use of dimensionally-corrected targets to obtain *a priori* information of targets of normal size has a history of success (*cf.* Cheville & Grischkowsky (1995), who used target at the 1/200th scale), and probably offers the most cost-effective way of obtaining such information in a relatively small, easily controlled laboratory environment.

APPENDIX

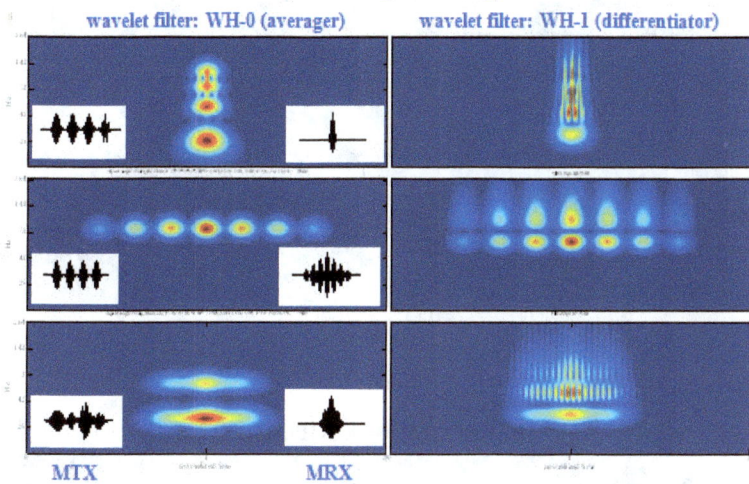

Aspect Angle : 00

wavelet filter: WH-0 (averager)   wavelet filter: WH-1 (differentiator)

MTX   MRX

MRXs

HUMVEE MRX; ASPECT 180; WH1 Q = 4

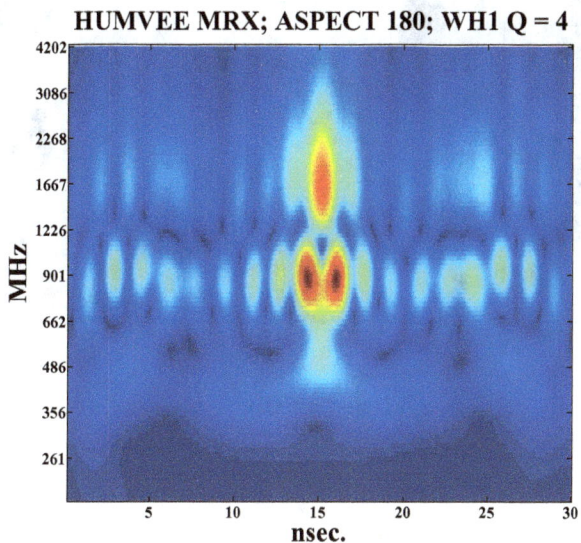

Figure 8A & 8C

BARREL UP MRX WH0:WH7 Q= 4

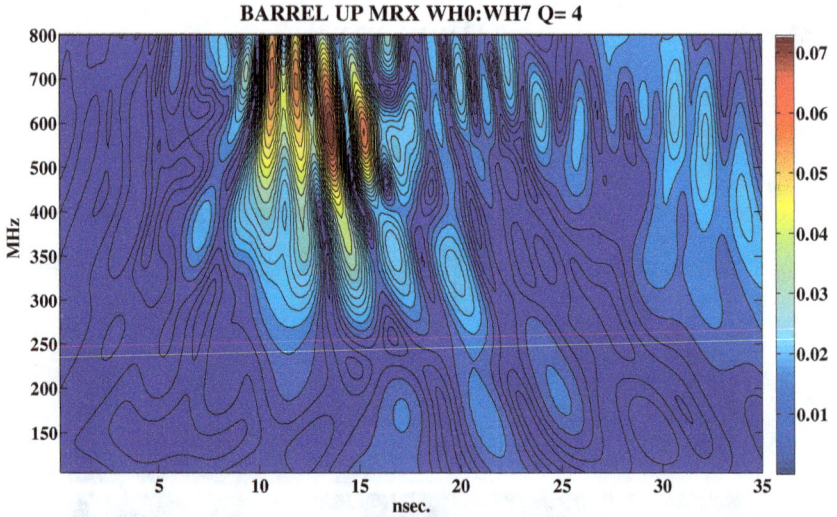

(A)

BARREL UP MRX WH0:WH7 Q = 4

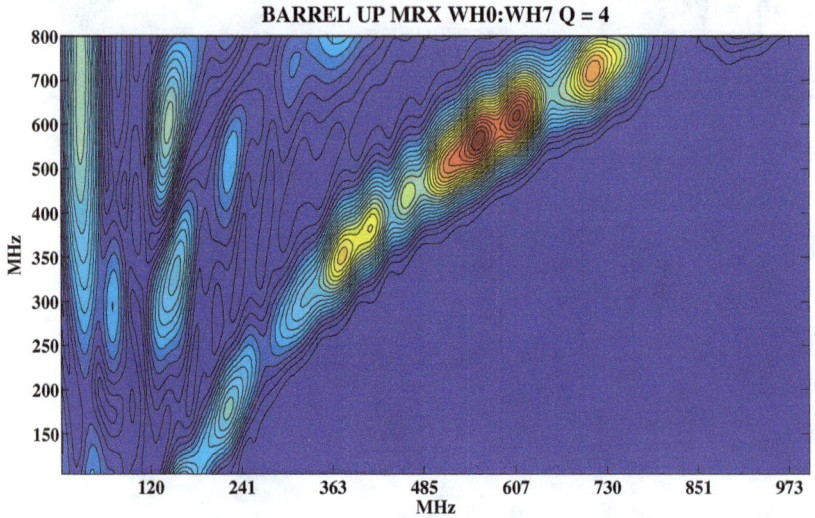

(B)

Figure 21

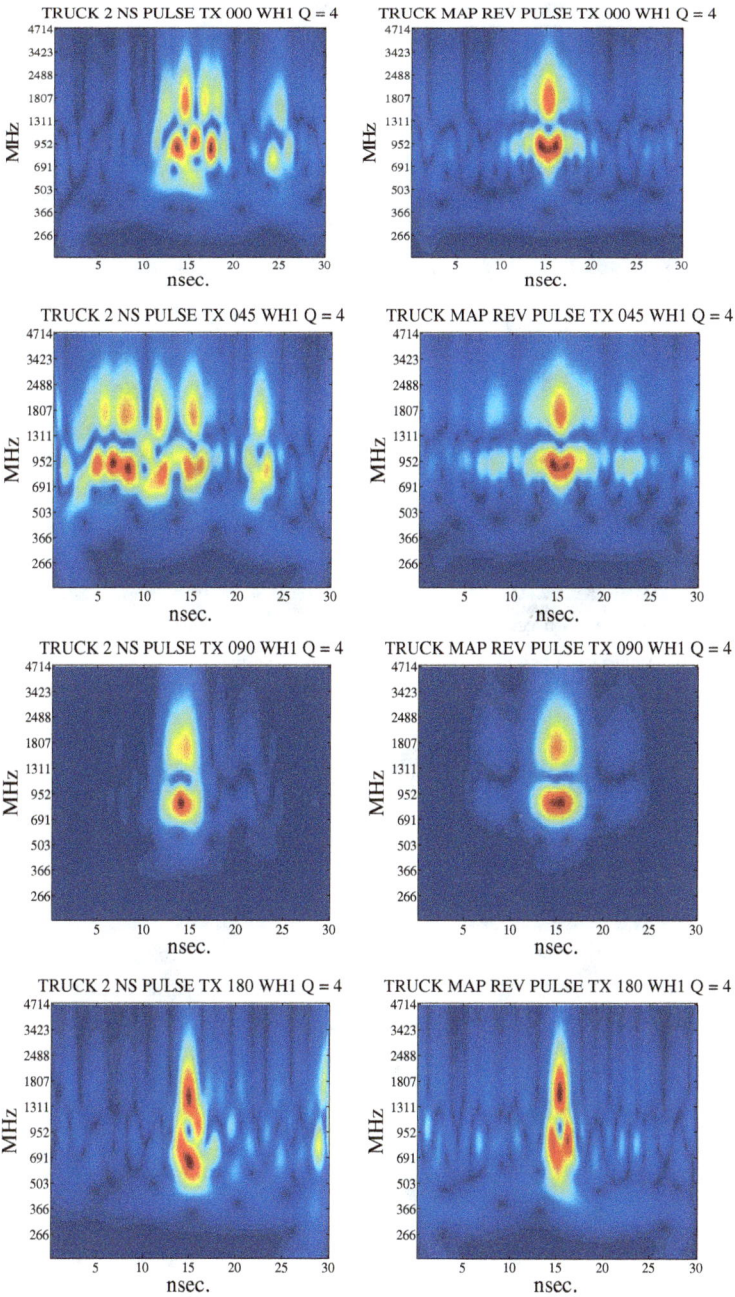

Figure 1.7.3

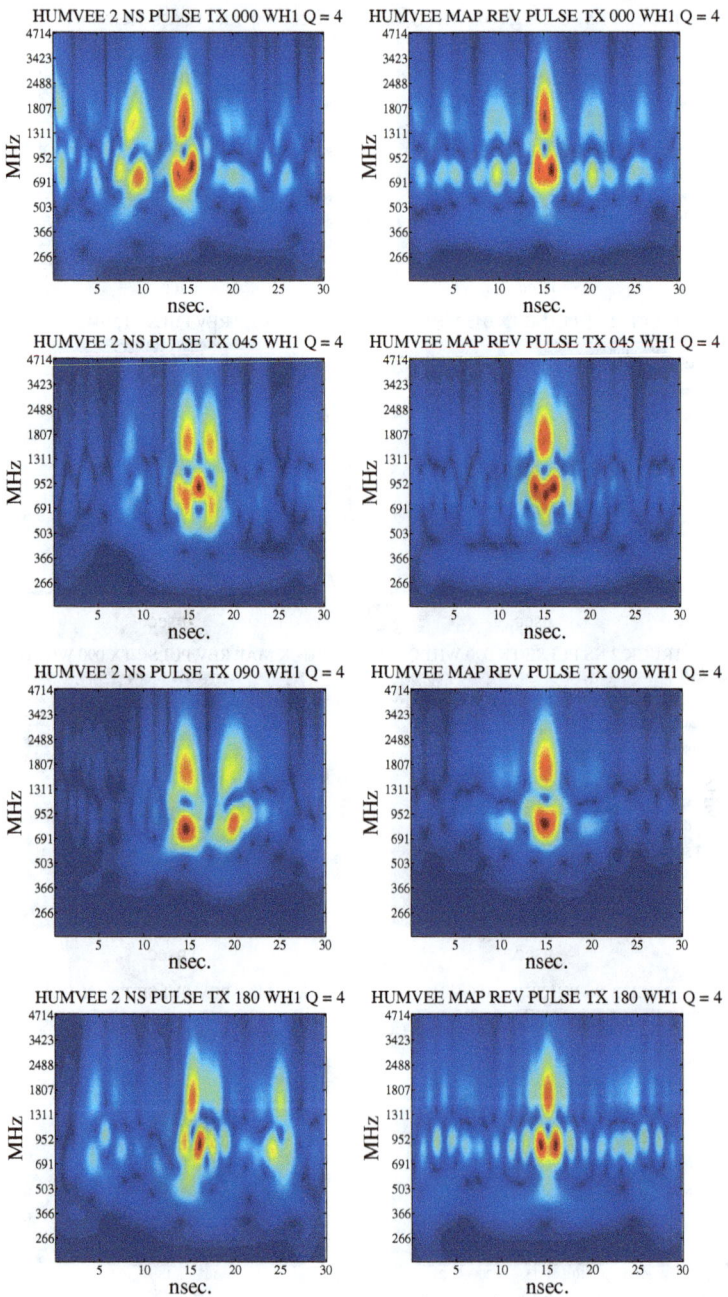

Figure 1.7.3   (*Continued*)

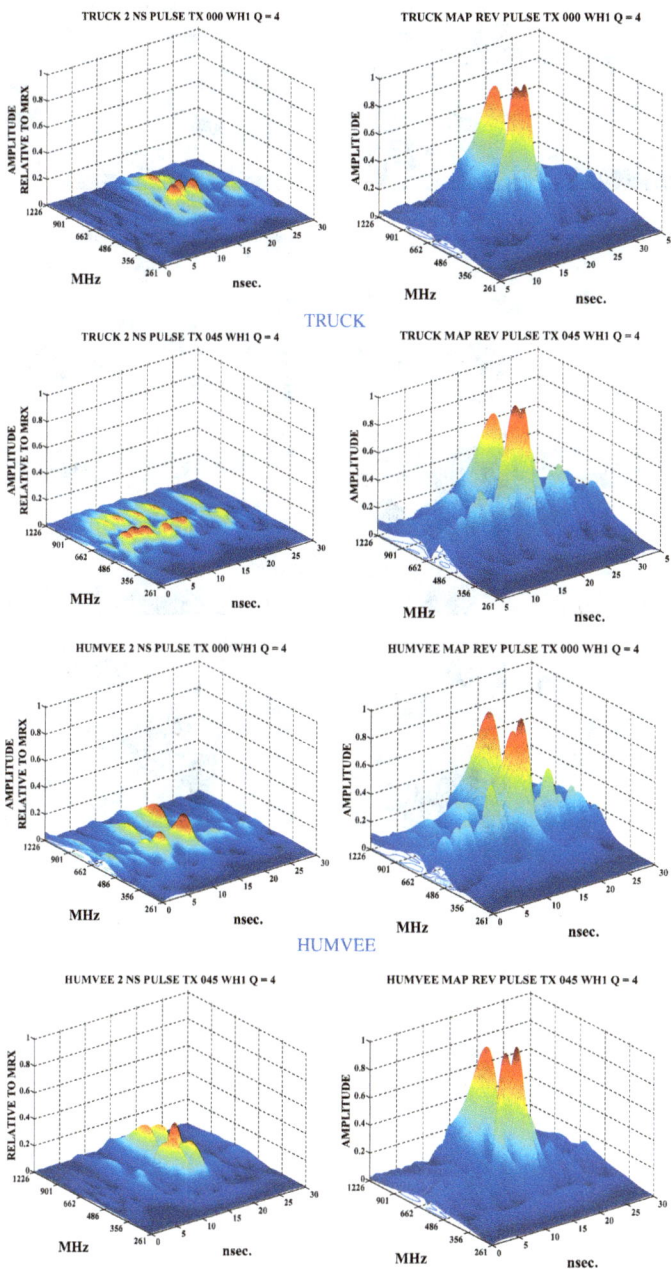

TRUCK 2 NS PULSE TX 000 WH1 Q = 4

TRUCK MAP REV PULSE TX 000 WH1 Q = 4

TRUCK

TRUCK 2 NS PULSE TX 045 WH1 Q = 4

TRUCK MAP REV PULSE TX 045 WH1 Q = 4

HUMVEE 2 NS PULSE TX 000 WH1 Q = 4

HUMVEE MAP REV PULSE TX 000 WH1 Q = 4

HUMVEE

HUMVEE 2 NS PULSE TX 045 WH1 Q = 4

HUMVEE MAP REV PULSE TX 045 WH1 Q = 4

Figure 1.7.4

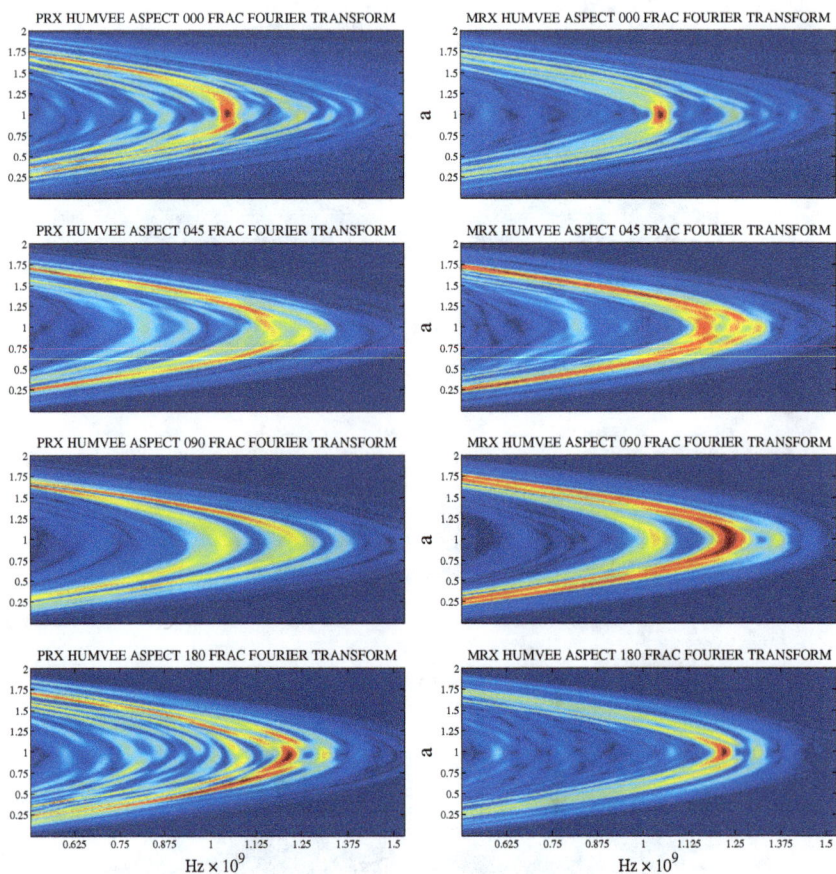

(b)

Figure 1.12.1B

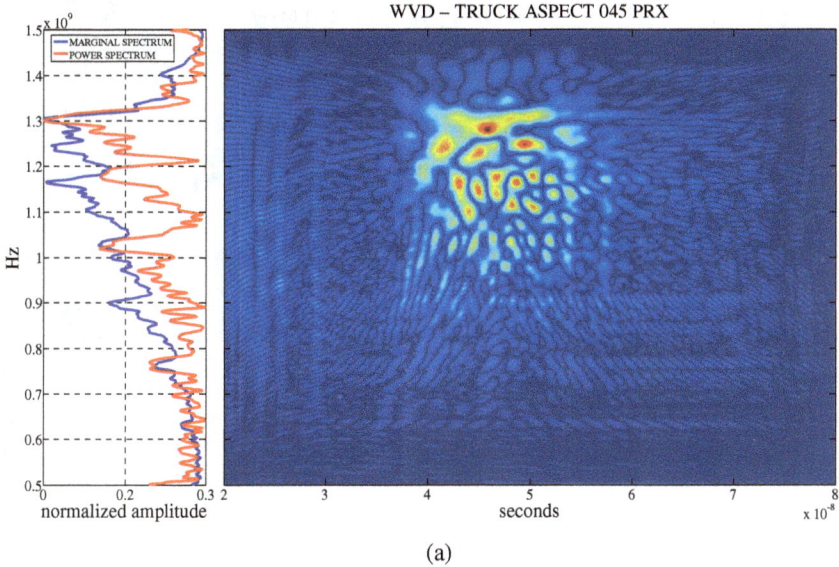

(a)

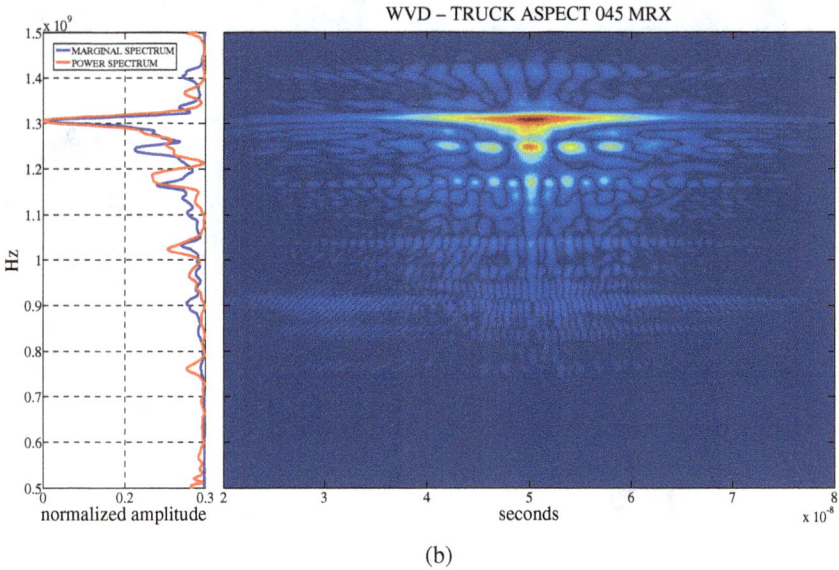

(b)

Figure 1.13.1.2A–B

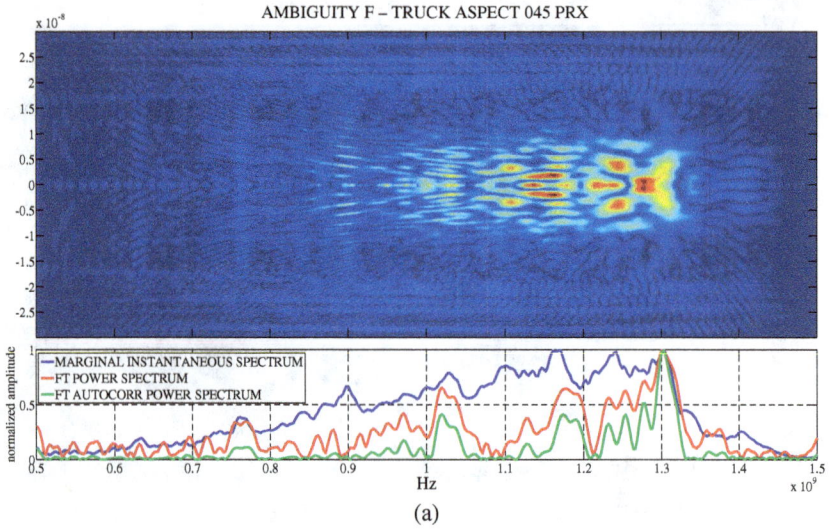

(a)

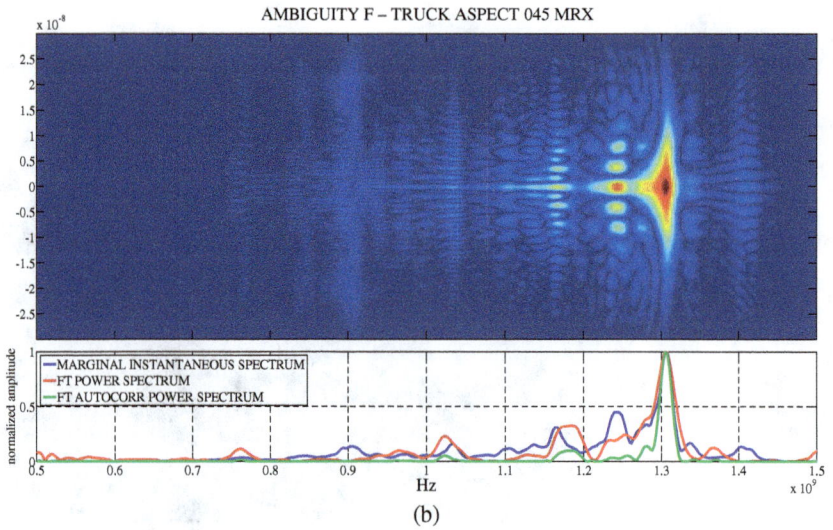

(b)

Figure 1.13.2.1(a)–(b)

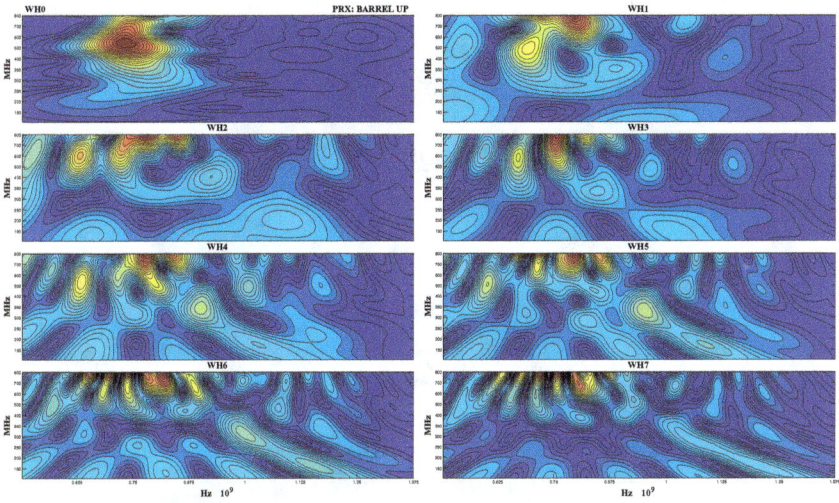

(a)

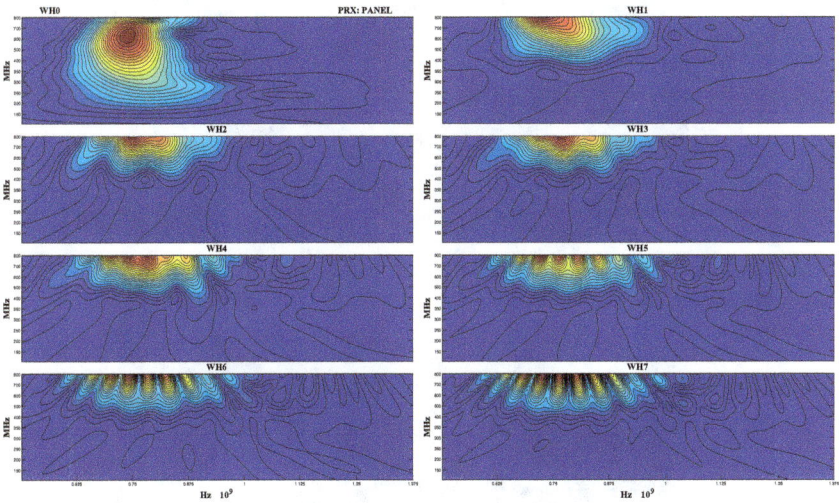

(b)

Figure 2.2.7.1(a)–(b)

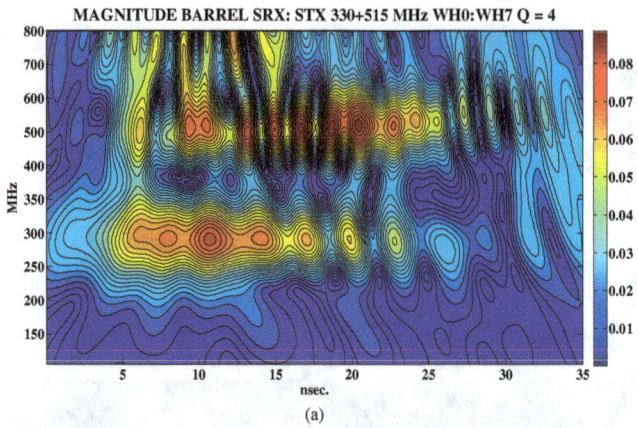

(a)

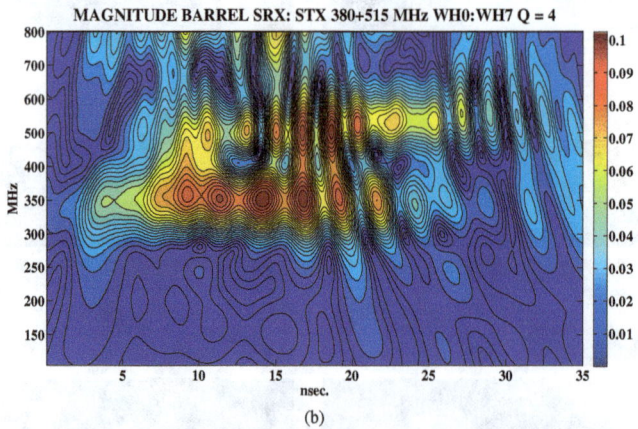

(b)

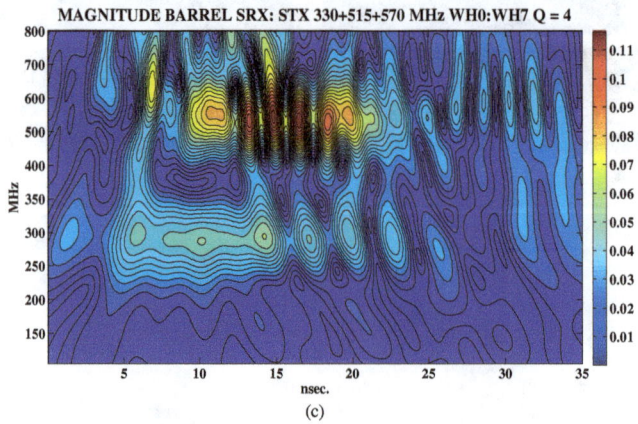

(c)

Figure 2.2.7.2(a)–(c)

MAGNITUDE BARREL SRX: STX 380+515+570 MHz WH0:WH7 Q = 4

(d)

MAGNITUDE BARREL SRX: STX 380+515+570+675 MHz WH0:WH7 Q = 4

(e)

MAGNITUDE BARREL SRX: STX 330+380+515+570+675 MHz WH0:WH7 Q = 4

(f)

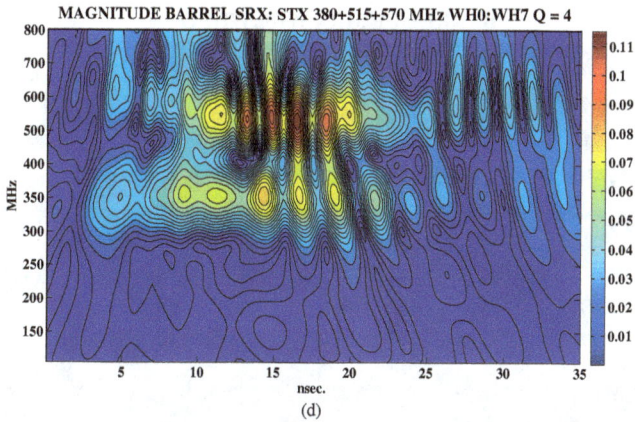

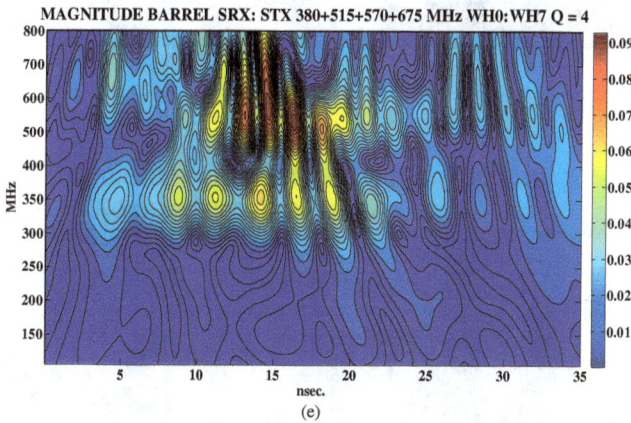

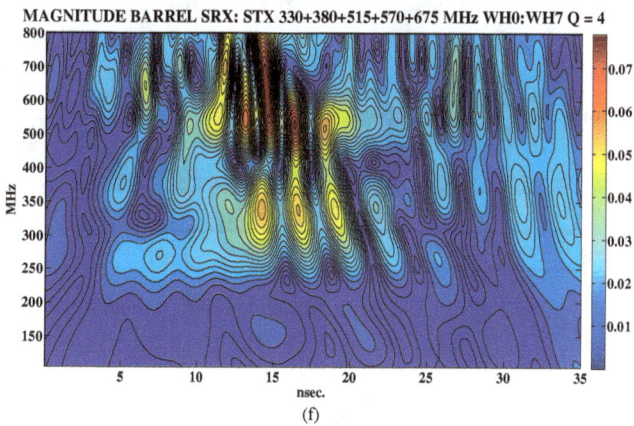

Figure 2.2.7.2(d)–(f)

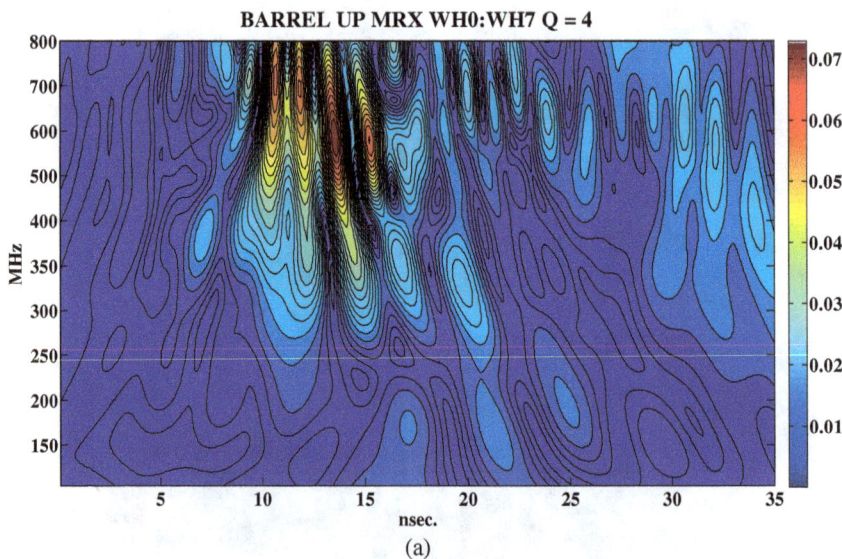

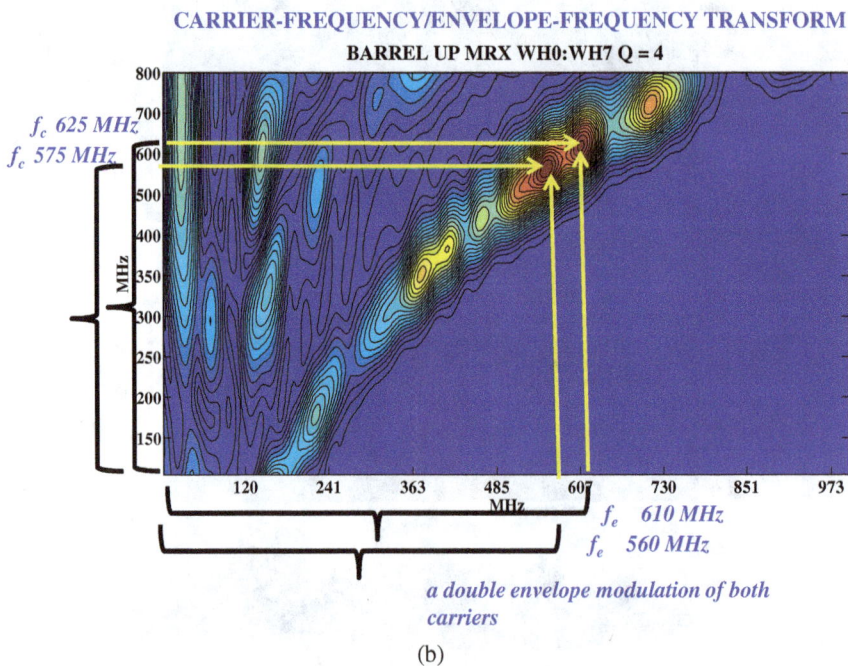

Figure 2.2.9.1(a)–(b)

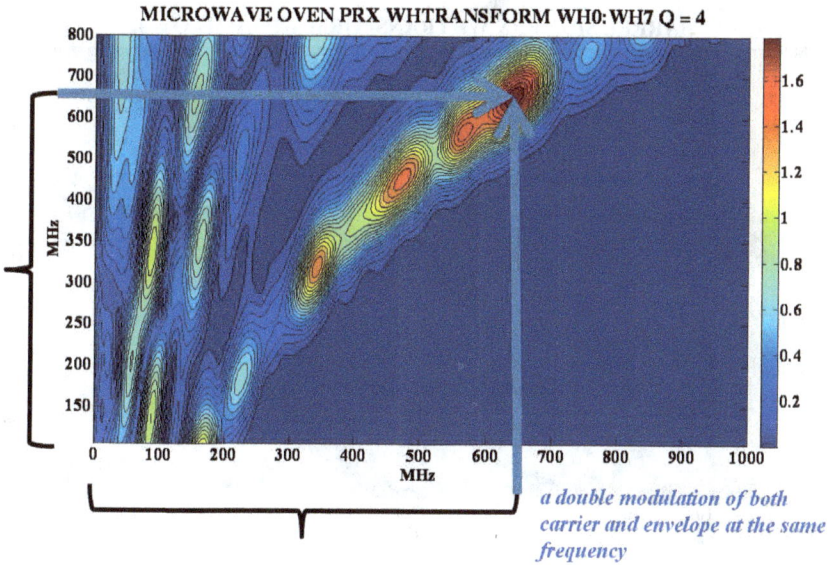

(a)

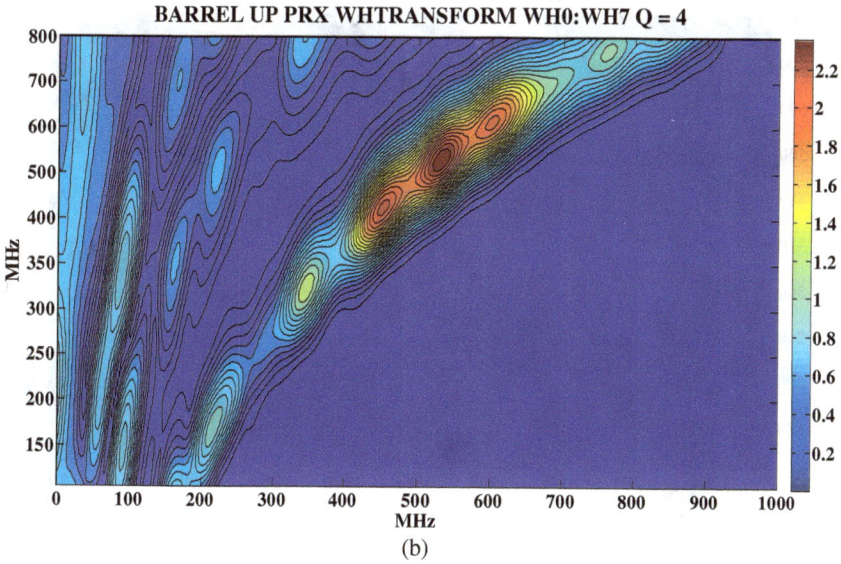

(b)

Figure 2.2.9.2(a)–(b)

BARREL SIDE PRX WHTRANSFORM WH0:WH7 Q = 4

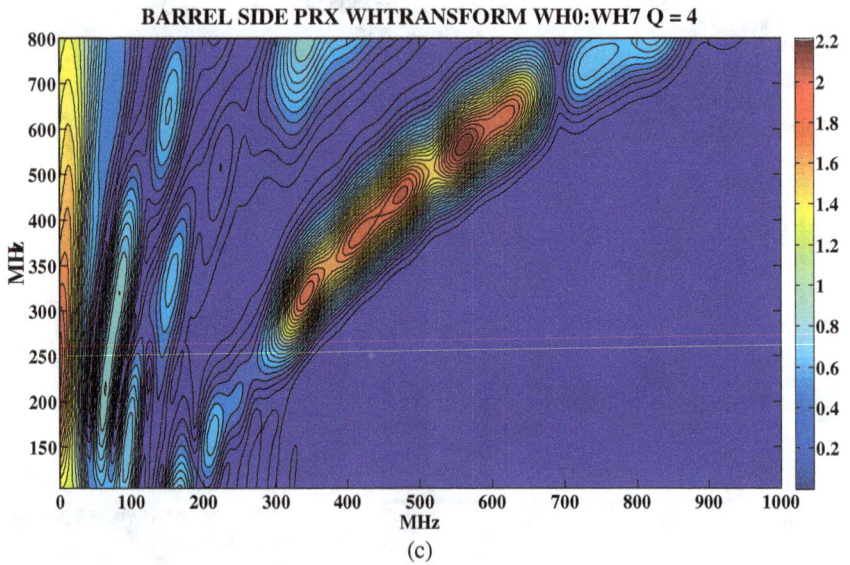

(c)

PANEL PRX WHTRANSFORM WH0:WH7 Q = 4

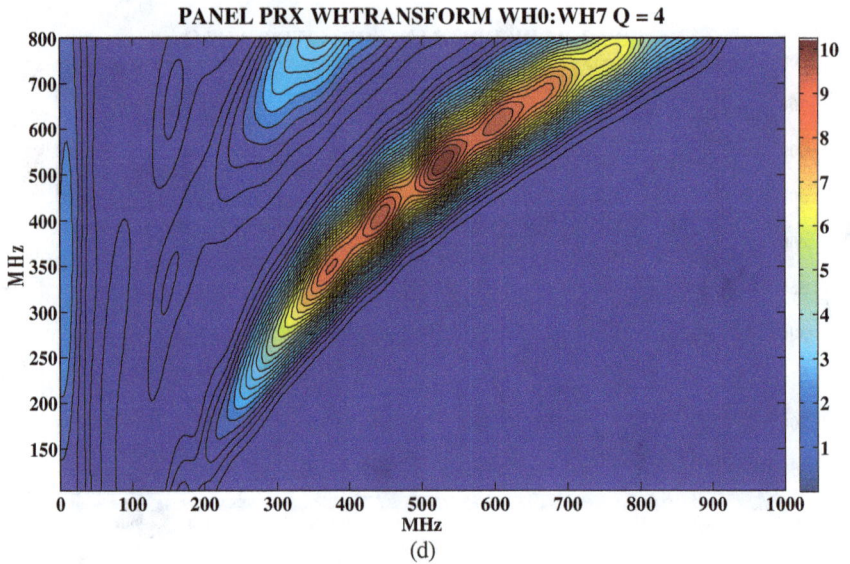

(d)

Figure 2.2.9.2(c)–(d)

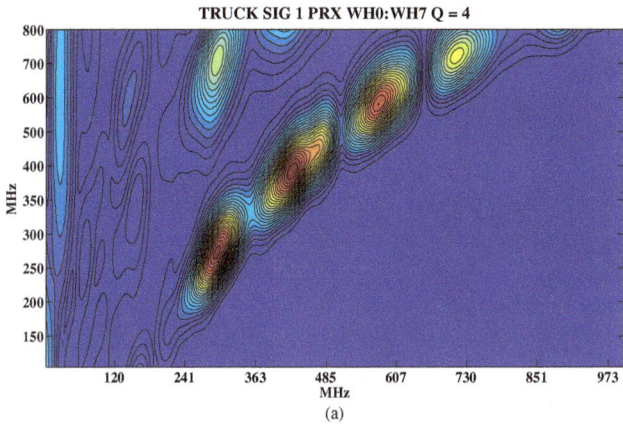

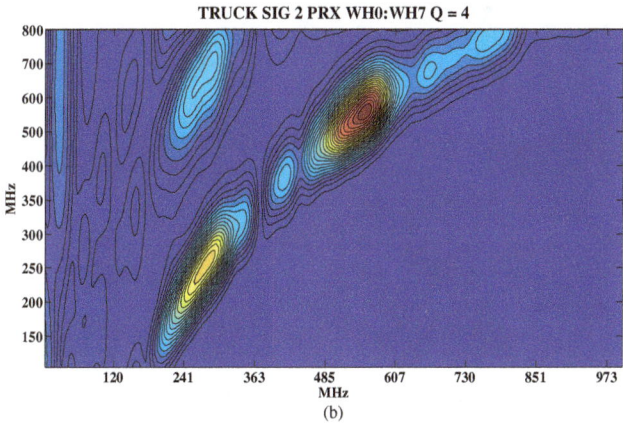

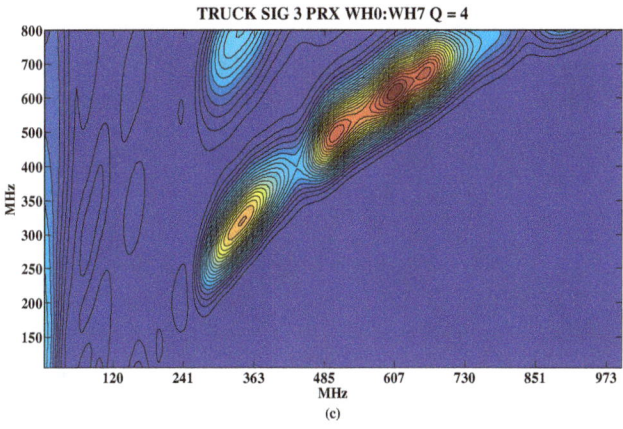

Figure 2.2.9.3(a)–(c)

REFERENCES

Abe, S. & Sheridan, J.T. (1994) Optical operations on wave functions as the Abelian subgroups of the special affine Fourier transformations, *Opt. Lett.*, 19, 1801–1803.

Abramowitz, M. & Stegun, C.A. (Eds.) (1972) Parabolic Cylinder Functions, Chapter 19 in *Handbook of Mathematical Functions with Formulas, Graphs and Mathematical Tables*, Dover, NY, pp. 685–700.

Almeida, L.B. (1994) The fractional fourier transform and time-frequency representations, *IEEE Trans. Signal Processing*, 42, 3084–3091.

Anderson, C.R., Bielawa, T., Davis, W.A., Licul, S., Noronha, J.A.N., Sweeney, D.J. (2003) Designing antennas for UWB systems: The Vivaldi antenna is an extremely broadband configuration that can be readily designed with modern CAD tools and fabricated with standard high-frequency substrate materials, *Microwave & RF Journal*, June 2003.

Apel, J.R. & Gjessig, A.T. (1989) Internal wave measurements in a Norwegian fjord using multifrequency radars, *Johns Hopkins APL Technical Digest*, 10(4), 295–306.

Ayasli, S., Dickens, T.A. & Fleischman, J.G. (1991) *Foliage Penetration Experiment with Three Frequency Synthetic Aperture Radars*, MIT Lincoln Laboratory Project Report STD-36, (23 July 1991).

Baez, J.C. (2002) The octonions, *Bull. American Math. Soc.*, 39, 145–205.

Baez, J.C. (2005) Review of: *On quaternions and octonions: Their geometry, arithmetic, and symmetry*, by John H. Conway and Derek K. Smith, A. K. Peters, Ltd., Natick, MA, 2003, *Bull. American Math. Society*, 42, 229–243.

Baez, J.C. & Huerta, J. (2011) The strangest numbers in string theory, *Scientific American*, 304, 60–65.

Barrett, H.H. (1984) The Radon transform and its applications. In *Progress in Optics XXI*, Chapter 3, pp. 217–286, Elsevier, Amsterdam.

Barrett, T.W. (1972) Conservation of information, *Acustica*, 27, 44–47.

Barrett, T.W. (1973a) Analytic information theory, *Acustica*, 29, 65–67.

Barrett, T.W. (1973b) Structural information theory, *J. Acoust. Soc. Am.*, 54, 1092–1098.

Barrett, T.W. (1975) Nonlinear analysis and structural information theory: A comparison of mathematical and physical derivations, *Acustica*, 33, 149–165.

Barrett, T.W. (1995a) Energy Transfer Through Media and Sensing of the Media, Chapter 7. Taylor, J.D. (ed) *Introduction to Ultra-Wideband Radar Systems*, CRC Press, Boca Raton, FL, pp. 365–434.

Barrett, T.W. (1995b) Performance Prediction and Modeling, Chapter 12. Taylor, J.D. (ed) *Introduction to Ultra-Wideband Radar Systems*, CRC Press, Boca Raton, FL, pp. 609–656.

Barrett, T.W. (1996) *Active Signalling Systems*, United States Patent No. 5,486,833, dated Jan. 23, 1996.

Barrett, T.W. (2008) *Method and application of applying filters to n-dimensional signals and images in signal projection space.* United States Patent No. 7,426,310, dated Sep. 16, 2008.

Bell, A.J. & Sejnowski, T.J. (1995) An information-maximization approach to blind separation and blind deconvolution, *Vision Research*, 37, 1129–1159.

Bell, M.R. (1988) Information theory and radar: Mutual information and the design and analysis of radar waveforms and systems. Ph.D. dissertation, California Institute of Technology, Pasadena, CA.

Bell, M.R. (1993) Information theory and radar waveform design, *IEEE Trans. Information Theory*, 39, 1578–1597.

Bendat, J.S. (1990) *Nonlinear System Analysis and Identification from Random Data*, Wiley, NY.

Bendat, J.S. & Piersol, A.G. (1993) *Engineering Applications of Correlation & Spectral Analysis*, 2nd edition, Wiley, NY.

Bendat, J.S. & Piersol, A.G. (2000) *Random Data Analysis and Measurement Procedures*, 3rd edition, Wiley, NY.

Bernardo, L.M. (1996) ABCD matrix formalism of fractional Fourier optics, *Opt. Eng.*, 35, 732–740.

Bienvenu, G. & Kopp, L. (1983) Optimality of high resolution array processing using the eigensystem approach, *IEEE Trans. Acoustics, Speech and Signal Processing*, 31, 1234–1248.

Born, M. & Wolf, E. (1999) *Principles of Optics*, 7th edition, Cambridge University Press.

Candan, C., Kutay, M.A. & Ozaktas, H.M. (2000) The discrete Fractional Fourier Transform, *IEEE Trans. Signal Processing*, 48, 1329–1337.

Cariolaro, G., Erseghe, T., Kraniauskas, P. & Laurenti, N. (1998) Unified framework for the Fractional Fourier Transform, *IEEE Trans. Signal Processing*, 46, 3206–3219.

Chambers, D.H., Candy, J.V., Lehman, S.K., Kallman, J.S., Poggio, A.J. & Meyer, A.W. (2004) Time reversal and the spatio-temporal matched filter, *J. Acoust. Soc. Am.*, 11, 1348–1350.

Chen, Q., Huang, N., Riemenschneider, S. & Xu, Y. (2006) A B-spline approach for empirical mode decomposition, *Advances in Computational Mathematics* 24, 171–195.

Cheville, R.A. and Grischkowsky, D. (1995) Time domain terahertz impulse ranging studies, *Appl. Phys. Lett.*, 67, 1960–1962.

Choi, H.I. & Williams, W.J. (1989) Improved time-frequency representation of multi-component signals using exponential kernels, *IEEE Trans. Acoust. Speech Signal Processing*, 37, 862–871.

Cichocki, A. & Amari, S. (2002) *Adaptive Blind Signal and Image Processing*, Wiley, NY.

Claasen, T.A.C.M. & Mecklenbräuker, W.E.G. (1980a) The Wigner distribution — A tool for time-frequency signal analysis — Part I: Continuous time signals, *Philips J. Research*, 35, 217–250.

Claasen, T.A.C.M. & Mecklenbräuker, W.E.G. (1980b) The Wigner distribution — A tool for time-frequency signal analysis — Part II: Discrete time signals, *Philips J. Research*, 35, 276–300.

Claasen, T.A.C.M. & Mecklenbräuker, W.E.G. (1980c) The Wigner distribution — A tool for time-frequency signal analysis — Part III: Relations with other time-frequency transformations, *Philips J. Research*, 35, 372–389.

Cohen, L. (1989) Time-frequency distributions — A review, *Proc. IEEE*, 77, 941–981.

Cohen, L. (1995) *Time-Frequency Analysis*, Prentice-Hall, Englewood Cliffs, New Jersey.

Conway, J.H. & Smith, D.K. (2003) *On quaternions and octonions: Their geometry, arithmetic, and symmetry*, A.K. Peters, Ltd., Natick, MA.

Cook, C.E. & Bernfeld, M. (1967) *Radar Signals: An Introduction to Theory and Application*, Artech, Boston, MA, USA.

Cover, T.M. & Thomas, J.A. (1991) *Elements of Information Theory*, Wiley, NY.

Crowne, F. & Fazi, C. (2009) Nonlinear radar signatures from metal surfaces, *Surveillance for a Safe World, Radar Conference*, Bordeaux, 12–16 October, 2009, pp. 1–6.

Daniels, D.J. (1996) *Surface Penetrating Radar*, IEEE Press, Piscataway, NJ.

Daniels, D.J. (2007) An assessment of the fundamental performance of GPR against buried landmines, *Proc. of SPIE*, 6553, 65530G, pp. 1–15.

Daniels, D.J., Gunton, D.J. & Scott, H.F. (1996) Introduction to subsurface radar, *IEE Proc. (London)*, H135, 278–320.

Datig, M. & Schlurmann, T. (2004) Performance and limitations of the Hilbert-Huang transformation (HHT) with an application to irregular water waves, *Ocean Engineering*, 31, 1783–1834.

Deans, S.R. (1981) Hough transform from the Radon transform, *IEEE Trans. Pattern Analysis and Machine Intelligence*, 3, 185–188.

Deans, S.R. (1983) *The Radon Transform and Some of Its Applications*, John Wiley, NY.

Deley, G.W. (1970) Waveform Design. Chapter 3, M.I. Skolnik (ed.) *Radar Handbook*, McGraw-Hill, NY.

De Visschere, P. (2009) Electromagnetic source transformations and scalarization in stratified gyrotropic media, *Progress in Electromagnetics Research*, 18, 165–183.

Dickinson, B.W. & Steiglitz, K. (1982) Eigenvectors and functions of the discrete Fourier transform, *IEEE Trans. Acoust. Speech Signal Process*, ASSP-30, 25–31.

Ding, Z. & Li, Y. (2001) *Blind Equalization and Identification*, Marcel Dekker, NY.

Dorme, C. & Fink, M. (1995) Focusing in transmit-receive mode through homogeneous media: The time reversal matched filter approach, *J. Acoust. Soc. Am.*, 98, 1155–1162.

Durka, P. (2007) *Matching Pursuit and Unification in EEG Analysis*, Artech House, MA.

Dym, H. & McKean, H.P. (1972) *Fourier Series and Integrals*, Academic Press, NY.

Erseghe, T., Kraniauskas, P. & Cariolaro, G. (1999) Unified fractional Fourier transform and sampling theorem, *IEEE Trans. Signal Processing*, 47, 3419–3423.

Chapou Fernández, J.L., Granados Samaniego, J., Vargas, C.A. & Velázques Arcos, J.M. (2009) Hertz tensor, current potentials and their norm transformations, *Progress in Electromagnetics Research Symposium, Proceedings*, Moscow, Russia, Aug 18–21, pp. 529–534.

Fink, M. (1992) Time reversal of ultrasonic fields — Part I: Basic principles, *IEEE Trans. Ultrason. Ferroelectr. Req. Control*, 39, 555–566.

Flandrin, P. (1998) *Time-Frequency and Time-Scale Analysis*, Academic Press. Volume 10 in the series: *Wavelet Analysis and Applications*.

Flandrin, P., Rilling, G. & Gonçalvès, P. (2003) Empirical mode decomposition as a filter bank, *IEEE Signal Processing Letters*, 10, 1–4.

Flandrin, P. & Gonçalvès, P. (2004) Empirical mode decompositions as data-driven wavelet-like expansions, *Int. J. of Wavelets, Multiresolution and Information Processing*, 2, 477–496.

Flandrin, P., Rilling, G. & Gonçalvès, P. (2004) Empirical mode decomposition as a filter bank, *IEEE Signal Processing Letters*, 11, 112–114.

Frazer, G. & Boashash, B. (1994) Multiple window spectrogram and time-frequency distributions, *Proc. IEEE Int. Conf. Acoustic, Speech, and Signal Processing — ICASSP'94, volume IV*, 293–296.

Gauss, C. F. (1866) Disquisitiones generales circa seriem infinitam $[\frac{\alpha\beta}{1-\gamma}\chi +$ $[\frac{\alpha(\alpha+1)\beta(\beta+1)}{1-2-\gamma(\gamma+1)}]]\chi^2 + [\frac{\alpha(\alpha+1)\alpha(\alpha+2)\beta(\beta+1)\beta(\beta+2)}{1-2-3-\gamma(\gamma+1)(\gamma+2)}]\chi^3 +$ etc. pars prior, *Commentationes Societiones Regiae Scientiarum Gottingensis Recentiores*, Vol. II. 1812. Reprinted in *Gesammelte Werke, Bd. 3*, pp. 123–163 & 207–229.

Gel'fand, I.M., Graev, M.I., & Vilenkin, N. Ya. (1966) *Generalized Functions. Volume 5, Integral Geometry and Representation Theory*, Academic Press, NY.

Georgieva, N.K. & Weiglhofer, W.S. (2002) Electromagnetic vector potentials and the scalarization of sources in a nonhomogeneous medium, *Phys. Rev. E*, 66, 046614-1-8.

Ghavami, M., Michael, L.B. & Kohno, R. (2007) *Ultrawideband Signals and Systems in Communication Engineering*, 2nd edition, Wiley, NY.

Ginkel, M. van, Hendricks, C.L.L. & Vliet, L.J. van (2004) *A short introduction to the Radon and Hough transforms and how they relate to each other*, Number QI-2004-01 in the Quantitative Imaging Group Technical Report Series, Delft University of Technology, Delft, The Netherlands.

Gjessig, D. (1978) *Remote Surveillance by Electromagnetic Waves for Air-Water-Land*, Ann Arbor Science Publishers.

Gjessig, D. (1981) *Adaptive Radar in Remote Sensing*, Ann Arbor Science Publishers.

Gjessig, D. (1986) *Target Adaptive Matched Illumination Radar: Principles & Applications*, Peter Peregrinus Ltd.

Guerci, J.R. (2010) *Cognitive Radar*, Artech House, MA.

Haykin, S. (2006). Cognitive radar: A way of the future, *IEEE Signal Processing Magazine*, 23, 30–40.

Hazewinkel, M. (Ed.) (2002), *Encyclopaedia of Mathematics*, Springer, New York.

Helgason, S. (1999) *The Radon Transform*, 2nd Edition, Birkhäuser, Boston.

Hertz, H. (1889) Die Krafte electrischer Schwingungen behandelt nach der Maxwell'schen Theorie, *Ann. d. Physik und Chemie*, 36, 1–22.

Hertz, H. (1893) The forces of electric oscillations, treated according to Maxwell's theory, *Wiedemann's Ann.* 36, 1–23, 1889. Reprinted in H. Hertz, *Electric Waves*, Macmillan, 1893, reprinted Dover Publications, 1962.

Hlawatsch, F. & Boudreaux-Bartels, G.F. (1992) Linear and quadratic time-frequency representations, *IEEE Signal Processing Magazine*, 9, 21–67.

Hough, P.V.C. (1962) Method and means for recognizing complex patterns. United States Patent No. 3,069,654 dated 1962.

Huang, N.E., Shen, Z., Long, S.R., Wu, M.C., Shih, H.H. & Zheng, Q. (1998) The empirical mode decomposition and the Hilbert spectrum for nonlinear and non-stationary time series analysis, *Proc. R. Soc. Lond.* A, 454, pp. 903–995.

Huang, N.E. & Attoh-Okine, N.O. (2005) *The Hilbert-Huang Transform in Engineering*, Taylor & Francis, Boca Raton, FL.

Huang, N.E. & Shen, S.S.P. (2005) *Hilbert-Huang Transform and Its Application*, World Scientific, Singapore.

Huang, N.E., Shen, Z. & Long, R.S. (1999) A new view of nonlinear waves — the Hilbert spectrum, *Ann. Rev. Fluid Mech.*, 31, 417–457.

Huang, N.E. & Wu, Z.H. (2008) A review on Hilbert-Huang transform: Method and its applications to geophysical studies, *Reviews of Geophysics*, 46, RG2006, doi:10.1029/2007RG000228.

Huang, N.E., Wu, M.L., Long, R.S., Shen, S.S., Qu, W.D., Gloersen, P. & Fan, K.L. (2003) A confidence limit for the empirical mode decomposition and Hilbert spectral analysis, *Proc. Roy. Soc. Lond.* A, 460, 1597–1611.

Hyvärinen, A. (2001) Complexity pursuit: Separating interesting components from time series, *Neural Computation*, 13, 883–898.

Hyvärinen, A., Karhunen, J. & Oja, E. (2001) *Independent Component Analysis*, Wiley, NY.

Illingworth, J. & Kittler, J. (1988) A survey of the Hough transform, *Computer Vision, Graphics and Image Processing*, 44(1):87–116.

Jackson, J.E. (2003) *A User's Guide to Principle Components*, Wiley, New York.

Janssen, A.J.E.M. (1981) Positivity of weighted Wigner distributions, *SIAM J. Mathematical Analysis*, 12, 752–758.

Janssen, A.J.E.M. (1982) On the locus and spread of pseudo-density functions in the time-frequency plane, *Philips J. Research*, 37, 79–110.

Janssen, A.J.E.M. (1984) Positivity properties of phase-plane distribution functions, *J. Math. Phys.*, 25, 2240–2252.

Jolliffe, I.T. (2002) *Principal Component Analysis*, 2nd edition, Springer, NY.

Jofre, L., Broquetas, A., Romeu, J., Blanch, S., Toda, A.P., Fàbregas, X. & Cardama, A. (2009) UWB tomographic radar imaging of penetrable and impenetrable objects, *Proc. IEEE*, 97, 451–463.

Jones, D.S. (1964) *The Theory of Electromagnetism*, Pergamon Press, New York.

Kak, A.C. & Slaney, M. (2001) *Principles of Computerized Tomographic Imaging*, Society for Industrial and Applied Mathematics, Philadelphia, 2001.

Kuperman, W.A., Hodgkiss, W.S., Song, H.C., Akal, T., Ferla, C. and Jackson, D.R. (1998) Phase conjugation in the ocean: Experimental demonstration of an acoustic time reversal mirror, *J. Acoust. Soc. Am.*, 103, 25–40.

Landau, H. J. & Pollak, H.O. (1961) Prolate spheroidal wave functions, Fourier analysis and uncertainty — II, *Bell Syst. Tech. J.*, 40, 65–84.

Landau, H.J. & Pollak, H.O. (1962) Prolate spheroidal wavefunctions, Fourier analysis and uncertainty — III: The dimension of the space of essentially time- and band-limited signals, *Bell Syst. Tech. J.*, 41, 1295–1336.

Li, B., Tao, R. & Wang, Y. (2007) New sampling formulae related to linear canonical transform, *Signal Processing*, 87, 983–990.

Li, Y. (2008) Wavelet-fractional Fourier transforms, *Chinese Physics* B, 17, 170–179.

Lohmann, A.W. (1993) Image rotation, Wigner rotation, and the Fourier transform, *J. Opt. Soc. Am.*, A10, 2181–2186.

Lohmann, A.W. & Soffer, B.H. (1993) Relationship between two transforms: Radon-Wigner and fractional Fourier, in *Annual Meeting, OSA Technical Digest Series* (Optical Society of America, Washington, D.C., 1993), Vol 16, p. 109.

Lohmann, A.W., Mendlovic, D., Zalevsky, Z. & Dorsch, R.G. (1996) Some important fractional transformations for signal processing, *Opt. Commun.*, 125, 18–20.

Mallat, S. (1999) *A Wavelet Tour of Signal Processing*, 2nd edition, Academic Press, NY.

Mallat, S. & Zhang, Z. (1993) Matching pursuit with time-frequency dictionaries, *IEEE Trans. Signal Processing*, 41, 3397–3415.

Marple, S.L. (1987) *Digital Spectral Analysis with Applications*, Prentice-Hall, Englewood Cliffs, NJ.

Martin, W. & Flandrin, P. (1985) Wigner-Ville spectral analysis of nonstationary processes, *IEEE Trans. Acoust., Speech, Signal Processing*, 33, 1461–1470.

McBride, A.C. & Kerr, F.H. (1987) On Namias' fractional Fourier transforms, *IMA J. Appl. Math.*, 39, 159–175.

McCrea (1957) Hertzian electromagnetic potentials, *Proc. Roy. Soc. Lond.*, A, 240, 447–457.

Meyer, Y. (1993) *Wavelets: Algorithms & Applications*, Society for Industrial & Applied Mathematics, Philadelphia.

Moll, J. & Fritzen, C. (2010) Time-Varying Inverse Filtering for High Resolution Imaging with Ultrasonic Guided Waves, *10th European Conference on Non-Destructive Testing*, Moscow, Russia, 1–10.

Morse, P.M. and Feshbach, H. (1953) *Methods of Theoretical Physics*, 2 Volumes, McGraw-Hill, NY.

Moshinsky, M. & Quesne, C. (1971) Linear canonical transformations and their unitary representations, *J. Math. Phys.*, 12, 1772–1783.

Moyal, J.E. (1949) Quantum mechanics as a statistical theory, *Proc. Camb. Phil. Soc.*, 45, 99–124.

Namias, V. (1980) The fractional order Fourier transform and its application to quantum mechanics, *J. Inst. Math. Appl.*, 25, 241–265.

Nisbet, A. (1955) Hertzian electromagnetic potentials and associated gauge transformations, *Proc. Roy. Soc. Lond. A*, 231, 250–262.

Onural, L. (1993) Diffraction from a wavelet point of view, *Opt. Lett.*, 18, 846–848.

Ozaktas, H.M. & Mendlovic, D. (1993) Fourier transforms of fractional order and their optical interpretation, *Optics Communications*, 101, 163–169.

Ozaktas, H.M., Barshan, B., Mendlovic, D. & Onural, L. (1994) Convolution, filtering and multiplexing in fractional domains and their relation to chirp and wavelet transforms. *J. Opt. Soc. Am.*, A11, 547–559.

Ozaktas, H.M., Zalevsky, Z. & Kutay, M.A. (2001) *The Fractional Fourier Transform with Applications in Optics and Signal Engineering*, Wiley, NY.

Panofky, W.G.H. & Phillips, M. (1962) *Classical Electricity and Magnetism*, 2nd edition, Addison-Wesley, Reading, MA.

Papoulis, A. (1962) *The Fourier Integral and Its Applications*, McGraw-Hill, NY.

Papoulis, A. (1968) *Systems and Transforms with Applications in Optics*, McGraw-Hill, N.Y.

Papoulis, A. (1974) Ambiguity function in Fourier optics, *J. Opt. Soc. Am.*, 64, 779–788.

Papoulis, A. (1977) *Signal Analysis*, McGraw-Hill, NY.

Pei, S.C. & Ding, J.J. (2000) Closed form discrete fractional and affine fractional transforms, *IEEE Trans. Signal Processing*, 48, 1338–1353.

Pei, S.C. & Ding, J.J. (2001) Relations between fractional operations and time-frequency distributions, and their applications, *IEEE Trans. Signal Processing*, 49, 1638–1655.

Pei, S.C. & Ding, J.J. (2002) Eigenfunctions of linear canonical transform, *IEEE Trans. Signal Processing*, 50, 11–26.

Radon, J. (1917) Über die Bestimmung von Funktionen durch ihre Integralwerte längs gewisser Mannigfaltigkeiten, *Berichte Sächsische Akademie der Wissenschaften, Leipzig, Mathematisch-Physikalische Klasse*, 69, 262–277.

Ramm, A.G. & Katsevich, A.I. (1996) *The Radon Transform and Local Tomography*, CRC Press, Boca Raton, FL.

Raveh, I. & Mendlovic, D. (1999) New properties of the Radon Transform of the Cross Wigner/Ambiguity Distribution Function, *IEEE Trans. Signal Processing*, 47, 2077–2080.

Righi, A. (1901) Sui campi elettromagnetiei e particolarmente su quelli create da cariche eletrriche o da poli magnetic in movimento, *Nuovo Cimento*, 2, 104–122.

Rihaczek, A.W. (1969) *Principles of High-Resolution Radar*, McGraw-Hill, NY.

Rihaczek, A.W. & Hershkowitz, S.J. (1996) *Radar Resolution and Complex-Image Analysis*, Artech House, MA.

Rihaczek, A.W. & Hershkowitz, S.J. (2000) *Theory and Practice of Radar Target Identification*, Artech House, MA.

Saxena, R. & Singh, K. (2005) Fractional Fourier transform: A novel tool for signal processing, *J. Indian Inst. Sci.*, 85, 11–26.

Schantz, H. (2005) *The Art and Science of Ultrawideband Antennas*, Artech House, MA.

Schmidt, R.O. (1986) Multiple emitter location and signal parameter estimation. *IEEE Trans. Antennas & Propagation*, AP-34, 276–280.

Shan, P.-W. & Li, M. (2010) Nonlinear time-varying spectral analysis: HHT and MODWPT, *Mathematical Problems in Engineering*, Volume 2010, Article ID 618231, doi:10.1155/2010/618231.

Skolnik, M.I. (2001) *Introduction to Radar Systems*, 3rd edition, McGraw-Hill, NY.

Slepian, D. (1964) Prolate spheroidal wave functions, Fourier analysis and uncertainty — IV: Extensions to many dimensions; generalized prolate spheroidal functions, *Bell System Tech. J.*, 43, pp. 3009–3057.

Slepian, D. (1978) Prolate spheroidal wave functions, Fourier analysis and uncertainty — V. The discrete case, *Bell System Technical J.*, 57, 1371–1430.

Slepian, D. & Pollak, H.O. (1961) Prolate spheroidal wave functions, Fourier analysis and uncertainty — I, *Bell System Tech. J.*, 40, 43–64.

Soumekh, M. (1999) *Synthetic Aperture Radar Signal Processing*, Wiley, NY.

Stone, J.V. (2004) *Independent Component Analysis: A Tutorial Introduction*, MIT Press, Cambridge, MA.

Stone, J.V. (2005) Independent Component Analysis, *Encyclopedia of Statistics in Behavioral Sciences*, B.S. Everitt & D.C. Howell, Editors, Volume 2, pp. 907–912, Wiley, NY.

Tanter, M., Thomas, J.-L. & Fink, M (2000) Time reversal and the inverse filter, *J. Acoust. Soc. Am.* 108, 223–234.

Thomson, D.J. (1982) Spectrum estimation and harmonic analysis, *Proc. IEEE*, 70, 1055–1096.

Toups, M.F., Ayasli, S. & Fleischman, J.G. (1993) *Analysis of Foliage-Induced Synthetic Pattern Distortions from the July 1990 Foliage Pentration Study*, MIT Lincoln Laboratory Project Report STD-50, (20 July 1993).

Trees, H.V.L. (1968) *Detection, Estimation, and Modulation Theory, Part I*, Wiley, NY.

Trees, H.V.L. (2002) *Optimum Array Processing*, Wiley, NY, 2002.

Vakman, D.E. (1968) *Sophisticated Signals and the Uncertainty Principle in Radar*, Springer-Verlag, NY.

Vickers, R.S., Lowry, R.T. & Schmidt, A.D. (1988) A VHF radar to make terrain elevation models through tropical jungle, *IEEE Radar-88 Symposium*, Ann Arbor, MI, USA.

Ville, J. (1948) Theorie et applications de la notion de signal analytique, *Cables et Transmission*, 2A, 61–74.

Weber, H. (1869) Über die Integration der partiellen Differentialgleichung: $\partial^2 u/\partial x^2 + \partial^2 u/\partial y^2 + k^2 = 0$, *Math. Ann.*, 1, pp. 1–36.

Weiglhofer, W.S. (2000) Scalar Hertzian potentials for nonhomogeneous uniaxial dielectric-magnetic mediums, *Int. J. Appl. Electrom.*, 11, 131–140.

Weiglhofer, W.S. & Georgieva, N. (2003) Vector potentials and scalarization for nonhomogeneous isotropic mediums, *Electromagnetics*, 23, 387–398.

Whittaker, E.T. (1902) On the functions associated with the parabolic cylinder in harmonic analysis, *Proc. Lond. Math. Soc.*, 35, 417–427.

Whittaker, E.T. (1903) On the partial differential equations of mathematical physics, *Math. Ann.*, 57, 333–355.

Whittaker, E.T. (1904) On an expression of the electromagnetic field due to electrons by means of two scalar potential functions, *Proc. Lond. Math. Soc.*, Series 2, 1, 367–372.

Whittaker, E.T. & Watson, G.N. (1927) *A Course of Modern Analysis*, 4th Edition, Cambridge University Press.

Whittaker, E.T. (1951) *History of the Theories of Aether and Electricity: Volume 1: The Classical Theories; Volume II: The Modern Theories 1900–1926*, Dover Publications, NY.

Wiener, N. (1933) *The Fourier Integral and Certain of Its Applications*, Cambridge University Press.

Wiener, N. (1958) *Nonlinear Problems in Random Theory*, MIT Press & Wiley.

Wigner, E.P. (1932) On the quantum correction for thermodynamic equilibrium. *Phys. Rev.*, 40, 749–759.

Wood, J.C. & Barry, D.T. (1994a) Tomographic time-frequency analysis and its application toward time-varying filtering and adaptive kernel design for multicomponent linear-FM signals, *IEEE Trans Signal Processing*, 42, 2094–2104.

Wood, J.C. & Barry, D.T. (1994b) Linear signal synthesis using the Radon-Wigner transform, *IEEE Trans Signal Processing*, 42, 2105–2111.

Wood, J.C. & Barry, D.T. (1994c) Radon transformation of time-frequency distributions for analysis of multicomponent signals, *IEEE Trans Signal Processing*, 42, 3166–3177.

Woodward, P.M. (1953) *Probability and Information Theory with Applications to Radar*, Artech, MA, USA.

Woodward, P.M. & Davies, I.L. (1950) A theory of radar information, *Phil. Mag.*, 41, 1001–1017.

Xie, S., He, Z. & Fu, Y. (2005) A note on Stone's conjecture of blind signal separation, *Neural Computation*, 17, 321–330.

Xu, Y., Haykin, S. & Racine, R.J. (1999) Multiple window time-frequency distribution and coherence of EEG using Slepian sequences, *IEEE Trans. Biomed. Eng.*, 49, 861–866.

Zadeh, L.A. (1950a) Frequency analysis of variable networks, *Proc. IRE*, 38, 291–299.

Zadeh, L.A. (1950b) Correlation functions and power spectra in variable networks, *Proc. IRE*, 38, 1342–1345.

GLOSSARY

AF: Ambiguity function.

BSS: Blind source separation.

CFEF spectrum: Carrier-Frequency-Envelope-Frequency spectrum

CRP: Corner Reflector Point

DTX: a PRX-derived transmitted (TX) arbitrarily long duration signal, modulating an arbitrary carrier, and which addresses some selected resonance or collection of resonances identified in the PRX spectrum.

DRX: the envelope of the received (RX) return target echo, modulating an arbitrary carrier, that is elicited by a DTX.

FRFT: Fractional Fourier transform.

ICA: Independent Component Analysis.

LFM: Linear frequency modulated signal or chirp.

LTI: Linear time-invariant.

LTV: Linear, time-varying.

MAP: (1) Matched Adaptive Time-Frequency Packet-Signal. (2) Maximum A Posteriori.

MAP signal: A target-matched adaptive time-frequency wave packet or signal using a priori information and permitting maximum a posteriori estimation detection according to Bayesian statistics.

ML: Maximum Likelihood.

MTX: a MAP transmitted (TX) short duration packet/signal, envelope or amplitude modulation of an arbitrary carrier. An MTX envelope is a PRX but time reversed resulting in a matching of the envelope of the MTX transmitted signal to the designated target.

MRX: the envelope of the received (RX) return target echo, modulating an arbitrary carrier, that is elicited by an MTX.

MUSIC: Multiple Signal Classification Algorithm.

MWTFA: Multiple Window Time-Frequency Analysis.

PCA: Principal Component Analysis.

PRX: the received (RX) return target echo, modulating an arbitrary carrier that is elicited by a PTX. A PRX is equivalent to an UWB return echo signal, and approximates the target impulse response at a specified target aspect to the transmitter.

PTX: a transmitted (TX) short duration packet/signal, modulating an arbitrary carrier, generally of 1 or 2 nanosecond duration, and that is used to obtain a priori information concerning the target in a specified aspect to the transmitter. A PTX is equivalent to an UWB transmitted pulse functions as a "δ function" surrogate or approximation,[9] and is usually a monocycle.

RAMAR: Resonance and Aspect Matched Adaptive Radar.

RX: Received signal.

SCR: signal-to-clutter ratio.

SNR: signal-to-noise ratio.

SRX: the received (RX) return target echo, modulating an arbitrary carrier, that is elicited by an STX.

STX: A transmitted (TX) arbitrarily long duration TX signal that is a collection or bundle of DTXs.

SVD: Singular Value Decomposition.

TBP: time-bandwidth product.

TX: transmitted signal.

UWB: ultrawideband.

WVD: Wigner-Ville Distribution Function.

---

[9]A true δ (delta) function is of infinitely short duration and of infinite bandwidth, and even an approximation could not be propagated. What is meant in this context is a very short pulse — usually a monocycle.

INDEX

273